W9-CBW-131

INNOVATING IN URBAN ECONOMIES

Economic Transformation in Canadian City-Regions

In a globalizing, knowledge-based economy, innovation and creative capacity lead to economic prosperity. Starting in 2006, the Innovation Systems Research Network began a six-year-long study on how city-regions in Canada were surviving and thriving in a globalized world. That study resulted in the "Innovation, Creativity, and Governance in Canadian City-Regions" series, which examines the impact of innovation, talent, and institutions on sixteen city-regions across Canada. This volume explores how the social dynamics that influence innovation and knowledge flows in Canadian city-regions contribute to transformation and long-term growth.

With case studies examining cities of all sizes, from Toronto to Moncton, *Innovating in Urban Economies* analyses the impact of size, location, and the regional economy on innovation and knowledge in Canada's cities.

(Innovation, Creativity, and Governance in Canadian City-Regions)

DAVID A. WOLFE is the Royal Bank Chair in Public and Economic Policy at the University of Toronto and the director of the Innovation Policy Lab at the Munk School of Global Affairs. He was National Coordinator of the Innovation Systems Research Network.

**Innovation, Creativity, and Governance
in Canadian City-Regions**

Series editor: David A. Wolfe

The series on *Innovation, Creativity, and Governance in Canadian City-Regions* presents the research results of a six-year, sixteen-city study of the social dynamics of innovation, creativity, and civic governance in Canadian cities. The first three volumes in the series provide detailed analyses of each of the three research themes carried out across a selection of large, medium-sized, and small Canadian cities, and the fourth one integrates the key findings across all themes for the individual cases.

While the cases covered are primarily Canadian, the volumes present the material in an international and comparative context that addresses ongoing intellectual and policy debates concerning urban economic development and civic governance. As such, the series offers important new insights that contribute to our contemporary understanding of the relationship between urban social dynamics and economic performance.

Innovating in Urban Economies

Economic Transformation in Canadian City-Regions

EDITED BY DAVID A. WOLFE

UNIVERSITY OF TORONTO PRESS
Toronto Buffalo London

© University of Toronto Press 2014
Toronto Buffalo London
www.utppublishing.com
Printed in Canada

ISBN 978-1-4426-4698-8 (cloth)
ISBN 978-1-4426-1476-5 (paper)

Printed on acid-free, 100% post-consumer recycled paper with vegetable-based inks.

Library and Archives Canada Cataloguing in Publication

Innovating in urban economies : economic transformation in Canadian city-regions / edited by David A. Wolfe.

(Innovation, creativity, and governance in Canadian city-regions)
Includes bibliographical references.
ISBN 978-1-4426-4698-8 (bound). – ISBN 978-1-4426-1476-5 (pbk.)

1. Cities and towns – Canada – Growth – Case studies.   2. Canada – Economic conditions – Case studies.   I. Wolfe, David A., editor of compilation   II. Series: Innovation, creativity, and governance in Canadian city-regions (Toronto, Ont.)

HC113.I55 2014   307.760971   C2014-900416-8

University of Toronto Press acknowledges the financial assistance to its publishing program of the Canada Council for the Arts and the Ontario Arts Council.

Canada Council   Conseil des Arts
for the Arts   du Canada

ONTARIO ARTS COUNCIL
CONSEIL DES ARTS DE L'ONTARIO
50 YEARS OF ONTARIO GOVERNMENT SUPPORT OF THE ARTS
50 ANS DE SOUTIEN DU GOUVERNEMENT DE L'ONTARIO AUX ARTS

University of Toronto Press acknowledges the financial support of the Government of Canada through the Canada Book Fund for its publishing activities.

Contents

**Part III: The Specialized Characteristics of Innovation
   in Medium-Sized Cities**

**Part IV: Innovation for Survival or Growth in Canada's
   Small Cities**

Tables

Figures

Foreword to the Series

Innovating in Urban Economies: Economic Transformation in Canadian City-Regions

Innovation and creative capacity are essential determinants of economic prosperity in a globalizing, knowledge-based economy. Although the process of globalization has led to numerous predictions of the "death of distance," growing evidence suggests that the contemporary global economy make cities more – not less – important as sites of production, distribution, and innovation. Over the past decade, recognition has grown that even the most global of economic activities remain fundamentally rooted in city-regions as critical sites for organizing economic activity. More significantly, the social dynamics of city-regions are crucial in shaping economic outcomes (Gertler 2001).

The interactive and social nature of the innovation process makes city-regions the appropriate scale at which social learning processes unfold. Knowledge transfer between highly skilled people happens more easily in cities, while talent flows within and between cities are a critical means for the dissemination of innovative ideas. In a country with diverse and strongly differentiated regional economies, relationships between economic actors, organizations, and institutions at the local and regional scale are crucial factors affecting national prosperity. The concentrations of talented people and skilled occupations in urban centres are seen as critical sources of the creative and innovative ideas which generate growth in city-regions. And leading urban regions are no longer prepared to be passive objects at the hands of globalizing forces associated with the spread of information technologies, but are taking control of their own economic future through efforts aimed at the "strategic management" of their own economies (Audretsch 2002). From this perspective, many of the key foundations of economic success in a globalizing, knowledge-based economy are the social qualities and properties of urban places.

The papers collected in this series report on the research results of a six-year study of the social dynamics of economic performance in Canadian city-regions by members of the Innovation Systems Research Network (ISRN). The ISRN itself was an innovative experiment in knowledge management that came into existence in 1998 through the leadership of the Social Sciences and Humanities Research Council of Canada and its partners, the National Research Council and Natural Sciences and Engineering Research Council. The ISRN is a collaborative, multidisciplinary network of university-based researchers analysing how innovation processes unfold in different regions and localities across the country. Since its inception in 1998, the network's research has focused on how the interaction among the major components of the regional innovation system shape the process of innovation and social learning that is critical for Canada's success in the knowledge-based economy. Its primary objectives have been: (1) to understand the process by which regional networks foster the production and circulation of knowledge that is critical to the innovation process, and (2) to deepen our understanding of the role of public policy in facilitating (or impeding) this process (Holbrook and Wolfe 2005).

The goal of the ISRN's first research program from 2001 to 2005 was to determine the prevalence and success of local industrial clusters across Canada and to analyse how the formation and growth of these clusters contribute to local economic growth and innovative capacity. Underlying this objective was a set of fundamental conceptual questions: how do local assets and relationships between economic actors enable firms – in any industry – to become more innovative? Under what circumstances does "the local" matter, and how important are local sources of knowledge and locally generated institutions (public and private) in strengthening the innovative capabilities of firms and industries? What is the relative importance of non-local actors, relationships, and flows of knowledge in shaping the development trajectories of localized innovation and growth?

The results of that five-year project made a distinctly Canadian contribution to the study of industrial clusters. The international literature on clusters and regional innovation systems recognizes that in a global marketplace, local input factors and inter-firm dynamics are critical to a firm's ability to innovate and thereby gain competitive advantage. But, as noted above, a critical question addressed in the ISRN study concerns the relative importance of local versus non-local factors in cluster performance and the relationship between the two sets of factors. The ISRN case studies provided important insights into these questions. They documented a balance between the relative impact of local and non-local relationships and knowledge flows – in other words, the dynamic tension between the local "buzz" and global "pipelines" that circulate knowledge between clusters (Bathelt, Malmberg, and Maskell 2004). They underlined the sectoral specificity of industrial clusters – clusters in different sectors draw upon different knowledge bases which influence both the innovation process within the clusters and the underlying relationship between the cluster and the research infrastructure which supports it. The case studies also highlighted the centrality of a strong, dynamic talent base, or "thick" labour market for the success of most clusters. The ability to draw upon a plentiful supply of labour with the skills required by cluster firms is often the most critical factor that attracts them to, and anchors them in, a specific geographic location (Wolfe 2009a).

Finally, the case studies suggested that some of the most successful clusters have profited from the development of strong social networks at the community level and the emergence of dedicated, community-based organizations. These entities link leaders in the individual clusters to a broader cross-section of the community. They appear to be supported by new institutions of civic governance that identify problems impeding the growth of the cluster and help mobilize support across the community for proposed solutions. The research also found some evidence to suggest that size is a critical variable in the success of civic engagement, with some of the larger urban centres encountering greater difficulty in achieving effective degrees of mobilization (Wolfe and Nelles 2008).

The results of the first ISRN project led to the overwhelming conclusion that many of the most significant factors which underlay the performance of individual clusters were not specific to the cluster itself. The research thus led to the conclusion that the social characteristics, dynamics, and relationships within the wider city-region are important determinants of economic performance. This led the members of the network to undertake a second Major Collaborative Research Initiative (MCRI), with funding from SSHRC and other partners to investigate the social dynamics of innovation and economic performance, with the city-region as the primary unit of analysis. The second MCRI project, from 2006 to 2011, set out to analyse what were believed to be three key determinants of economic performance at the local level: the broader dynamics of innovation and knowledge flows, the role of creativity and talent, and the contribution made by new forms of civic governance in Canadian city-regions (Wolfe 2009b).

The study involved detailed investigations of all three of these themes – innovation, talent, and governance – in a sample of sixteen large, medium-sized, and small Canadian cities across eight provinces, resulting in more than fifty case studies in all. Each case was examined using a common research methodology, primarily based on in-depth interviews with key participants in critical sectors in the local economy using common interview guides. The key questions for each theme were largely the same, but were structured to reflect the researchers' detailed knowledge of their own local economy and institutions.

This series of four volumes presents the results of a selection of the case studies conducted under each of the themes. The present volume examines how the dynamics which influence innovation and knowledge flows in Canadian city-regions contribute to the process of economic transformation and long-term growth in their local econo-mies. The individual case studies feature reports from Canada's largest city, Toronto (Ontario), to much small cities, such as Moncton (New Brunswick). The chapters illuminate the contribution that urban size, relative location, the nature of the regional economy, the underlying industrial structure, and the strength of local research institutions all make to the dynamics of innovation and knowledge flows in the local economy. The volume provides a rich and nuanced analysis of how the interplay of these various factors influences innovation in a broad cross-section of Canadian city-regions.

David A. Wolfe, Series editor

REFERENCES

Audretsch, D.B. 2002. "The innovative advantage of US cities." *European Plan-ning Studies* 10 (2): 165–76. http://dx.doi.org/10.1080/09654310120114472.

Bathelt, H., A. Malmberg, and P. Maskell. 2004. "Clusters and knowledge: Lo-cal Buzz, Global Pipelines and the Process of Knowledge Creation." *Progress in Human Geography* 28 (1): 31–56. http://dx.doi.org/10.1191/0309132504 ph469oa.

Gertler, M.S. 2001. "Urban economy and society in Canada: Flows of people, capital and ideas." *Isuma: Canadian Journal of Policy Research* 2 (3): 119–30.

Holbrook, J.A., and D.A. Wolfe. 2005. "The Innovation Systems Research Network: A Canadian experiment in knowledge management." *Science & Public Policy* 32 (2): 109–18. http://dx.doi.org/10.3152/147154305781779623.

Wolfe, D.A. 2009a. "Introduction: Embedded clusters in a global economy." *European Planning Studies* 17 (2): 179–87. http://dx.doi.org/10.1080/09654310802553407.

Wolfe, D.A. 2009b. *21st century cities in Canada: The geography of innovation.* Ottawa: Conference Board of Canada.

Wolfe, D.A., and J. Nelles. 2008. "The role of civic capital and civic associations in cluster policies." In *Handbook of research on innovations and cluster policies,* ed. Charlie Karlsson, 374–92. Cheltenham, UK: Edward Elgar Publishers.

Acknowledgments

A national research project inevitably ends with an army of people and institutions whose contributions merit gratitude. First and foremost, the team acknowledges the vital financial and advisory inputs provided by funding agencies and partners in the research. The primary source of funding for the work came from the Social Sciences and Humanities Research Council of Canada through the Major Collaborative Research Initiatives program #412-2005-1001. Throughout the work SSHRC proved immensely supportive of the team's activities.

In addition, we appreciate the support of our research partners: Atlantic Canada Opportunities Agency, Health Technology Exchange, National Research Council, Nova Scotia Economic Development, the Ontario ministries of Economic Development and Trade, Research and Innovation, and Economic Development and Innovation, Queen's University, Ryerson University, Simon Fraser University, Statistics Canada, Université Laval, and University of Toronto, as well as in-kind support from THECIS in Calgary.

The team's annual meetings (King City, Ontario, 2006; Vancouver and Bowen Island, 2007; Montreal, 2008; Halifax, 2009; Toronto, 2010) would not have proved productive without the contributions of local organizing groups who helped with logistics, policymakers who participated in the knowledge exchange opportunities of the policy events, and the local partners mentioned above who provided additional funding.

Several people played pivotal roles in running the research program. This project has been a true partnership from the outset. Without the intellectual guidance and supportive friendship of Meric Gertler, the successful conclusion of this project would not have been possible. Deborah Huntley, who has been an indispensible member of the team over two major research projects, provided excellent administrative skills that kept the project on time, on track, and on budget. These two key individuals were joined on the project management committee by Charles Davis, Jill Grant, Adam Holbrook, and Réjean Landry. At and between the annual meetings, the team received guidance and oversight from the project's Research Advisory Board: Bjørn Asheim, Susan Christopherson, Susan Clarke, Philip Cooke, Hervey Gibson, Gernot Grabher, Anders Malmberg, Peter Maskell, Kevin Morgan, Claire Nauwelaers, Tod Rutherford, and Allen Scott. Their willingness to freely devote their time to our endeavours is a testament to their collegiality and dedication to scholarship.

Teams in each of the city-regions studied, coordinated, and conducted the case study work and many city-region teams involved several junior colleagues and students – too numerous to name – who made invaluable contributions to the work. Many began working on the project as doctoral students, graduated to post-doctoral fellows and ultimately became colleagues and collaborators.

Without the participation of community members from across the country who donated their time and knowledge to the team in agreeing to sit for interviews or complete surveys, none of this work would have been possible. The team sincerely appreciates their willingness to share their wisdom and experience, and hopes that they find the results of the research useful in their efforts to keep their communities vital and progressive.

Final thanks go to Daniel Quinlan and his team at the University of Toronto Press for believing in the importance of bringing this work to a wider audience and for providing assistance along the way to make it possible.

PART I

Dynamics of Innovation in City-Regions – Diversity, Specialization, and Variety

1 Introduction

DAVID A. WOLFE

Innovation and creative capacity are essential determinants of economic prosperity in a globalizing, knowledge-based economy. The creation and diffusion of new knowledge drives innovation in knowledge-intensive production and service activities, which in turn drives economic performance and growth. Although these processes are strongly shaped by national institutions and global knowledge flows, innovation and creativity overwhelmingly occur in the geographic context of city-regions, which are consequently critical sites for determining economic performance. Many aspects of the trend towards globalization makes cities *more*, not less, important as principal sites for innovation, creativity, and the production of knowledge-intensive goods and services. In somewhat paradoxical fashion, the global revolution in transportation and communications technologies has accentuated the concentration of ideas, innovation, and economic growth in city-regions. "This then is the age of cities – and increasingly, of large city regions, both in Canada and in other developed nations" (Bourne, Hutton, Shearmur, et al. 2011, 8).

Most Canadians live in city-regions, defined simply as a continuous network of urban communities, whose "boundaries move outward – or halt – only as city economic energy dictates" (Jacobs 1984, 45). More precisely, a city-region can be defined as "the presence of a core city linked by functional ties to a hinterland" (Rodríguez-Pose 2008, 1027). According to the 2011 census, more than twenty-three million people live in one of Canada's thirty-three largest city-regions (census metropolitan areas), which were the only areas where population growth exceeded the national average.[1] Within this broad trend, however, the greatest concentration of population and economic activity lies within

Canada's six largest, which accounted for two-thirds of the growth (Bourne, Brunelle, Polèse, et al. 2011, 45).

The growing concentration of our population in city-regions high-lights the fact that the social dynamics of urban regions are particularly significant in shaping economic outcomes. This raises the first key question for our analysis: why do large spatial disparities persist among regions and cities? While standard neoclassical theory suggests that initial differences among urban regions should gradually be levelled away through the effects of market forces, the evidence overwhelming points to the opposite result (Venables 2006). As the world economy becomes more globalized, the pressure increases for city-regions to create distinct advantages across a diverse range of sectors and industries. The introduction of new ideas in knowledge-intensive production and services is the source of economic growth in the industrial countries. The social nature of the innovation process means that city-regions provide the locale in which the transmission and diffusion of knowledge necessary for innovation occurs. But as Edward Glaeser perceptively notes, "The urban ability to create collaborative brilliance isn't new. For centuries, innovations have spread from person to person across crowded city streets" (Glaeser 2011, 8; Hall 1998).The dense concentration of economic actors in cities offers multiple opportunities for contact, interaction, and the exchange of ideas among highly skilled people. Canada's city-regions are not just the dominant sites of economic activity; they are also the leading edges of innovation that will generate the new ideas, new products, and new industries that will drive the economy in the future.

Despite this recognition of the importance of city-regions as the loci of knowledge creation and innovation, a number of critical questions remain about how these processes occur within the urban economy. First, there is considerable debate about how the economic structure and specific characteristics of urban economies affect innovation and knowledge circulation. Some analysts maintain that the most important dynamics are those generated by the advantages that accrue to firms located in dense clusters of similar and related firms – in other words, dynamics generated by a greater degree of specialization within city-regions. Conversely, others emphasize the innovation potential that arises when new forms of knowledge circulate among a wide range of sectors within a city – in other words, from the dynamics associated with greater diversity in the economic structure of a city-region. In this latter view, ideas that are commonplace or widely accepted within a particular sector may have novel value in another. The possibility of cross-fertilization arising from an economic structure with greater variety enhances the potential for the generation of new ideas and innovation within the local economy. The question of whether industrial specialization or diversity has the greater potential for innovation and growth has critical implications for how city-regions adapt in a knowledge-based and globalizing economy.

A second question concerns the relative advantages or disadvantages associated with city size or urban agglomeration. There is widespread agreement that the largest city-regions enjoy certain advantages as centres of innovation and creativity, but they are also subject to greater costs that result from rising land prices and the congestion associated with larger size. Recent surveys of the literature on this question suggest that the doubling of city size can produce a noticeable increase in productivity across a wide range of different city sizes. At the same time, however, the advantages derived from their more diversified economies, strong research institutions, and a deeper talent pool may be offset by negative consequences of urban agglomeration linked to rising land costs and the increased costs associated with urban congestion (OECD 2006). However, the critical issue here concerns the specific way in which these contradictory economic forces making for urban concentration and dispersion play out across different economic sectors in specific city-regions (Venables 2006).

At the same time, there is less agreement on the prospects for midsized and small cities. Small and medium-sized cities often operate from a more specialized industrial base concentrated in a narrower range of more traditional economic sectors. That offers some advantages, but the future of these cities is closely tied to the specific industries in which they are specialized. The industrial structure of such cities increases the risk that they will be locked into declining sectors or obsolete technologies that may be supplanted by newer ones. At the same time, the speed or flexibility with which their economies can shift from declining sectors to emerging ones may have a disproportionate effect on their economic future. The ability of city-regions to adapt to, and absorb, rapid changes in technology, and the competitiveness of their industrial mix, are critical to their economic future.

This volume presents the results of recent empirical research and conceptual analysis on the nature of innovation and creative processes in a wide range of Canadian cities and their contribution to the economic future of those cities. This introduction sets out a broad framework for understanding the results in terms of the implications of city size, the relative degree of specialization and diversity, and the contribution of a talented and creative workforce to the economic performance of city-regions. The framework developed in this chapter provides the context for examining how these trends are currently playing out across a range of Canadian cities presented as case studies in the rest of the volume. The research presented in each chapter is based on a common research methodology followed in each case study, based on in-depth interviews using a common interview guide and quantitative overviews of the economic structure of the individual cities. In each case, the study was undertaken by a member of the research team with a deep knowledge of their own city-region. Field research was conducted in the cities between 2008 and 2010, with the chapters written in 2011. As such, the case studies provide snapshots of how urban social and political dynamics shaped economic trajectories during a turbulent period in Canada's economic development.

Industrial Evolution and the Life Cycle of City-Regions

One of the defining features of contemporary economies is the central role of knowledge and learning in creating economic value and determining competitive success. The literature on this point is abundant and compelling: innovation is a socially organized process that depends on interactive, social learning among individuals and firms (Lundvall 1992). The critical issue is how the socially embedded nature of the innovation process affects economic growth and development in an urban setting. A number of theories have been advanced to explain the relative pace of industrial growth and decline in city-regions. Traditional explanations of the factors that affect the economic performance of city-regions have been framed in terms of the origins and growth of urban centres; the relationship between the concentration of firms in an urban economy and the growth of local labour markets; the relative degree of specialization or diversity that characterizes the economic structure of individual cities; and the relative importance of lifestyle amenities and the quality of place in increasing the attractiveness of particular cities (Wolfe and Bramwell 2008). Recent research suggests that these issues need to be considered within the context of a broader set of changes that influence the economic performance of city-regions. The additional factors include the relative size of the individual city, the city's point of insertion into an evolving global hierarchy of urban centres, and the evolution of the city's industrial structure towards the growth of higher-level business services associated with a more knowledge-intensive economy.

Conventional theories of urban economic growth emphasize the historical decisions made by firms regarding where to locate. These decisions were often influenced by physical factors, such as the practical necessity of locating near traditional modes of transportation such as rivers, coastal harbours, or railway lines, or the advantage of operating near rich endowments of natural resources. As transportation systems improved in the post-war period, especially with the introduction of inter-city highway networks in the 1950s and 1960s and the deregulation of airline travel in the 1970s, and as trading relations were increasingly liberalized on both a continental and global basis, the relative influence of these historical factors on urban location was reduced. Some researchers point to the impact that the mobility of highly skilled and educated workers has on the growth of specific locations, while others focus on the pull exerted by existing concentrations of industry. At the heart of this debate is a question about how the locational choices of individuals intersect with and reinforce those made by firms to contribute to the growth and dynamism of modern cities: Do people choose to locate where they find the greatest number of job opportunities, or do businesses locate in city-regions with the largest potential labour market (Glaeser and Gottlieb 2006; Storper and Scott 2009)?

A related issue concerns the advantages that firms derive from locating in cities with firms in similar or different industries, and which type of structure contributes most to industrial innovation and economic growth. One perspective emphasizes the benefits for urban growth of specialization in similar or closely related industries, while the alternative focuses on the advantages that flow from a diverse and variegated urban environment. The first approach argues that the advantages created by a dense network of suppliers, a deep pool of skilled labour, and the knowledge spillovers that occur among geographically concentrated groups of firms in related industries make the most significant contribution to growth. These advantages are associated with a greater degree of specialization in an urban economy; they are derived from forces that lie outside the individual firm, but are embedded in the industrial sector of a particular city.

This research draws upon a perspective dating back to the late nineteenth century, which suggests that once a region or city establishes itself in a particular set of production activities, its chances for continued growth tend to be high. Paul Krugman, building on the work of Alfred

Marshall in the early twentieth century, argues that three types of benefits are created for firms located in the same city-region that specialize in similar technologies or production techniques. The first is the deep pool of specialized labour created by the concentration of firms within similar industries, which makes it easier to hire people with the specialized skills the firms require and attracts more workers to the city-region because of the resulting employment opportunities. The second benefit is the fact that a local concentration of firms in the same industry can support a larger number of specialized providers of intermediate inputs or services, thus enabling firms to concentrate on their areas of competence through the specialized division of labour in the local economy. Finally, knowledge is transferred more easily among firms located close to each other in a city-region than it is over longer distances. The transfer of knowledge between firms in the same industrial sector in the same city-region occurs through the mobility of specialized labour among them, through the tendency of serial entrepreneurs to establish a succession of firms in the same city, and through the "learning-by-observing" effects of densely concentrated industries (Krugman 1991).

The contrasting perspective, associated with that of Jane Jacobs, suggests that new and innovative ideas derive from the cross-fertilization that occurs across different industrial sectors. Therefore, city-regions that are endowed with a diverse range of industries, rather than those that are specialized in a smaller number of industrial sectors, enjoy the conditions most conducive to innovation and growth. Innovative ideas are derived by applying knowledge that may be considered standard in one sector to help solve problems or develop new products in another sector of the local economy. From this perspective, larger city-regions, with a broader cross-section of industries, have a greater potential for generating innovative new ideas and, as a result, enjoy faster rates of growth and higher levels of innovation than do smaller ones. Jacobs also suggested that competition among alternative sets of ideas embodied in a diverse set of economic actors is more conducive to generating new knowledge than the local monopoly over ideas that exists in a more specialized urban economy (Jacobs 1969).

These competing views on the sources of urban economic growth have stimulated a substantial body of academic research that provides some evidence to support both perspectives: that the presence of both specialized *and* diversified urban economies contributes to the overall performance of city-regions, but the two factors may act in different ways in cities of different size. A recent survey suggests a possible explanation for the apparently contradictory results. Part of the confusion may arise from the omission of time as a factor in determining whether the economic benefits associated with specialization or diversity exert a more important influence on the economic performance of firms in city-regions. In effect, "the role of externalities varies according to the maturity of the industry. Jacobs externalities predominate in the early stages of the industry life cycle, whereas Marshall externalities enter at a later point, and in the end, specialization will in fact hinder economic growth" (Beaudry and Schiffauerova 2009, 334)

This conclusion is consistent with other findings that the introduction of an industry life-cycle perspective, as well as a better appreciation of the relationship between city size and diversity or specialization, help explain the relative contribution made by industrial specialization and diversification to urban economic growth. The economic benefits associated with a more diversified local economy play different roles in the innovation process at various stages in the maturity of the industry, while differences in population also affect cities' ability to create and diffuse new knowledge. Duranton and Puga (2000) suggest that firms often develop new products in the diversified, creative environment found in larger urban centres, but as the technology and industry mature, there is a strong incentive for them to relocate to more specialized cities in the mass production phase of the industry's life cycle in order to exploit urban cost advantages. Larger city-regions tend to be more diversified and knowledge-intensive than medium-sized and small cities. Where large cities have multiple specializations, medium-sized cities have significantly fewer. Related findings reveal that levels of innovative activity are also strongly linked to city size, with R&D, patenting, and major product innovations much more concentrated in large urban areas (Audretsch 2002).

This insight is reinforced by an argument linking industrial activity, economic fortunes, and city size. Large cities with a diversified industrial base are more insulated from the impacts of economic change, while smaller ones with a narrower industrial base are more subject to a life cycle of growth and decline (Brezis and Krugman 1997). While a greater degree of specialization does stimulate the growth of some medium-sized cities, the outlook for those cities is linked to the economic prospects of the specific sectors in which they are specialized. Once the sectors lose their competitive edge, the cities may lack the knowledge base or the quality of place to diversify their local economy into newer and expanding industries. They are often confronted with the challenge of regenerating their local basis for economic development without the institutional capacity that can furnish a fresh supply of ideas and new sources of growth.

This suggests that ultimately the source of innovation and economic growth for a city-region rests not just on the degree of specialization or diversification in its industrial structure, but more importantly on the resilience of the city-region in mobilizing its economic assets in the pursuit of a new basis for growth. The key issue concerns the ability of firms, industries, and institutions in a specific city or region to adapt their existing knowledge base and localized capabilities to the generation and exploitation of new commercially valuable sources of knowledge. "New paths do not emerge in a vacuum, but always in the contexts of existing structures and paths of technology, industry and institutional arrangements" (Martin and Simmie 2008, 186). Resilient regions tend to be those which prove adept at making the transition out of declining industries, while simultaneously being able to exploit the local knowledge infrastructure to cultivate new, potential growth fields. However, future patterns of development are strongly influenced by the existing industrial structure, as well as the broader set of institutions that have supported that structure. Those sectors in which the city-region has historically been specialized will constrain its future ability to grow, or create opportunities for new sectors to emerge. The basis on which those sectors can emerge will be influenced in turn by the capacity of firms and institutions within the region to develop and exploit new sources of knowledge and their existing knowledge infrastructure, as well as the talents and skills of the workforce (Wolfe 2010).

Variations in the ability of cities to create and diffuse new knowledge appear to be important for the cities' long-term growth prospects, as well as their ability to adjust to changing economic conditions. Indeed, as John Montgomery argues, "At the end of the day, there are only a handful of means by which city and urban regional economies can grow. One is the introduction of new production processes and services to create new work and new divisions of labour … More important than this even is the extension of new technologies to create new products and therefore economic sectors" (Montgomery 2007, 29). This introduces a distinctly Schumpeterian dimension into the analysis of urban economics that underlines the impact of the capacity to innovate. New economy sectors are sustained by the continuous pace of innovation and learning needed to keep abreast of the rapidly moving knowledge frontier in their industries.

This need for continuous innovation extends well beyond the manufacturing sector of the economy. In an innovative economy where the knowledge frontier is moving rapidly, dynamic cities are those able to draw on their local knowledge assets and research infrastructure to re-invent themselves by moving from one field of specialization to another. In this transition, existing industries may provide the essential building blocks for the emergence of a new innovative industry, because the skills and talents that have accumulated over time may furnish critical inputs needed by the emerging industry. The ability of individual city-regions to marshal their local knowledge assets and develop local concentrations of expertise in emerging technology areas may be a good indicator of their prospects for resurgence and growth. As two experts put it, "The important question may not be specialization versus diversity but whether a city has specialized in the right thing at the right time" (Storper and Manville 2006, 1250).

The capacity of individual cities to make the transition to more information-intensive forms of industrial activity depends not just on the functioning of autonomous market-based processes; it is also affected by their social and political structures. The emergence of new technologies and industrial sectors is often associated with a corresponding set of changes in the industrial geography of cities and regions. It is not surprising that the new information technologies or biotechnologies tend to be associated with names such as Silicon Valley, San Diego, Austin (Texas), or Research Triangle Park (North Carolina) in the United States or Helsinki and Bangalore in Europe and Asia – places that scarcely registered on the industrial map before 1970. These cities, which were among the most dynamic of the late twentieth century, have been home to many of the technologies that anchor the expanding knowledge economy (Montgomery 2007, 22).

At the same time, this transition poses significant economic challenges for older, established metropolitan areas in the United States, Europe, and Canada that dominated the previous industrial era. These cities were once the pinnacles of economic growth and prosperity to which other urban regions aspired: "The Silicon Valleys of the Second Industrial Revolution had names like Akron, Detroit, Pittsburgh, and Rochester" (Safford 2004, 16; Moretti, 2012, 20). While some older industrial cities in the United States have experienced a recent economic resurgence due to their ability to shift to knowledge-intensive activities, many have not. Cities and regions that remain locked into traditional specializations in mature manufacturing and are unable to capitalize on existing knowledge assets or to mobilize their local endowments of human capital face greater challenges in effecting this transition (Glaeser 2011; Moretti 2012).

Diversity and the Emergence of an International Hierarchy of Cities

The relation between the size of a city-region and its economic prospects is determined not just by its place within the national economy, but also by its relative standing within a hierarchy of global or world cities. In a seminal paper, Peter Hall identified the importance of urban size and the degree of specialization as important signifiers for an emerging international hierarchy of urban centres (Hall 1966). Around the same time, the Canadian economist Stephen Hymer suggested that the growing predominance of multinational corporations in the global economy was likely to increase the stratification of world cities, creating a new global division of labour between geographic areas corresponding to the vertical one within the firm. This would result in a concentration of those occupations with responsibility for corporate decision-making within a few of the world's major cities – such as New York, London, Paris, Frankfurt, and Tokyo – supported by a larger number of regional hubs. As a consequence, the structure of income and consumption in those major cities would match the distribution of status and authority, with the citizens of these global capitals enjoying the best jobs and the highest rates of remuneration (Hymer 1972).

In the decades since, this provocative insight has been expanded upon in the growing literature on the stature and importance of "world cities" or "global cities." According to Saskia Sassen, the fundamental dynamic at work "is that the more globalized the economy becomes, the higher the agglomeration of central functions in a relatively few sites … the global cities" (Sassen 2001, 5). The position of these cities in the global hierarchy is determined by the role they play both in coordinating the processes of production and distribution of goods around the world and in providing the specialized services that large, complex firms require to manage a spatially distributed network of offices, factories, and distribution centres. Increasingly, these global cities have also become the key sites for innovation in the financial services industry and the development of new financial instruments, two of the most notable features of the global economy since the 1980s. Rather than being randomly distributed around the world, as some enthusiastic supporters of the digital revolution maintained, these high-end financial and information services have in fact become ever more concentrated in the central business districts of a few leading cities that are specialized in the production of producer and financial services. International gateway cities – such as Paris, London, Tokyo, New York, and Los Angeles – located at the peak of the urban hierarchy achieve a higher level of economic performance because of their access to large pools of highly specialized and technical workers and the wide range of innovative firms located in the cities (Simmie and Wood 2002; McCann 2008).

Below this top tier of truly global cities is a wider range of urban centres with a concentration of diverse knowledge-intensive production and service activities that act as the economic hubs of their respective national economies and serve as the principal nodes linking those economies into the global economy (Simmie 2003). They play a key role as knowledge hubs – both for their countries and for the wider world – due to their ability to attract exceptional talent and to capitalize on global and local sources of knowledge, much of which flows through their local companies. Knowledge flows more easily in big urban centres, which are advantaged in their abilities to draw on both local and global sources of knowledge and attract the best "talent," thereby insulating themselves from the consequences of population and industrial change (Florida 2002; Scott 2008; Glaeser 2011).

However, these leading cities are not the only sources of innovation. Regional hubs, such as Montreal, Toronto, Boston, and Milan, are also highly competitive and play an important role in the global hierarchy. These national and regional hub cities contain high proportions of elite business and political leaders with the authority to make local investment decisions, which gives these urban areas a greater degree of autonomy. With their large populations, these cities also enjoy the benefits of economic agglomeration due to the concentration of a wide range of knowledgeable collaborators from different disciplines who can contribute to the innovation process (Simmie 2002). Medium-sized cities that are specialized in a narrower range of industrial activities can serve as hubs for their regional economies, but they have more limited access to global knowledge flows and trade. The most dynamic medium-sized cities, which make the most effective use of local institutional research supports (universities) and social networks, are able to specialize successfully in knowledge-intensive industrial activities (Safford 2004).

Urban Agglomeration and the Concentration of Talent and Creativity

Another view of the underlying determinants of urban dynamism and economic growth focuses on the role of talent and creativity. The critical link between innovation, personalized knowledge exchanges, and economic growth makes the most important locational asset a dense labour market of highly educated and creative workers – what Cooke calls "regional talent pools of global significance" – with the potential to attract and embed globally mobile investment, and generate innovative growth (Cooke 2007). This view suggests that the local attributes that attract talented workers are of paramount importance in determining local economic prosperity. Such talent is attracted to and retained by cities, but not just *any* cities; those that offer rich employment opportunities, a high quality of life, a critical mass of cultural and entertainment activity, and social diversity are said to exert the strongest pull (Glaeser and Gottlieb 2006).

A different line of reasoning maintains that the causal link between concentrations of creative and talented people and regional economic growth may in fact be reversed; instead of skills driving economic growth, the preference of firms for locating in regions with large, diversified economies may be the primary factor in attracting and retaining large concentrations of creative workers, thus stimulating urban growth and innovation (Scott 2006, 2007; Storper and Manville 2006). "Though person-embodied talent remains a critical input into innovation, it needs to be considered in the context of ... other factors ... such as city size, industry specialization, local institutional infrastructure, and knowledge flows" (Wolfe and Bramwell 2008, 177).

In addition, the more concentrated the talent, the more innovative the output. One of the key advantages of cities in a globalized economy is that they reduce the cost of knowledge transfer, and act as centres of idea creation and diffusion where talent clusters. There is also a strong correlation between population densities in general, the density of creative workers in particular, and metropolitan patenting activity, all of which suggests that population density is critical for generating knowledge spillovers and innovation (Carlino, Chatterjee, and Hunt 2007).

Cities as "Schumpeterian Hubs" of Innovation

While a significant body of research points to the positive correlation between urban concentrations of talent and human capital in urban centres and the prospects for economic growth, some scholars suggest that the high concentrations of human capital may be the result of other positive externalities found in these cities. This perspective maintains that skills-led explanations of economic growth confuse the nature of the relationship between the locational decisions of individual workers – creative or otherwise – and those of firms. These critiques argue that the primary determinant of urban growth is not just the locational preferences of highly skilled and creative workers, but the concentration of firms that generate a dense labour market in the first place. Pointing to the fact that economic resurgence has occurred not just in Sunbelt cities, but also in "old, cold, dense city-regions" such as Boston, Chicago, and New York, Storper and Manville argue that recent population growth in cities – both older, northern ones and new, southern ones – is linked to shifts in regional economic geography and industrial activity. Workers are drawn to urban centres where employment opportunities are the greatest, not just to those with lifestyle amenities such as shopping and entertainment, which are ubiquitous and readily available in most cities of a certain size: "Jacobs, Florida, and Glaeser are all on to something in claiming that skills and amenities go together, but they may have got their causality reversed. It is the fact that these skilled workers are congregated in certain places that leads to the presence of amenities and, in some cases, makes the places tolerant and bohemian as well" (Storper and Manville 2006, 1254).

The argument that workers are attracted by employment opportunities more than by consumer, lifestyle, and social amenities suggests that explanations of urban economic growth need to be more nuanced. A related body of research indicates that human capital levels are becoming more unequally distributed across urban centres, as cities with a higher concentration of innovation-related jobs generate greater employment opportunities in other sectors of the urban economy. Recent research points to the fact that knowledge-related jobs in innovation-based industries have a much higher multiplier effect than jobs in other industries. According to Enrico Moretti, "For each new high tech job in a metropolitan area, five additional local jobs are created outside of high tech in the long run" (2012, 60). Most economic sectors have a multiplier effect on related employment, but few sectors have as great an effect as high tech with its concentration of innovation-related jobs. The reason for this is that high-tech firms tend to co-locate due to advantages created by potential knowledge spillovers and access to a larger labour market. The implication of this is that urban centres with a higher concentration of jobs in these sectors are likely to grow faster than other cities due to the spin-off of related jobs. This trend is reinforcing the bifurcation of cities into larger ones with greater concentrations of innovation-based and knowledge-related jobs and more traditional manufacturing or resources centres that face increased difficulty in maintaining their employment base. In the middle may be a few cities that are able to shift their base of economic activity from more traditional resource and manufacturing activities to knowledge-based ones (Moretti 2012).

At the same time, the boundaries between traditional sectors of the economy is shifting as the impact of innovation is blurring the distinction between traditional manufacturing and service-oriented activity. While many industrial activities still occur in identifiable sectors staffed by industry-specific occupations, many of the knowledge-intensive activities associated with new and emerging sectors of the economy are less easily categorized. Changing patterns of urban development are similarly ambiguous. Allen Scott describes shifts in the nature of economic activity in terms of an emerging "cognitive-cultural economy" where leading-edge economic growth and innovation are driven by "technology-intensive manufacturing, diverse services, 'fashion-oriented neo-artisanal production,' and cultural products industries" (Scott 2007, 1466; Scott 2008). The shift to this new form of production is facilitated by the steady adoption of digital technologies for creating more customized goods through a less routinized organization of the production process. In this view, the location choices of the creative class and related concentrations of human capital result from the broader economic transformation to a knowledge-based economy that is under way. The larger forces at work are the outcome of historically conditioned trajectories of urban economic growth, where the supply of and demand for labour evolve in a mutually reinforcing fashion.

The interaction between the location decisions of firms and those of highly skilled workers is strongly influenced, though not completely determined, by city size. Scott argues that the transition to a "cognitive-cultural economy" is most pronounced in the major metropolitan centres of the advanced economies, where "the internal production spaces ... are being remade in the image of the core sectors of the new economy, such as technology-intensive industry, finance and business services, fashion and media" (Scott 2012, 34). Consequently, the presence of the "cognitive-cultural economy" is most apparent in large metropolitan areas or "flagship hubs," such as New York, London, Paris, Amsterdam, and Tokyo, where production activities are densely concentrated in firms with global market reach. However, intermediate cities, such as Austin, Texas, or Calgary in Canada, have also developed large concentrations of technology-intensive activity and higher-end business service. New information technologies permit the simultaneous dispersion and concentration of economic activity, allowing producers in various urban centres to benefit from the knowledge flows in a specific location, as well as to access global knowledge flows and markets. Virtuous cycles of growth result as the number of producers increases and local growth accelerates, leading to a deepening of localized returns and the intensification of economic benefits.

The emphasis on growth driven by the virtuous interaction of skilled labour and firm preferences characterizes large metropolitan cities as environments where value chains underlying production, and the associated networks of economic actors, can adapt rapidly because of their efficiency at coordinating and managing the processes that are the basis of innovation and growth. In this sense, cities are acting like giant "Schumpeterian hubs" of innovative activity, or "switchboards which permit the constant creation and reshaping of the chains linking producers, consumers, and different kinds of indirect players of the economy" (Veltz 2004). Signs of this developmental dynamic are evident in large metropolitan areas, both in rapidly growing "cognitive-cultural sectors" and in the formation of "intra-urban industrial districts devoted to specialized facets of cognitive-cultural production," such as high-technology industries and software in the San Francisco Bay area, movies in Hollywood, business and financial services in New York and London, and fashion in Paris and Milan (Scott 2007, 1470). These emerging areas of cognitive-cultural production tend to be located in, or close to, the central business district and often take advantage of low-cost space available in abandoned industrial warehouses or factories. The conversion of existing physical spaces associated with the older industrial economy to new uses for the emerging cognitive-cultural economy illustrates the critical way in which the spatial landscape of an inner city is reconfigured in dynamic urban regions (Hutton 2008, 11).

Despite the appeal of a talent-based approach to urban economic growth and development, the uneven distribution of creative occupations and highly skilled labour and the resulting increase in social disparities in some of the most successful city-regions suggests there are inherent limits to this approach to urban growth strategies. The rise of creative cities in North America and Europe is a product of their role as places with the ability to generate a high level of technological innovations and economically useful knowledge. Further confirmation for this insight lies in the fact that scientific and creative occupations tend to cluster together in urban centres (Beckstead, Brown, and Gellatly 2008).

Cities in the Canadian Urban System

Cities do not exist in isolation from each other, but as part of the regional economy in which they are located, as well as in relation to other cities in the national economy. The economic standing of an individual city and its long-term prospects are influenced by its relative position within the urban system – which is defined by its spatial relations with other cities. The urban system is the organizational dynamic that underlies the economic and human geography of the country. Different-sized cities play different roles in Canada's urban system. Each city in the urban system "represents a unique combination of population size, demographic structure, economic specialization and rate of growth that together define the opportunities open to each resident" (Bourne and Simmons 2003, 24).

Canada's urban system is distinguished by its relatively large number of smaller cities with only a few large ones, and the substantial differences between those larger cities. Large cities within the urban system are clearly differentiated in turn from medium- and smaller-sized cities in terms of their more diversified economic base. The ten largest cities – Halifax, Montreal, Ottawa-Gatineau, Toronto, Winnipeg, Regina, Saskatoon, Calgary, Edmonton, and Vancouver – generate a disproportionate share of the national wealth and dominate the regional hinterlands in their respective parts of the country (Brender, Cappe, and Golden 2007). Despite the privileged status of these cities within the Canadian urban system, none is substantial enough to have attained a significant position in the global urban hierarchy, with only one city – Toronto – in the ten-member "beta" group of world cities (no Canadian city ranked in the top-most "alpha" group) and the twelve-member "well-rounded global cities" category (ibid.). Smaller cities, in contrast, which are more dependent on relatively few areas of industrial concentration, may be more vulnerable to external economic shocks, which can adversely affect their economic status.

The increasing role of the largest cities in the Canadian urban system, and the challenges this poses for the country as a whole, is highlighted in a recent analyses of trends in the urban system over the past four decades. Pierre Filion notes that the traditional distinction between the core industrial cities and the more distant hinterland cities, referred to as the "heartland-hinterland" distinction, is on the wane. Over this period it has been displaced by a growing differentiation between Canada's largest cities and its medium- and smaller-sized ones. Consistent with Moretti's argument above, he postulates that as this disparity continues to grow, the urban system will become more bifurcated, with a growing concentration of population and economic opportunities in the largest cities at the expense of the rest. Their increasing size and economic influence will afford them greater weight in our political system as well, to the disadvantage of smaller and medium-sized cities (Filion 2010). At the same time, there is a growing distinction in the reconfiguration of those sectors of economic activity concentrated in Canada's largest urban centres compared to small and medium-sized ones. The knowledge inputs, even for Canada's traditional resources industries, are increasingly concentrated in the larger cities, while manufacturing activity continues its migration to cities below the one million mark, which is consistent with the arguments noted above. "Mid- and low-technology manufacturing is deserting the large metropolitan areas for smaller places where land prices and wages are lower. This industrial exodus ... is the corollary of the concentration of rapidly growing information- and knowledge-rich services in large metropolitan areas" (Bourne, Brunelle, Polèse, and Simmons 2011, 56). The key challenge this poses for the smaller and medium-sized cities concerns the degree to which they will be able to maintain and expand the employment opportunities for their residents. This will increasingly depend on the extent to which they are able to leverage their existing industrial capabilities and knowledge assets to create jobs in the expanding sectors of the economy.

Outline of the Volume

The ensuing chapters in this volume examine how the changing nature of innovation and knowledge flows are playing out across a wide range of Canadian cities – large, medium-sized, and small. The cases were selected in order to ensure that different-sized cities in various regions of the country, each with a distinctive industrial structure and areas of economic selection, were represented. The following chapters provide a rich set of insights into the way in which these cities that make up the Canadian urban system are changing in the context of the underlying economic factors described above.

The next chapter, by Gregory Spencer, provides a broad statistical overview of the nature of the knowledge bases that support the innovation process in those cities. The chapter starts from the proposition that as the social processes of knowledge production differ, each will tend to thrive in certain social environments which vary between city-regions. In this way, the social characteristics of a city-region impact the social processes of knowledge production and over time these impacts contribute to the spatial patterns of industrial location. The chapter uses a comprehensive database of statistical indicators for Canadian city-regions to find connections between the characteristics of city-regions and the predominant knowledge bases that drive their economies.

The next sections of the volume organize the individual case studies largely on the basis of where they fall in terms of urban size and their concentration of economic activities. The cases have been grouped into the three broad categories discussed above. The first includes Canada's larger cities, which are increasingly becoming concentrations of innovation-based activity and expanding their reach in terms of research and high-end business services. The second includes a number of cities whose economic strength has historically been based on traditional resource or manufacturing activity, but are beginning to make the transition to more knowledge-intensive activities. The final category includes a number of Canada's smaller urban centres whose economies have been based on traditional manufacturing, or in one case administrative, activities and who face greater challenges in making the transition to the knowledge-based economy. While the cases present evidence that some of these centres are beginning to forge successful strategies, others are still confronted by significant challenges.

Part II of the volume explores the nature of innovation and knowledge flows in four of Canada's largest metropolitan economies. The chapter on Toronto by Charles Davis and Nicholas Mills focuses on the concept of the cognitive-cultural economy, which emphasizes the significance of a group of industries, occupations, work tasks, and functions with high cognitive and/or cultural intensity in the growth of core metropolitan regions. The City of Toronto's diverse cultural, creative, and knowledge-based industries generate extraordinary potential for knowledge and innovation spillovers within the cultural and media cluster. As Davis and Mills note, there are currently considerable spillovers between the information and communications technology sector, the traditional media clusters, and the rapidly growing digital-media sector which take the form of product spillovers where a growing number of firms are developing highly specialized capabilities to create cross-platform content for film, television, and mobile screens, as well as more diverse network spillovers where the co-location of a diverse range of cultural and media industries is creating the kind of milieu conducive to innovation and new product development.

The next chapter, on Vancouver, by Brian Wixted and Adam Holbrook, takes up the issue of the role of knowledge specialization and diversity in industry innovation within the context of the influence of the larger physical geographic setting. To do this, it focuses on the knowledge specialization and diversity dynamic in an inter-city (multi-spatial) framework. Given the absence of a strong manufacturing base in Vancouver's urban economy, the chapter selects five human capital–based segments of the economy in order to study the knowledge dynamics in an inter-urban setting – fuel cells, wireless (mobile phones and related technologies), new media, bio-pharma, and Hollywood North (feature movies) in metropolitan Vancouver. For the most part, the selected sectors rely on knowledge flows from other activities and, significantly, knowledge flows from beyond Vancouver. Apart from fuel cells, all of the sectors relied upon significant knowledge exchanges with organizations in cities in the eastern North American mega-region or in southern California. Some inter-city relations were based in specialization (bio-pharma), while others (new media) were based in diversity locally and inter-regionally. The authors conclude that a critical factor in Vancouver's ability to sustain its current level of creative activity has been its attractiveness as a desirable location for creative talent and its strategic location on international transportation routes.

The case study of Montreal by Diane-Gabrielle Tremblay and Juan-Luis Klein adopts a different perspective to examine some of the most distinctive features of the innovative process in Canada's second-largest city. It starts out with a review of recent economic strategies in response to the crisis in its traditional role as a manufacturing and financial services centre that laid the base for Montreal's economic renewal and reorientation towards a more knowledge-based economy.

The chapter then goes on to analyse the innovation process in three of the city's design, creative, and high-technology sectors – the fashion sector, the TOHU cultural district, and the aeronautics sector – which demonstrates the beginning of some cross-sectoral patterns of knowledge flows, while also recognizing that sectoral specialization tends to be the more common pattern for innovation. The authors find that a distinctive feature of Montreal's economy is the extent to which intermediate organizations and business associations play an important role in the acceleration of knowledge flows, by increasing the quality and the intensity of these flows. Overall, however, they conclude that a sectoral logic still dominates productive interactions and knowledge flows, even within those sectors more closely associated with the cognitive-cultural economy.

In stark contrast to Canada's three largest cities, Calgary is one of the most specialized in the country. Since the discovery of oil at Leduc in 1947, Calgary has emerged as the command centre of the oil and gas industry. The case study by Cooper Langford, Ben Li, and Cami Ryan notes that despite the fact there is no actual oil and gas production within the boundaries of the city, mining and oil and gas extraction dominate the city's economy. However, they also observe that closely tied to oil and gas is a set of related industries, particularly in construction, business, and financial and ICT services. Knowledge is embodied in the people working in the oil and gas industry and innovation within Calgary's leading sectors is primarily based on local knowledge sources. Knowledge flows occur through the movement of personnel within specific sectors, but also occur through the assembly of networks of firms to meet the needs of individual clients, as well as through the broader social networks within the city. Economic sectors that have grown up in close proximity to the oil and gas industry, and have developed specialized knowledge and technological capabilities in that role, can apply their knowledge to other opportunities in new markets in different industries, such as software, global information systems, and wireless communications. These new industries are thus drawing upon a platform of diverse related knowledge, including managerial, technical, and financial knowledge that emerged out of the primary resource-based economy. This related-knowledge platform is now being applied to new commercial products and economic opportunities in a manner that suggests the possibility of moving beyond the confines of the specialization-versus-diversity model of innovation in urban city-regions. The result of this cross-sectoral application of the knowledge base developed for a highly specialized sector suggests the possibility of a different form of innovation, labelled related-knowledge diversity, where the existing knowledge base is deployed to work on a related target problem, drawing upon a diversity of knowledge inputs from an existing sector.

Part III of the book examines the nature of innovation in a selection of Canada's medium-sized cities, which tend to be characterized by a higher degree of industrial specialization. These case studies look at the way some of these cities are beginning to exploit their existing knowledge base and dynamic research infrastructure to develop new areas of specialization or exploit new opportunities emerging in the knowledge-based economy. The case study of London, Ontario, by Neil Bradford and Jen Nelles deals with the nature of the innovation process in what are termed "ordinary cities" – those without world-class industrial clusters or a hip cultural and creative economy. They argue that innovation dynamics in London exhibit a pattern of strong external connectivity. Many of its leading firms are innovative despite the fact that they are not embedded in an identifiable local industrial cluster; rather, they are globally networked through their "Global Niche Markets" or "Custom Services and Solutions" strategies. Both of these business orientations rely on strategic interface with, and position in, global value chains to access the ideas and information that drive innovation. The London case study found that global orientations and connections were often the reason why interaction with other co-located firms and the area's universities, community colleges, and laboratories remained a low priority. The evidence from this case suggests that the key to success for innovative firms in Canada's medium-sized cities lies in their ability to find and exploit niche opportunities in dynamic and rapidly changing global markets.

Peter Warrian's case study of Hamilton provides another example of an older industrial city with a traditional concentration in one particular sector, namely, steel. Hamilton has a strong, unitary industrial identity that marks the city as a classic in specialization, displaying all the potential gains and dangers of economic development tied to the life cycle of one industry. However, with the emergence of the health sciences sector in the west end of the city, based around the McMaster University Health Sciences complex, the other side of Hamilton's innovative economy has introduced a new degree of diversity into the urban economy with significant, but somewhat surprising linkages between the two sectors. Research into the pattern of knowledge flows in the

Hamilton economy reveals an unexpected linkage between the older industrial manufacturing sector that has traditionally dominated the city and the emerging health sciences one. Steel company executives have traditionally played key leadership roles in the governance structures of the Health Sciences complex; and the well-endowed health-benefit plans of the local unions have provided the financial basis for the new innovative firms in health services. This is a surprising twist on the conventional innovation model, where the cross-sectoral linkages that are proving influential in stimulating the development of innovative products and serves lie within the consumer or demand side of the local economy, rather than the supply side.

The next chapter, by Andrew Munro and Harald Bathelt, examines the region around the cities of Cambridge, Guelph, Kitchener, and Waterloo – more commonly referred to as the Waterloo region. The region, which has been a strong centre for manufacturing firms since the nineteenth century, is often portrayed as a Canadian success story thanks to its transformation to a regional economy driven by knowledge-based information technology (IT) firms and the central role of the research-intensive University of Waterloo. This chapter explores the nature of the innovation dynamics in the Waterloo region and examines the degree to which regional innovation networks and trans-regional knowledge pipelines underlie the success of innovative firms in both the IT and traditional manufacturing sectors. The authors also investigate the extent to which cross-sectoral innovation networks have formed across firms in the two segments. The results indicate that the economic success of the region can best be explained by the presence of a distinct entrepreneurial culture and international linkages in production and innovation, but that both sectorally based linkages among regional firms and cross-sectoral knowledge flows are weakly developed.

The final chapter in this section, by Jill Grant, explores the nature of innovation and knowledge flows in three sectors in the city of Halifax. Halifax is strategically located as the regional metropolis for much of Atlantic Canada, not just the province of Nova Scotia. As a consequence, it is large enough to attract a critical mass of creative workers. The chapter investigates the extent to which this strategic advantage is helping to transform Halifax into a cognitive-cultural economy. The findings are equivocal. While the cognitive and creative sectors are increasing in strength in Halifax and contributing to innovation and growth, their capacity to become dominant in the economy of the region remains unproved. Despite the relative strength and diversity of

Halifax's cultural sectors and knowledge-intensive business services, the research found that knowledge flows across these sectors remain relatively undeveloped. Advertising and design firms are drawing on other sectors, such as music, for cultural product inputs, and built-environment consultants sometimes include public art in their designs, but the local economy appears to provide few mechanisms for mutual learning or exchange between sectors. As a relatively small city on the rim of a national network of cities, Halifax faces a greater challenge in achieving the level of innovation which may be necessary to support its future economic growth.

The next section of the volume explores the nature of the innovation process in Canada's smaller cities, which face their own set of distinctive challenges in adapting to the restructuring of traditional manufacturing sectors in their urban economies. The evidence from these cases suggests that the challenge of developing more innovative, knowledge-based sectors is not insurmountable for these cities, but depends considerably on their ability to mobilize the assets of their local research infrastructure and develop more specialized, niche products for international markets. Peter Phillips and Graeme Webb's study of Saskatoon is instructive in this respect. The Saskatoon case offers some interesting perspectives on the role of social dynamics, diversity, and infrastructure in mid-sized creative and innovative cities. First, the case offers some (but not total) support for the assertion that knowledge flows within sectors and clusters, linked to specialized structures and processes, matter for innovation. Both firms and employees indicated that thick, specialized labour markets, formal knowledge-transfer and informal knowledge systems are important for knowledge-intensive economic development, but it is not clear that these matter for less knowledge-intensive industries. Second, they did not find much evidence of any acknowledged externalities between parallel innovative clusters or between innovative clusters and the rest of the economy – if anything, they seem to be solitudes. In the case of Saskatoon, innovation both within specific clusters and in the broader economy is supported by a general focus on innovation and by the capacity to leverage capital in support of the foundational infrastructure generated. Nevertheless, much of this capacity seems to be narrowly focused in specific sectors, rather than widely distributed. This finding may simply be the result of the relatively small population in the city and narrower base of its economic structure.

The Moncton study by Yves Bourgeois explores the question of whether smaller urban centres can reposition their local economies as centres for innovation in emerging technologies. Moncton was a critical transportation and retail hub for the Maritime economy, but has undergone a slow but steady process of restructuring to a business-service economy over the past two decades. During this period, the city has supported the growth of information technology firms and knowledge-intensive business services. In examining the innovation process in these sectors, the study investigates the extent to which the use of information technologies can help to level the playing field for smaller cities located in more peripheral regions of the country. In the case of Moncton, which lacks a major research institution on the scale of Fredericton or Saint John, firms depend on knowledge flows into the region from the outside, primarily through the mobility of their skilled labour force and, to a lesser extent, through the use of information technology. This intensifies the competition for talent in these cities, especially when they must compete with larger regional centres, such as Halifax. Innovative firms tend to adopt two different competitive strategies. Those identifying their major strength as an intimate knowledge of the local market tended to customize solutions to local clients from globally accessible technologies. Those identifying the technical knowledge they developed (products or processes) as their competitive advantage tended to be global exporters. Despite the innovative success of a number of firms, the greatest deficiency overall was the weak innovation linkages among community, business, government, post-secondary education, and other stakeholders, certainly not to the level found in other centres. The region has done relatively well at addressing important economic challenges since the 1980s, but the study concludes that continued prosperity requires the creation of a more innovative culture.

The last case study in this section deals with the city of Trois-Rivières and examines a cross-section of three industries in older industrial sectors of the economy, as well as two emerging sectors. It explores the question of whether the deficiencies observed in the renewal of the economy of the city-region are linked not only to its size or degree of specialization (or lack of variety), but also to the specific characteristics/dynamics of the sectors in which it is specialized or is trying to be. The study shows that the "traditional" sectors of the regional economy are historically aligned with the local infrastructure of research, transfer, and support institutions which was put in place from the 1970s onward. The result is a regular collaboration between firms and support organizations, in other words, a regional embeddedness of the innovation activities. In light of these findings, it is difficult to see the characteristics of the city-region (size and diversity) as the explanatory factors of their weak economic performance and challenges for renewal. Instead, the answer lies with the internal characteristics of the individual sectors and their firms. In two of the city's emerging sectors, the patterns of interaction that lead to innovation provide evidence of the importance of external knowledge flows and extra-regional assets. Although the geographical and social proximity of the partner is also important for them, the companies of these sectors collaborate outside the region because they do not find what they need for their innovation projects in the local environment. The study concludes that both the general characteristics of the city-region and the specific characteristics of individual sectors play a significant role in explaining the economic performance of the city-region and also as inhibitors of the renewal and diversification that can potentially emerge from the new knowledge produced and diffused within each sector.

In summary, the findings of the three case studies of smaller Canadian city-regions confirm some of the broader points made in the first part of this introduction. Knowledge-flows and innovation patterns tend to occur overwhelmingly along sectoral lines, rather than across sectoral boundaries in the manner that is occurring in the larger and some of the medium-sized cities. Those cities that enjoy the presence of a strong local research institution and technology-transfer and support institutions are able to draw upon these institutions for innovative ideas, technical support, and skilled labour. Frequently, the most innovative firms, even in these smaller cities, tie into global knowledge networks as further sources of innovative ideas and practices. Furthermore, those cities which are located in relative proximity to larger hubs in the Canadian urban system face further challenges in attracting talent and retaining firms in their regional economy. There is widespread recognition of the need to shift the economies in these city-regions to a more knowledge-intensive and innovative basis, but the resources for doing so are often more limited. In this instance, Saskatoon, which has benefited from considerable investment in its research infrastructure, provides a highly instructive counter-example.

The final part of the book contains a chapter by Philip Cooke, a long-time member of the research advisory committee of the Innovation Systems Research Network (ISRN), which locates the key findings and insights of our national study of innovation and knowledge flows in

Canadian city-regions in the broader context of international debates about the relative importance of specialization versus diversity for growth in city-regions. Cooke argues that while the ISRN case studies have added considerably to our knowledge of the innovation dynamics in city-regions of widely varying size and competences, they don't fully address how these cities and regions can adapt to the challenge of a post-recession global economy. He suggests the need to go beyond the limits of the specialization-versus-diversity debate and bring into the discussion key insights derived from recent European research on the importance of related variety. A key insight is that economic change occurs as entrepreneurs take knowledge from their own and their firm's path-dependent evolution in one sector and finds ways, in combination with network partners from related but distinctive industry clusters, to form a new or emergent cluster built from these knowledge convergences. Thus, the presence of related variety speeds up the lateral absorptive capacity between neighbouring sectors and facilitates the process of innovation through knowledge spillovers across related sectors. Cooke goes on to argue that the policy implication of this perspective is that regional development "platforms" need to be facilitated to reproduce these conditions. He concludes that a rebalancing of urban economies towards new kinds of manufacturing requires both creativity and innovation, in a range of new and emerging industries, where cities as Schumpeterian hubs may play a key role through a judicious mix of governance and policies.

The Dynamics of Innovation and Creativity in Cities: Final Observations

The case studies gathered together in this volume provide a rich body of research into the nature of the innovation process in different-sized cities spread across the diverse regions of the country. They provide detailed evidence on the relative virtues of economic specialization versus diversity for economic growth in cities of different sizes; a discussion of where Canadian cities are fitting into the emerging hierarchy of global cities around the world with ever more differentiated economic roles; the relative importance of greater concentrations of highly skilled and creative workers as attractors for firms and industries; and the potential for Canadian cities to emerge as "Schumpeterian hubs" in the cognitive-cultural economy. While there is considerable variation in the innovation processes that we have uncovered in Canadian cities, some interesting patterns do emerge.

First, the evidence from the case studies confirms that cities of different sizes play different roles in the broader urban system of both their own national economy and the global economy. Larger cities with a more diverse range of industries and more extensive research infrastructure are frequently, though not exclusively, the locations where new ideas leading to new products and industries are developed. Medium-sized cities with traditional cost advantages have tended to be the home to more manufacturing and distribution industries, although there is a strong indication that many of these are moving to more innovative and knowledge-intensive activities. The challenge of making this transition remains much greater for the smaller cities in the Canadian urban system, especially those which lack an indigenous or fully developed research infrastructure.

Second, the spread of a globalized economy has also led to the concentration of higher-order financial and business services in larger cities around the world, but a clear division has emerged between those cities that operate on a truly global scale and those that act as national or regional hubs for their hinterland economies. At the same time, there is contrasting evidence of the extent to which different cities across the country are generating greater concentrations of both the knowledge-intensive business services and the design-intensive creative industries that are most closely associated with the cognitive-cultural economy.

Finally, what is clear is the growing importance of knowledge-based activities, design, and cultural industries to future economic growth across virtually all sectors in cities of all sizes. While the evidence suggests that large cities have certain advantages in this respect, the successful development of new dynamic regions in the United States and elsewhere, with respect to both the production of new technologies and cultural activities, suggests that is not a foregone conclusion. This trend poses a serious challenge for small and medium-sized cities that have developed historically as more specialized manufacturing centres. Addressing that challenge will undoubtedly require major policy support across all levels of government in Canada.

NOTE

1 Statistics Canada defines a census metropolitan area as an urban core with a population of at least 100,000 and adjacent urban or rural areas that have a high level of economic and social integration with the core.

REFERENCES

Audretsch, D.B. 2002. "The innovative advantage of US cities." *European Planning Studies* 10 (2): 165–76. http://dx.doi.org/10.1080/09654310120114472.

Beaudry, C., and A. Schiffauerova. 2009. "Who's right, Marshall or Jacobs? The localization versus urbanization debate." *Research Policy* 38 (2): 318–37. http://dx.doi.org/10.1016/j.respol.2008.11.010.

Beckstead, D., W.M. Brown, and G. Gellatly. 2008. *Cities and growth: The left brain of North American Cities: Scientists and engineers and urban growth.* Catalogue no. 11-622-MIE, no. 017. Occasional paper. Ottawa: Statistics Canada.

Bourne, L.S., C. Brunelle, M. Polèse, and J. Simmons. 2011. "Growth and change in the Canadian urban system." In *Canadian urban regions: Trajectories of growth and change,* ed. L.S. Bourne, T. Hutton, R.G. Shearmur, and J. Simmons, 43–80. Toronto: Oxford University Press.

Bourne, L.S., T. Hutton, R. Shearmur, and J. Simmons. 2011. "Introduction and overview: Growth and change in Canadian cities." In *Canadian urban regions: Trajectories of growth and change,* ed. L.S. Bourne, T. Hutton, R.G. Shearmur, and J. Simmons, 2–18. Toronto: Oxford University Press.

Bourne, L.S., and J. Simmons. 2003. "New fault line? Recent trends in the Canadian urban system and their implications for planning and public policy." *Canadian Journal of Urban Research* 12: 22–47.

Brender, N., M. Cappe, and A. Golden. 2007. *Mission possible: Successful Canadian cities.* The Canada Project Final Report. Ottawa: Conference Board of Canada.

Brezis, E.S., and P. Krugman. 1997. "Technology and the life cycle of cities." *Journal of Economic Growth* 2 (4): 369–83. http://dx.doi.org/10.1023/A:1009754704364.

Carlino, G.A., S. Chatterjee, and R.M. Hunt. 2007. "Urban density and the rate of invention." *Journal of Urban Economics* 61 (3): 389–419. http://dx.doi.org/10.1016/j.jue.2006.08.003.

Cooke, P. 2007. "Regional innovation, entrepreneurship and talent systems." *International Journal of Entrepreneurship and Innovation Management* 7 (2/3/4/5): 117–39. http://dx.doi.org/10.1504/IJEIM.2007.012878.

Duranton, G., and D. Puga. 2000. "Diversity and specialization in cities: Why, where and when does it matter?" *Urban Studies* (Edinburgh, Scotland) 37 (3): 533–55. http://dx.doi.org/10.1080/0042098002104.

Filion, P. 2010. "Growth and decline in the Canadian urban system: The impact of emerging economic, policy and demographic trends." *GeoJournal* 75 (6): 517–38. http://dx.doi.org/10.1007/s10708-009-9275-8.

Florida, R. 2002. *The rise of the creative class: And how it's transforming work, leisure, community and everyday life.* New York: Basis Books.

Glaeser, E.L. 2011. *Triumph of the city: How our greatest invention makes us richer, smarter, greener, healthier, and happier.* New York: Penguin Press.

Glaeser, E.L., and J.D. Gottlieb. 2006. "Urban resurgence and the consumer city." *Urban Studies* (Edinburgh, Scotland) 43 (8): 1275–99. http://dx.doi.org/10.1080/00420980600775683.

Hall, P. 1966. *The world cities*. London: Weidenfeld and Nicolson.

Hall, P. 1998. *Cities in civilization*. New York: Pantheon Books.

Hutton, T.A. 2008. *The new economy of the inner city*. London, New York: Routledge.

Hymer, S. 1972. "The multinational corporation and the law of uneven development." In *Economics and world order from the 1970s to the 1990s*, ed. J.N. Bhagwati, 113–140. London: Collier-Macmillan.

Jacobs, J. 1969. *The economy of cities*. New York: Random House.

Jacobs, J. 1984. *Cities and the wealth of nations: Principles of economic life*. New York: Random House.

Krugman, P. 1991. *Geography and trade*. Cambridge, MA: MIT Press.

Lundvall, B.-Å. 1992. "Introduction." In *National systems of innovation: Towards a theory of innovation and interactive learning*, ed. B.-Å. Lundvall, 1–19. London: Pinter Publishers.

Martin, R., and J. Simmie. 2008. "Path dependence and local innovation systems in city-regions." *Innovation: Management, Policy & Practice* 10 (2–3): 183–96. http://dx.doi.org/10.5172/impp.453.10.2-3.183.

McCann, P. 2008. "Globalization and economic geography: The world is curved, not flat." *Cambridge Journal of Regions, Economy and Society* 1 (3): 351–70. http://dx.doi.org/10.1093/cjres/rsn002.

Montgomery, J. 2007. *The new wealth of cities: City dynamics and the fifth wave*. Burlington, VT: Ashgate.

Moretti, Enrico. 2012. *The new geography of jobs*. Boston: Houghton Mifflin Harcourt.

OECD. 2006. *Territorial reviews: Competitive cities in the global economy*. Paris: Organisation for Economic Co-operation and Development.

Rodríguez-Pose, A. 2008. "The rise of the 'city-region' concept and its development policy implications." *European Planning Studies* 16 (8): 1025–46. http://dx.doi.org/10.1080/09654310802315567.

Safford, S. 2004. *Searching for Silicon Valley in the Rustbelt: The evolution of knowledge networks in Akron and Rochester*. Working paper. Cambridge, MA: MIT Industrial Performance Center.

Sassen, S. 2001. *The global city: New York, London, Tokyo*. Princeton, NJ: Princeton University Press.

Scott, A. 2006. "Entrepreneurship, innovation and industrial development: Geography and the creative field revisited." *Small Business Economics* 26 (1): 1–24. http://dx.doi.org/10.1007/s11187-004-6493-9.

Scott, A.J. 2007. "Capitalism and urbanization in a new key? The cognitive-cultural dimension." *Social Forces* 85 (4): 1465–82. http://dx.doi.org/10.1353/sof.2007.0078.

Scott, A.J. 2008. *Social economy of the metropolis: Cognitive-cultural capitalism and the global resurgence of cities*. Oxford, New York: Oxford University Press. http://dx.doi.org/10.1093/acprof:oso/9780199549306.001.0001.

Scott, A.J. 2012. "A world in emergence: Notes towards a resynthesis of urban-economic geography for the twenty-first century." In *Reframing Regional Development: Evolution, Innovation and Transition*, ed. Philip Cooke, 29–53. London, New York: Routledge.

Simmie, J. 2002. "Knowledge spillovers and reasons for the concentration of innovative SMEs." *Urban Studies* (Edinburgh, Scotland) 39 (5–6): 885–902. http://dx.doi.org/10.1080/00420980220128363.

Simmie, J. 2003. "Innovation and urban regions as national and international nodes for the transfer and sharing of knowledge." *Regional Studies* 37 (6–7): 607–20. http://dx.doi.org/10.1080/0034340032000108714.

Simmie, J., and P. Wood. 2002. "Innovation and competitive cities in the global economy: Introduction to the special issue." *European Planning Studies* 10 (2): 149–51. http://dx.doi.org/10.1080/09654310120114454.

Storper, M., and M. Manville. 2006. "Behaviour, preferences and cities: Urban theory and urban resurgence." *Urban Studies* (Edinburgh, Scotland) 43 (8): 1247–74. http://dx.doi.org/10.1080/00420980600775642.

Storper, M., and A.J. Scott. 2009. "Rethinking human capital, creativity and urban growth." *Journal of Economic Geography* 9 (1): 147–67.

Veltz, P. 2004. "The rationale for a resurgence in the major cities of advanced economies." Opening plenary address to Levehulme International Symposium on The Resurgent City. London School of Economics, London.

Venables, Anthony J. 2006. "Shifts in economic geography and their causes." Economic Review, Federal Reserve Bank of Kansas City, 4th quarter: 61–85.

Wolfe, D.A. 2010. "The strategic management of core cities: Path dependence and economic adjustment in resilient regions." *Cambridge Journal of Regions, Economy and Society* 3 (1): 139–52.

Wolfe, D.A., and A. Bramwell. 2008. "Innovation, creativity and governance: Social dynamics of economic performance in city-regions." *Innovation: Management, Policy & Practice* 10 (2–3): 170–82.

2 Systems of Innovation and Contexts of Creativity: An Assessment of the Knowledge Bases of Canadian City-Regions

GREGORY M. SPENCER

Models of local and regional economic development have generally evolved from neo-classically based cost reduction strategies to high value-added knowledge-based strategies. This typically involves identifying local strengths and building on them through cooperative governance models that seek to transcend divisions between competing firms and municipalities in order to build collective solutions that benefit the local economy as a whole. In this regard a tricky balance must be struck between maintaining existing strengths while remaining open to new opportunities. Jurisdictions that fail in this respect can suffer decline from being "locked in" to a particular technology or industry and as a result are unable to adapt to changing global realities, or they can decline from being able to identify and build on any discernible strength. In short, places must be able to have specializations while excelling at innovation. This paper presents a theoretical argument about the role of local industrial structure in maintaining a balance between "lock-in" and a lack of specialization and then provides data analysis on Canadian city-regions in order to support the hypothesis.

The theoretical reasoning of this paper seeks to connect three main suppositions in order to present a systematic accounting of the role of local industrial structure in knowledge production processes and ultimately economic outcomes. The first supposition is that knowledge production is a social process and not simply about individual acumen. While there are no doubt people with special abilities they are only able to use the full extent of their talent if they are embedded in supportive and stimulating environments full of other individuals who share, circulate, and evaluate knowledge and ideas. The second supposition is that as there are different types of knowledge there are also different types of social processes that produce them. The epistemology of arts and cultural knowledge differs from that of the natural sciences and mathematics, and thus artistic creativity, technological innovation, and scientific discovery each have distinctive characteristics. The third supposition is that as there are different types of knowledge production processes, the environments in which they thrive also have specific qualities. In linking these ideas together literatures from the social psychology of creativity and social network analysis are injected into the economic geography literature in an attempt to specify details of the social dynamics involved in knowledge production processes and how they relate to regional economies. Ultimately, it is argued that traditional debates around local diversity versus specialization are overly binary and simplistic in their framing, and require a much more nuanced and differentiated set of questions if there is to be further progress made in theory development and policy implementation. Basic descriptive data on Canadian city-regions is presented in order to bolster this case.

Knowledge Production as a Social Process

Tales of heroic creativity and innovation are often framed around a cult of personality. There are many stories of lone geniuses toiling away by themselves in their garage or college dorm room and inventing the next great thing virtually by themselves. These stories, however, sell short all of the many people and institutions that assisted and influenced those individuals along the way. We know that the production of knowledge and learning in general is fundamentally a social process (Foray and Lundvall 1996) rather than one that is primarily about the special intellect of particular people. The notion that economic activity, and especially knowledge-driven economic activity, is embedded in social relations (Granovetter 1985) has been a major focus of social science literature over the past three decades. It has also featured in the economic geography literature vis-à-vis the "relational turn" (Bathelt and Gluckler 2003; Boggs and Rantisi 2003) and conceptions of "local embeddedness" (Gertler 2003). The main shift in perspective is that instead of primarily observing the traits and characteristics of individuals and groups, the emphasis is to try and understand the patterns of relations *between* individuals and groups.

This theoretical framework has been supported by rapid methodological developments in social network analysis (SNA). Specifically, the "strength of weak ties" (Granovetter 1973), "structural holes" (Burt

1992), and "networks of learning" (Powell et al. 1996) are leading examples of applications of SNA in the economic sociology literature that have had significant impacts across the social sciences. These works exemplify how a systematic understanding of the patterns of relationships between individuals and other actors is an essential step in understanding how knowledge is produced in the context of economic activity. A consistent finding of these works involves how previously unconnected ideas are brought together via the bridging of previously unconnected people, groups, or organizations. One of the central tenets of knowledge production theory is that new ideas are produced from the formation of novel connections between existing sets of knowledge (Jacobs 1969; Weitzman 1998). This, however, is one of the major barriers to knowledge production, as social networks have a tendency to be self-reinforcing and exclusionary (Christopherson and Clark 2007). The sociological term for this is "homophily" and is explained as follows by MacPherson, Smith-Lovin, and Cook (2001):

> Contact between similar people occurs at a higher rate than among dissimilar people. The pervasive fact of homophily means that cultural behavioral, genetic, or material information that flows through networks will tend to be localized. Homophily implies that distance in terms of social characteristics translates into network distance, the number of relationships through which a piece of information must travel to connect two individuals. It also implies that any social entity that depends to a substantial degree on networks for its transmission will tend to be localized in social space and will obey certain fundamental dynamics as it interacts with other social entities in an ecology of social forms.

The concept of homophily presents a dilemma in that it represents a pervasive constraint on the bridging of divergent networks which is an essential component of the knowledge production process. Organizational and institutional factors are an important element in this respect as they can affect patterns of relationships, but also increase trust and reduce the risks associated with making unfamiliar connections. Macro structural factors are also influential, as contexts that contain greater amounts of diversity also offer a greater range of bridging opportunities (Spencer 2012).

While structural factors that affect rates of knowledge production are of vital importance, this should not be interpreted as a complete dismissal of the role of agency. Ultimately, new ideas are formed in the minds of individuals, and so gaining an understanding as to what drives people to produce knowledge is imperative. On this score, the social psychology literature on creativity not only tackles questions of agency but does so with the wider social environment in mind. Notably, Amabile's (1996) *Creativity in context* is one of the key pieces of literature on the subject of situating human creativity within workplace settings. Her insights into motivation are particularly revealing, as she finds string evidence to suggest that the desire to produce new ideas is inherently intrinsic. In other words, people are not generally motivated by external pressures via the offers of rewards or by threat of punishment. Instead, Amabile argues convincingly that certain environmental conditions must be produced that enable people to produce knowledge. Specifically, this means ensuring that environments are open and tolerant and that there is a free flow of information between actors.

Others in the social psychology field have also written extensively on similar tracks. Gardner (1993), Csikszentmihalyi (1996), and Sternberg and Lubart (1999) all discuss the importance of social context to the creative process. One aspect that this literature covers in greater detail is the notion that the social process of knowledge production is as much about how new knowledge is evaluated by networks and communities as it is about providing inputs. This literature has stressed that new ideas are only truly "creative" if they provide some sense of value. Thus, knowledge production does not end with the formulation of some novel insight, but rather this represents a midpoint of sorts.[1] The social evaluation of novelty is fraught with risk. As Schumpeter (1976) pointed out, new creations tend to displace and destroy existing beliefs and assumptions; there will always be some who are not ready to let go of existing norms and so will resist change. Research on the creativity in small group settings demonstrates how the social mix can have a significant impact on such an outcome. For example, more diversity/less conformity in groups tends to reduce the social risks associated with making novel suggestions (De Dreu and West 2001). Conversely, higher degrees of social uniformity tend to increase the prevalence of "groupthink" and the pressure to conform (Nemeth and Nemeth-Brown 2003).

A wide variety of literatures on knowledge production highlight various aspects of the social dynamics of the process. They also tend to stress the importance of context and explain ways in which it influences levels of creativity. In particular, environments that offer a wide array of possible inputs in the form of divergent knowledge sets are demonstrated to be more effective at enabling people to generate novel ideas. Furthermore, contexts that mitigate the social risks associated with creativity are also beneficial to the process. A great deal of the research that supports these theories, especially in sociology and social psychology, is centred on small groups, typically within either firms or laboratory settings, while not explicitly accounting for the influences of larger contextual settings. Conversely, the economic geography literature takes into account the impacts of structural variations of the larger environment in the production of knowledge, but does a poorer job of providing a systematic description of social processes. The following section attempts to reconcile these two different bodies of literature.

Knowledge Production as an Embedded Local Process

The aim of the preceding section is to firmly establish knowledge production as a fundamentally social one. The aim of this section is to establish that the production of knowledge is heavily localized. At first inspection, the question of establishing a "geography of knowledge production" may seem like a fool's errand, as rapidly evolving information and communications technologies have drastically altered how people interact with one another. Location, it may appear, has lost its relevance. To wit, distance has been declared "dead" (Cairncross 1997) and the world "flat" (Friedman 2005) by some observers. This hypothesis has elicited a spirited response from geographers (Martin and Sunley 1998; Morgan 2004) who counter that information and communications technologies have inherent limitations and therefore physical proximity will continue to matter. This position has typically been framed by the tacit versus codified knowledge paradigm. The basic argument is that information and communications technologies are effective at facilitating the transmission of codified knowledge, but are less effective when it comes to tacit knowledge. For this reason the importance of face-to-face contact has been emphasized when examining the transmission of tacit knowledge (Storper and Venables 2004).

The sociology and social psychology literatures on knowledge production tend to be situated within organizational frameworks which act to constrain the boundaries of networks. A sense of geography is either implicit or absent in these studies. The geography literature on the other hand is explicit about spatial constraints, but often lacks comprehensive data on social interaction and networks that would allow for conclusive and systematic findings about the physical locations of relationships. Instead, geography has a tradition of using proxy measures in order to describe and define socio-spatial phenomena. One highly relevant example of this is the use of commuting-pattern data in order to identify the functional boundaries of a city-region. With the knowledge of where people live and where they work, statistical agencies in both Canada and the United States use thresholds to define which municipalities overlap to such a degree that they can be considered a single unit of economic activity based on the typical day-to-day movements of their constituent populations. Researchers performing regional analysis then typically adopt these definitions as "labour market areas" that act as containers of locally transmitted tacit knowledge that (re)produces local culture and practices. While it is possible to produce case studies that map out local networks, collecting multiple region-wide comprehensive data on the spatial nature of social interaction is too daunting to undertake due to reasons of scale and complexity. Instead, what is left is the tradition of using measures of association between observed phenomena (i.e., knowledge production) and traits of regions (i.e., educational attainment) in order to make empirically driven assumptions about the role of place in the production of knowledge.

This raises the question of whether or not the theoretical aspects of the sociology and social psychology literatures can be applied on a regional scale and how various concepts may be operationalized. The theme that may be most congruent is the notion that diverse groups and the ability to bridge them are key factors in the creative process. From a theoretical perspective, the presence of localized diversity should provide two general benefits to the creative process: one is that a wider variety of knowledge present exponentially increases the possible number of novel combinations; and the second is that novel ideas are less likely to be rejected due to groupthink. In short, there should more potential and fewer risks in diverse places. This is essentially the argument put forth by Jacobs (1969) and supported by a host of others (Weitzman 1998; Desrochers 2001), that diversity begets creativity which ultimately drives growth. The counterview, mainly associated with the works of Marshall (1890), Arrow (1962), and Romer (1986), is that greater specialization is a more effective local structure in generating knowledge-based growth. In this case, the argument is centred on efficiency, both in terms of ability to communicate and the focusing of shared infrastructure. This fits with the idea of "homophily" in that familiarity breeds similarity, which improves the flow of tacit knowledge. In addition, a high degree of related activity requires similar infrastructure, both hard and soft, thus providing that these become more efficient. The debate over the relative merits of local diversity versus specialization has existed for at least a century and is showing few signs of abating. There is strong evidence from meta-analysis that much of the reason behind mixed empirical results is mainly due to inconsistent methods that vary by length of study, how diversity is measured, and the choice of dependent variables (Beaudry and Schiffauerova 2009). More recently, a middle ground has been proposed with the concept of "related variety" (Frenken et al. 2007). This view incorporates elements from each side of the debate by proposing that possessing a diverse set of similar industries is the most effective industrial structure for growth. A dose of relatedness aids communication and efficiency, while a degree of variety boosts dynamism and reduces risk.

The general lack of consensus on the local diversity versus specialization debate can be ascribed to not only methodological issues but conceptual ones as well – this is where the social psychology and SNA literatures can clarify some of the issues present in the economic geography literature that are generally characterized as a lack of specificity. The three main areas in which these literatures may help are: (a) a broadening of the concept of diversity from industrial diversity to cognitive diversity; (b) a focus on actual rather than implied bridging relationships and interaction; and (c) a differentiation of knowledge production processes along an arts and culture/natural science spectrum. In return, the (economic) geography literature can instil a sense of location and proximity into these other conceptual approaches to provide a sense of why place matters to the production of knowledge. On the first point, an issue with the typical economic geography approach to operationalizing the concept of diversity is that it almost always relies on a narrow definition based on industrial classification. The social psychology and sociology literatures are not similarly bounded and offer greater detail and specificity when speaking to the importance of "cognitive" diversity. With the second point, the main issue is the use of proxy measures in regional analysis that assume underlying social processes without actual observation. Specifically, levels of diversity are employed without any actual observations of flows of knowledge between groups (read industries). While the first two issues are fairly well acknowledged, the third point of contention is only beginning to be addressed in the geography literature. The recognition that there are different types of knowledge and that producing them involves somewhat different social process is the focus of the remainder of this chapter.

Different Types of Knowledge Production (Processes)

Studying knowledge production at the regional scale necessitates a certain amount of generalization in order to present functional arguments. The danger is that these generalizations can go too far and skim over important details that undermine the analysis. Many of the assumptions built into the local diversity versus specialization debate are particularly vulnerable in this sense, especially when it comes down to the social dynamics of the production of knowledge. A recent exception to this in the economic geography literature is the work on knowledge bases (Asheim and Gertler 2005; Asheim and Hansen 2009). This body of literature identifies three distinct types of knowledge: symbolic, synthetic, and analytic. Symbolic knowledge is associated with the arts and culture, synthetic knowledge is linked to technology and engineering, while analytic knowledge is connected to the natural sciences. From a research perspective these knowledge types are mapped onto particular industries in relation to the core products that they produce (and the knowledge required to produce them). As industries are unevenly distributed among places, regions can be characterized as having a particular knowledge base depending on the local mix of economic activity. This matters from a policy perspective insofar as different knowledge bases involve different processes and subsequent institutional support.

While the knowledge-base literature is still somewhat fresh in the economic geography literature, differentiating between types of knowledge production has a much longer and richer tradition in the social psychology literature on creativity. There is general agreement that artistic creativity differs from the production of scientific knowledge. These differences relate in some way to either how new ideas are arrived at, or how new ideas are evaluated by a community. Novelty and value form the basic building blocks of the definition of creativity. For instance, Sternberg and Lubart (1999) define creativity as "the ability to produce work that is both novel (i.e., original, unexpected) and appropriate (i.e., useful, adaptive concerning task restraints)." Similarly, Unsworth (2001) states that the usual definition of creativity is akin to "the production of novel ideas that are useful and appropriate to the situation." Other terms and phrases can be substituted for novelty and value, but the core ideas remain fairly constant. Creativity is a process of producing new knowledge that is deemed to have worth by some group or audience.

This definition is quite broad in scope in that it encapsulates all knowledge production. While this satisfies some, others take a more nuanced view in the belief that creativity is only one type of knowledge production. Santagata (2004), for instance, claims that creativity is "non-utilitarian" as opposed to innovation, which is driven by a logic of specific and measurable improvement. Sternberg and Lubart (1996) make a distinction between "open" and "closed" problem solving whereby "closed" problems have specific answers while "open" ones do not. Amabile (1996) makes a similar distinction between heuristic knowledge production and algorithmic knowledge production, in which the former can only be truly considered "creativity." The key difference is that heuristic knowledge production is based on the subjective values of both the creator and the evaluators, while algorithmic knowledge production is observable and repeatable by others who should come to the exact same conclusions. Little if any knowledge production can be purely subjective or objective, but these two ideal types can prove useful as ends of a conceptual spectrum of knowledge production. There are some parallels between the heuristic/algorithmic dichotomy and the tacit/codified distinctions (Dosi 1988; Gertler 2003; Amin and Cohendet 2004) that are more familiar in the innovation literature. By definition, tacit knowledge cannot be articulated in a formulaic manner, while codified knowledge is precisely that. These parallels are particularly useful when it comes to introducing physical proximity to the equation in terms of how well types of knowledge "travel."

There have been many models developed that seek to describe the basic elements of creativity. Santagata (2004) proposes that one of the key distinctions between the creative process and science-based innovation is that the former mode of knowledge production is "non-cumulative" in nature. This suggests that while all knowledge production can be thought of as path dependent in some way, different types may follow different patterns.

While there are many examples in the literature as to how the production of arts and culturally based knowledge differs from that of science based knowledge, there are no definitive models. The three knowledge-base typologies, symbolic, synthetic, and analytic, offer the clearest way forward, but can be aided by elements from the social psychology and SNA literatures, particularly in terms of how they may relate to the local diversity versus specialization debate. These typologies are linked to specific industries according to the core products that they produce and the knowledge that is required to produce them. The underlying social dynamics of each type may differ due to the nature of the subject matter. Symbolic knowledge is associated with the artistic creativity end of the spectrum wherein the subject matter (and related products) is based on human experience. One could say that human interaction and how people relate to one another *is* the fundamental core of symbolic knowledge. Analytic knowledge is located at the other end of the spectrum in that it is an attempt to understand the material world. While there is certainly a degree of human subjectivity involved which is limited by our own physical nature and ability to sense the world around us, there remains a certain amount to quantify and formally model what we perceive that can be observed and replicated in the same way in a culturally neutral manner. The key message is that there are some fundamental differences between types of knowledge and, more to the point, the social processes that generate them. Synthetic knowledge is the third type and can be imagined as occupying space on the spectrum between symbolic and analytic knowledge. In this sense, synthetic knowledge is related to technological innovation, which requires both an understanding of the material world as well as knowledge of the human experience. In other words, technology can be described as applying knowledge of the material world in a way that betters the human experience.

When thought of in terms of social process the three types can be associated with artistic creativity (symbolic), technological innovation (synthetic), and scientific discovery (analytic) (figure 2.1). These connections bring the social psychology and SNA literatures on knowledge production in line with the economic geography literature on knowledge bases. This is an important step, especially insofar as these processes relate to the wider social environment. For example, artistic creativity draws inputs from social activity for novel ideas and interpretations and also relies heavily on the social arena for the evaluation of novel ideas. Scientific discovery differs somewhat in this respect, in that knowledge at the frontiers of enquiry is undertaken by groups of highly specialized experts who typically learn from a much narrower segment of society. These differences suggest that local diversity may be more important to artistic creativity, while specialization may be more important to scientific discovery. In addition, the fundamental tacit nature of symbolic knowledge suggests that it may not travel as well as analytic knowledge, and thus the wider local social environment may matter more to the creative process than it does to the

Figure 2.1 Three types of knowledge production

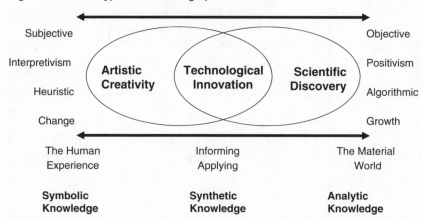

scientific discovery process. While this chapter does not seek to fully answer these questions, the remainder of the discussion is devoted to laying out a basic description of the knowledge-base landscape of city-regions in Canada and what it may mean for further enquiry and possible policy interventions.

Knowledge Bases of Canadian City-regions

The goal here is to develop a systematic method for assessing what the knowledge bases of various city-regions are, and to find common traits and characteristics that are associated with each type. The potential benefit of such an approach is the ability to further understand the knowledge production processes within city-regions and thereby be able to build more focused policy initiatives and more effectively allocate public resources.

Standard classification systems of industrial activity such as the North American Industrial Classification System (NAICS) and the Standard Industrial Classification (SIC) system are organized around the types of goods and services that firms produce. While to some extent they can reflect the underlying knowledge bases that are applied within the day-to-day operations of companies, they do not fully capture the extent of the knowledge production processes that are occurring. Examining the constituent workforces of firms and industries gets us somewhat closer to this end goal. A common approach (e.g., Florida

Figure 2.2 Classifying regions according to knowledge base

		Subject area of post-secondary qualifications		
		Fine Arts, Humanities & Social Sciences	Engineering & Applied Sciences	Natural Sciences & Mathematics
Per cent of labour force with post-secondary qualifications	Above national average	Symbolic Knowledge	Synthetic Knowledge	Analytic Knowledge
	Below national average	N/A	N/A	N/A

2002) in this regard is to analyse occupational structures using occupational classification systems such as the Standard Occupation Classification (SOC) system or National Occupational Classification (NOC) system, which categorize people by the jobs that they perform. Assessing the specific functions that individuals perform moves us one step closer to mapping the knowledge production systems in firms, industries, and entire city-regions. Cross-referencing occupations by industry gets us even nearer. An additional mode of measuring knowledge production is examining the educational attainment of the labour force (Glaeser 2003). While this can help illuminate the amount of knowledge production present in any given unit of study, it does not say anything about the types of knowledge being produced. Each of these traditional methods of quantifying knowledge and knowledge production is useful in its own way, but they do not directly gauge the concept of knowledge bases as articulated by Asheim and Gertler (2005).

The remedy proposed here is to incorporate the field of study of higher education qualifications into the analysis (figure 2.2). We divide educational learning into types of knowledge (subjects), each with distinct pedagogical characteristics, and so it makes sense that these same categories can be used to assess the knowledge bases of firms, industries, and city-regions. By cross-tabulating both the amount of post-secondary qualifications and the fields of study in which the degrees were granted, we can more closely assess the relative knowledge bases at various scales of analysis. Certainly there is a long lag time between the education of some workers and their current employment, but in aggregate this measure can provide a generalized account as to which types of knowledge are desired by firms and industries.

Measuring the Knowledge Bases of City-Regions

Most measurements of human capital in regions tend to assess the amount of knowledge while not saying very much about the types of knowledge present. The Canadian Census of Population provides a detailed picture not only of the number of individuals with post-secondary education qualifications, but also of the specific field of study in which the qualifications were obtained. This information can be used to generate a more specific assessment of the knowledge characteristics of city-regions. Fields of study can be aggregated in order to represent the concepts of knowledge bases (symbolic, synthetic, and analytic) as well as the types of knowledge production (artistic creativity, technological innovation, and scientific discovery). To this end the basic categories of field of study used by Statistics Canada are assigned to one of each of the three types of knowledge production. Arts, humanities, and social sciences are associated with creativity as they deal primarily with human experience and more subjective forms of knowledge production and learning. Engineering and applied sciences are related to technological innovation as they require both an understanding of physical science as well as of human needs. Mathematics and natural sciences are placed with the discovery form of knowledge production as they are mainly based on more objective forms of learning. Degrees in education, health care, and protective services are omitted as they are primarily held by workers in the public sector. Because the goal here is to better understand the geographic patterns of private-sector economic activities, these workers are not included in the analysis. Location quotients (LQ) are calculated for each of the three categories of individuals for all census metropolitan areas (CMAs) and census agglomerations (CAs) in Canada. The location quotients show where there is a higher

Figure 2.3 Distribution of Canadian city-regions by knowledge base

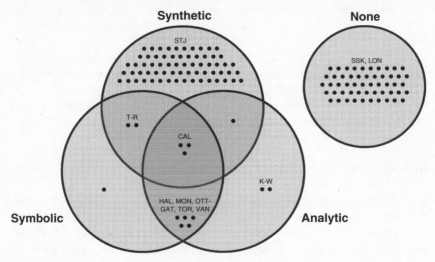

than expected concentration of adults with post-secondary qualifications in each category of the fields of study. When an LQ is greater than one, it is interpreted to mean that a city-region has a significant knowledge base of a particular type. It is of course possible that a city-region can have multiple types of knowledge bases, or otherwise none at all. Figure 2.3 shows the distribution of 144 city-regions according to their LQs across three types of knowledge production.

This distribution shows that there are, broadly speaking, three types of city-regions that are prevalent according to the knowledge production structures. City-regions in Canada with strong knowledge bases have an abundance of people trained in engineering and applied sciences (synthetic) related fields ($N = 67$) or they have some combination of all three types of knowledge production ($N = 18$). City-regions with only a symbolic knowledge base ($N = 1$) or only an analytic knowledge base ($N = 3$) are so rare that they can be considered outliers. There are also a large number ($N = 55$) of city-regions that do not have an above-average number of people who are trained in any of the three knowledge categories. Also indicated on figure 2.3 are the case studies (by city) presented in subsequent chapters of this volume. The cases cover most of the spectrum, although they most commonly represent the symbolic-analytic typology found in larger urban regions.

Table 2.1 Characteristics of city-regions according to knowledge base

	N	Average no. in labour force, 2005	CAGR, 2000–5 (%)	Average employment income
Multiple bases	18	702,623	2.1	$39,574
Synthetic base	67	50,922	2.5	$36,008
No base	55	66,110	2.2	$33,729

Source: Statistics Canada, Census of Population 2001, 2006

The Geography of Knowledge Production in Canadian City-Regions

The three broad types of city-regions display widely divergence traits and outcomes (see table 2.1). Understanding these characteristics is essential to providing effective policy interventions derived from an awareness of the local context. While the synthetic knowledge–driven city-regions and the city-regions which lack a clear knowledge driver are much greater in number, the eighteen city-regions with multiple knowledge bases represent almost twice the population. This reaffirms the connection made between scale and diversity as noted by Beckstead and Brown (2003). The converse is that smaller city-regions tend to be either specialized (almost exclusively in technological innovation) or do not possess a clear critical mass of any form of knowledge production. Both highly creative economic activities and ones based on scientific discovery are mainly concentrated in large urban centres, although the corresponding rationales for these phenomena may differ. As Spencer (2011) shows, localized cognitive diversity is a strong predictor of the location of creative economic activity. As the knowledge production processes differ with scientific discovery, the same environmental influences are not likely the cause of the co-location with creative endeavours. More probable is the spatial concentration of science-producing institutions such as universities, research hospitals, and government laboratories in large urban areas. People with technology- and engineering-related qualifications are more prevalent in smaller Canadian city-regions, as they are more likely to work in industries such as manufacturing, construction, and resource extraction, which are more prevalent in smaller city-regions. This distribution is consistent with the argument made in the introduction to this volume.

Figure 2.4 Industrial structure by knowledge base of region

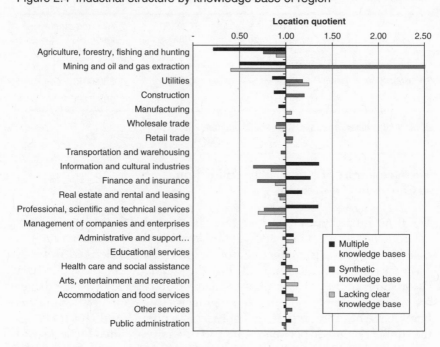

The city-regions with multiple knowledge bases have the highest average annual employment incomes, at $39,574. The city-regions specialized in innovation have an average income roughly $3500 lower, while the places without a clear knowledge base have average incomes over $2000, lower still. In terms of compound annual growth rates of employment, the innovative-driven regions show the highest level between 2001 and 2006, at 2.5 per cent, while the diverse regions and those lacking a clear knowledge base have rates of 2.1 per cent and 2.2 per cent, respectively. Although the synthetic knowledge–driven regions have a higher growth rate, it must be noted that in terms of the overall amount, growth is mainly found in the diverse regions due to their much larger size.

The key difference between the diverse regions with multiple knowledge bases, those with a synthetic knowledge base, and those regions lacking a clear knowledge base, is the local industrial structure. Figure 2.4 shows the location quotients for employment by industry (2-digit NAICS) by groups of regions. A clear pattern in the specializations of these types of places is evident whereby the diverse regions are heavily endowed with producer services ranging from information and cultural industries to professional, scientific, and technical services. The synthetic knowledge regions are specialized in mining, oil, and gas, utilities, and construction, while the industrial structure of regions lacking a significant knowledge base are skewed towards education, health, and consumer services. These patterns offer a great deal of explanation to the relative income disparities noted in table 2.1, as the types of economic activities found in each generally provide divergent wage levels.

While the reasons that drive industrial location patterns are complex and varied, there is a clear relationship between large, diverse city-regions and high concentrations of economic activities that rely on symbolic and analytic knowledge. In addition, economic activities that are driven by synthetic knowledge tend to be focused in smaller, more homogeneous city-regions. This conforms to the basic theoretical suppositions that (a) diversity increases creativity, as there are more possible combinations of divergent knowledge clustered in close proximity; and (b) specialization fosters innovation via efficient communication and infrastructures. This suggests that framing questions around the notion of diversity versus specialization is a poor starting point, as they both offer different types of regional advantages, albeit to different types of knowledge production, and subsequently to different types of economic activity. Instead, regional policymaking should start with the recognition that certain types of knowledge production are better suited for certain types of local environments. More to the point, policymakers can help shape diverse local contexts in which the creative process can thrive, in which the main issues may be helping to bridge divergent groups. And within homogeneous environments, policymakers can help reinforce local systems of innovation that capture the advantages of specialization. With the latter cases it is important that policy tackles potential problems associated with "lock-in" and helps to build information pipelines to new and current knowledge production centres in other parts of the world (Bathelt et al. 2004).

Discussion and Conclusion

The bigger picture that emerges from these figures shows a significantly divergent pattern of regional economic activities and outcomes. The largest urban areas and regional centres tend to have diverse knowledge bases driven mainly by symbolic and analytic-based economic activities, which tend to produce high value-added services resulting in higher average incomes for the local labour force. A second type of city-region is characterized by smaller urban areas that are specialized in resource extraction, utilities, and construction. These regions have middling incomes and higher rates of growth. The third type of city-region is places that lack clear advantage in any type of knowledge production and are specialized in non-basic (non-exporting) economic activities such as education, health care, and consumer services. They tend to be smaller, with significantly lower average incomes. This seems to suggest a three-speed national economy whereby local economic structures have developed self-reinforcing trajectories. As Jacobs (1969) noted, diverse environments tend to produce more novelty, which leads to greater and greater divisions of labour, which by definition means greater diversity. The opposite has also been identified, often in cautionary terms, whereby (over)specialization can lead to "lock-in" (Arthur 1989) that limits diversity and ultimately can lead to economic decline. Efficiencies in terms of shared infrastructure and better communication are examples of positive outcomes of specialization that are sources of advantage. The third type of city-region poses a conundrum in that it begs the question as to what can be done by city-regions lacking a significant knowledge specialization in order to close the income gap with their more endowed counterparts.

There are policy-centric theories in the economic geography literature that offer possible pathways to more prosperous local outcomes. Florida (2002) espouses a labour-driven strategy whereby a highly talented "creative class" can be attracted to places that develop certain type of living environments. A problem with this approach, however, is that the very characteristics which Florida identifies as being important, such as open, tolerant societies, are found in greater abundance in the larger urban regions that already possess a creative advantage vis-à-vis higher levels of diversity and its related benefits. Storper and Scott (2009) are critical of labour-led economic development theories and contend that the broad patterns of industrial location are primarily the result of decisions taken at the level of the firm. Furthermore, Scott (2008) believes that the emergent "cognitive-cultural economy" is one that is mainly present in urban areas, and that they are co-constituted in that dense networks of relationships are essential to the functioning of the contemporary economic system. Whichever point of view one adopts in this debate, it is undeniable that both clearly include various·evolutionary processes. Though there has always been a degree of evolutionary thought within these literatures, the call by Boschma and Frenken (2006) has created a more formalized approach to developing an evolutionary economic geography. This has occurred roughly in parallel with the development of the knowledge-base literature led by Asheim and Gertler (2005). If we fuse these two bodies of literature together, a number of interesting and important questions arise. Principal among them is how have the knowledge bases of city-regions come to be?

Many of the Innovation Systems Research Network case studies presented in this volume address this question in greater detail, but more generally it is an essential question especially for the regions that do not exhibit a clear knowledge base of any kind. A great deal of literature and case studies exist on the more "successful" examples of industrial clusters and dynamic knowledge driven regions, but there are precious few exemplars of places that have gone from the bottom to the top of the knowledge economy league tables. What needs to be understood most is where the initial sparks come from that set regions on evolutionary trajectories that generate high value-added, knowledge-driven economic activity. From there we also need to understand how momentum is maintained, whether it be through either labour or firm action or, more likely, a system of innovation that includes both public- and private-sector partners. In order to identify such cases and understand them, it may be that we need to look specifically for anomalous places that defy our expectations. Regions that have been counted out and yet somehow managed to develop strong knowledge-driven economies when it appeared that they were on a path of terminal decline are most useful in this regard. Thus, the growing literature on regional resiliency has a lot to offer, especially to the policymaking community as economic development challenges continue to mount.

NOTE

1 It is important to keep in mind that knowledge production is not often a simply linear process, but involves a great deal of iteration.

REFERENCES

Amabile, T. 1996. *Creativity in context*. Boulder, CO: Westview Press.
Amin, A., and P. Cohendet. 2004. *Architectures of knowledge: Firms, capabilities and communities*. New York: Oxford University Press. http://dx.doi.org/10.1093/acprof:oso/9780199253326.001.0001.

Arrow, K.J. 1962. "The economic implications of learning by doing." *Review of Economic Studies* 29 (3): 155–73. http://dx.doi.org/10.2307/2295952.

Arthur, W.B. 1989. "Competing technologies, increasing returns, and lock-in by historical events." *Economic Journal* 99 (394): 116–31. http://dx.doi.org/10.2307/2234208.

Asheim, B.T., and M.S. Gertler. 2005. "The geography of innovation: Regional innovation systems." In *The Oxford handbook of innovation*, ed. J. Fagerberg, D.C. Mowery, and R.R. Nelson, 291–317. Oxford: Oxford University Press.

Asheim, B., and H.K. Hansen. 2009. "Knowledge bases, talents, and contexts: On the usefulness of the creative class approach in Sweden." *Economic Geography* 85 (4): 425–42. http://dx.doi.org/10.1111/j.1944-8287.2009.01051.x.

Bathelt, H., and J. Gluckler. 2003. "Toward a relational economic geography." *Journal of Economic Geography* 3 (2): 117–44. http://dx.doi.org/10.1093/jeg/3.2.117.

Bathelt, H., A. Malmberg, and A. Maskell. 2004. "Clusters and knowledge: Local buzz, global pipelines and the process of knowledge creation." *Progress in Human Geography* 28 (1): 31–56. http://dx.doi.org/10.1191/0309132504ph469oa.

Beaudry, C., and A. Schiffauerova. 2009. "Who's right, Marshall or Jacobs? The localization versus urbanization debate." *Research Policy* 38 (2): 318–37. http://dx.doi.org/10.1016/j.respol.2008.11.010.

Beckstead, D., and M. Brown. 2003. *From Labrador City to Toronto: The industrial diversity of Canadian cities, 1992–2002*. Ottawa: Statistics Canada.

Boggs, J.S., and N.M. Rantisi. 2003. "The 'relational turn' in economic geography." *Journal of Economic Geography* 3 (2): 109–16. http://dx.doi.org/10.1093/jeg/3.2.109.

Boschma, R.A., and K. Frenken. 2006. "Why is economic geography not an evolutionary science? Towards an evolutionary economic geography." *Journal of Economic Geography* 6 (3): 273–302. http://dx.doi.org/10.1093/jeg/lbi022.

Britton, J. 2004. "High technology localization and extra-regional networks." *Entrepreneurship and Regional Development* 16 (5): 369–90. http://dx.doi.org/10.1080/08985620410001674351.

Burt, R.S. 1992. *Structural holes*. Cambridge, MA: Harvard University Press.

Cairncross, F. 1997. *The death of distance: How the communications revolution will change our lives*. Boston, MA: Harvard Business School Press.

Christopherson, S., and J. Clark. 2007. "Power in firm networks: What it means for regional innovation systems." *Regional Studies* 41 (9): 1223–36. http://dx.doi.org/10.1080/00343400701543330.

Csikszentmihalyi, M. 1996. *Creativity: Flow and the psychology of discovery and invention*. New York: Harper.

De Dreu, C.K.W., and M.A. West. Dec 2001. "Minority dissent and team innovation: The importance of participation in decision making." *Journal of Applied Psychology* 86 (6): 1191–201. http://dx.doi.org/10.1037/0021-9010.86.6.1191. Medline:11768061.

Desrochers, P. 2001. "Local diversity, human creativity, and technological in-novation." *Growth and Change* 32 (3): 369–94. http://dx.doi.org/10.1111/0017-4815.00164.

Dosi, G. 1988. "The nature of the innovation process." In *Technical change and economic theory*, ed. G. Dosi et al., 221–38. New York: Pinter.

Florida, R. 2002. *The rise of the creative class*. New York: Basic Books.

Foray, D., and B.A. Lundvall. 1996. "The knowledge-based economy: From the economics of knowledge to the learning economy." In *Employment and growth in the knowledge-based economy*, ed. D. Foray and B.A. Lundvall, 11–32. Paris: OECD.

Frenken, K., F. Van Oort, and T. Verburg. 2007. "Related variety, unrelated variety and regional economic growth." *Regional Studies* 41 (5): 685–97. http://dx.doi.org/10.1080/00343400601120296.

Friedman, T. 2005. *The world is flat: A brief history of the twenty-first century*. New York: Farrar, Straus & Giroux.

Gardner, H. 1993. *The creators of the modern era*. New York: Basic Books.

Gertler, M.S. 1995. "Being there: Proximity, organization, and culture in the development and adoption of advanced manufacturing technologies." *Economic Geography* 71 (1): 1–26. http://dx.doi.org/10.2307/144433.

Gertler, M.S. 2003. "Tacit knowledge and the economic geography of context, or the undefinable tacitness of being (there)." *Journal of Economic Geography* 3 (1): 75–99. http://dx.doi.org/10.1093/jeg/3.1.75.

Glaeser, E.L. 2003. "The new economics of urban and regional growth." In *The Oxford handbook of economic geography*, ed. G.L. Clark, M.P. Feldman, and M.S. Gertler, 83–98. Oxford: Oxford University Press.

Granovetter, M. 1973. "The strength of weak ties." *American Journal of Sociology* 78 (6): 1360–80. http://dx.doi.org/10.1086/225469.

Granovetter, M. 1985. "Economic action and social structure: The problem of embeddedness." *American Journal of Sociology* 91 (3): 481–510. http://dx.doi.org/10.1086/228311.

Grotz, R., and B. Braun. 1997. "Territorial or trans-territorial networking: Spa-tial aspects of technology-oriented cooperation within the German mechani-cal engineering industry." *Regional Studies* 31 (6): 545–57. http://dx.doi.org/10.1080/00343409750131686.

Guilford, J.P. 1950. "Creativity." *American Psychologist* 5 (9): 444–54. http://dx.doi.org/10.1037/h0063487. Medline:14771441.

Jacobs, J. 1969. *The economy of cities*. New York: Random House.

Kulkarni, D., and H.A. Simon. 1988. "The processes of scientific discovery: The strategy of experimentation." *Cognitive Science* 12 (2): 139–75. http://dx.doi.org/10.1207/s15516709cog1202_1.

Lubart, T.I. 2001. "Models of the creative process: Past, present, and future." *Creativity Research Journal* 13 (3–4): 295–308. http://dx.doi.org/10.1207/S15326934CRJ1334_07.

Marshall, A. 1890. *Principles of economics*. London: Macmillan.

Martin, R., and P. Sunley. 1998. "Slow convergence? The new endogenous growth theory and regional development." *Economic Geography* 74 (3): 201–27. http://dx.doi.org/10.2307/144374.

Maskell, P., H. Bathelt, and A. Malmberg. 2006. "Building global knowledge pipelines: The role of temporary clusters." *European Planning Studies* 14 (8): 997–1013. http://dx.doi.org/10.1080/09654310600852332.

McPherson, J.M., L. Smith-Lovin, and J. Cook. 2001. "Birds of a feather: Homophily in social networks." *Annual Review of Sociology* 27 (1): 415–44. http://dx.doi.org/10.1146/annurev.soc.27.1.415.

Morgan, K. 2004. "The exaggerated death of geography: Learning, proximity and territorial innovation systems." *Journal of Economic Geography* 4 (1): 3–21. http://dx.doi.org/10.1093/jeg/4.1.3.

Nemeth, C.J., and B. Nemeth-Brown. 2003. "Better than individuals? The potential benefits of dissent and diversity for group creativity." In *Group creativity*, ed. P.B. Paulus and B.A. Nijstad, 63–84. New York: Oxford University Press. http://dx.doi.org/10.1093/acprof:oso/9780195147308.003.0004.

Powell, W.W., K.W. Koput, and L. Smith-Doerr. 1996. "Interorganizational collaboration and the locus of innovation: Networks of learning in biotechnology." *Administrative Science Quarterly* 41 (1): 116–45. http://dx.doi.org/10.2307/2393988.

Romer, P.M. 1986. "Increasing returns and long-run growth." *Journal of Political Economy* 94 (5): 1002–37. http://dx.doi.org/10.1086/261420.

Rothwell, R. 1994. "Towards the fifth-generation innovation process." *International Marketing Review* 11 (1): 7–31. http://dx.doi.org/10.1108/02651339410057491.

Santagata, W. 2004. "Creativity, fashion, and market behaviour." In *Cultural industries and the production of culture*, 75–90. New York: Routledge.

Schumpeter, J.A. 1976. *Capitalism, socialism and democracy.* 5th ed. London: Taylor and Francis.

Scott, A.J. 2008. *Social economy of the metropolis: Cognitive-cultural capitalism and the global resurgence of cities.* Oxford: Oxford University Press. http://dx.doi.org/10.1093/acprof:oso/9780199549306.001.0001.

Spencer, G.M. 2011. "Local diversity and economic activity in Canadian cityregions." In *Beyond territory: Dynamic geographies of knowledge creation, diffusion, and innovation*, ed. H. Bathelt, M.P. Feldman, and D.F. Kogler, 46–63. New York: Routledge.

Spencer, G.M. 2012. "Creative economies of scale: An agent-based model of creativity and agglomeration." *Journal of Economic Geography* 12 (1): 247–71. http://dx.doi.org/10.1093/jeg/lbr002.

Sternberg, R., and T. Lubart. 1996. "Investing in creativity." *American Psychologist* 51 (7): 677–88. http://dx.doi.org/10.1037/0003-066X.51.7.677.

Sternberg, R.J., and T.I. Lubart. 1999. "The concepts of creativity: Prospects and paradigms." In *Handbook of creativity*, ed. R.J. Sternberg and T.I. Lubart, 3–15. Cambridge: Cambridge University Press.

Storper, M., and A.J. Scott. 2009. "Rethinking human capital, creativity and
    urban growth." *Journal of Economic Geography* 9 (2): 147–67. http://dx.doi.
    org/10.1093/jeg/lbn052.
Storper, M., and A.J. Venables. 2004. "Buzz: Face-to-face contact and the urban
    economy." *Journal of Economic Geography* 4: 351–70.
Unsworth, K. 2001. "Unpacking creativity." *Academy of Management Review*
    26: 289–97.
Weitzman, M. 1998. "Recombinant growth." *Quarterly Journal of Economics*
    113 (2): 331–61. http://dx.doi.org/10.1162/003355398555595.

PART II

Diversity, Variety, and the Cognitive-Cultural Economy in Large Cities

3 Innovation and Toronto's Cognitive-Cultural Economy

CHARLES H. DAVIS AND NICHOLAS MILLS

The term "creative" is increasingly employed to characterize a broad sweep of novelty-producing industries, regions, cities, occupations, workers, and tasks, but the multiple meanings attributed to the word have caused some confusion. In this chapter we investigate the concept of the *cognitive-cultural economy* (CCE), a promising framework for apprehending the key novelty-producing and wealth-generating features of the contemporary urban economic and social order, and apply it to the case of Toronto. As laid out by Alan Scott in *Social economy of the metropolis* and other recent publications (Scott 2010; 2008a, 2008b, 2008c; 2007), the concept of the cognitive-cultural economy emphasizes the economic and social significance of a group of industries, occupations, work tasks, and functions with high *cognitive and/or cultural intensity* in the growth of core metropolitan regions.

The cognitive-cultural economy is closely associated with contemporary processes of urbanization, globalization, and "worlding" (Roy 2009). Urban regions have become the economic motors of the world economy (Berube et al. 2010; Scott et al. 2001) as well as the principal sites of cognitive-cultural production and consumption (Scott 2008b). The predominance in cognitive-cultural production of global urban hubs such as New York, London, Los Angeles, Paris, and Tokyo suggests that the contemporary CCE "is both firmly anchored in specific places and a very insistent element of the global economic order at large" (Lorenzen, Scott, and Vang 2008).

The cognitive-cultural economy has emerged in different configurations and manifests different degrees of local value-creating capability in different places (D'Ovidio, 2010; Kloosterman, 2010). What accounts for the greater growth and vitality of cognitive-cultural economic activity in some places rather than others? An important debate is currently taking place among geographers, demographers, regional economists, and urban planners about the relative contributions of various cognitive and cultural occupations and industries to economic growth, the significance of urban amenities and services in the attraction and retention of talent and investment, the modalities and processes of effective planning and governance of metropolitan cognitive-cultural economies, and the consequences of cognitive-cultural economic development for income distribution (Bontje and Musterd 2009; Donegan et al. 2008; Indergaard 2009; Scott 2011, 2010, 2008b, 2004; Storper and Scott 2009; Wolfe 2009).

The division of cognitive and cultural labour within the global economic system has important implications for Toronto, Canada's largest metropolitan area and its principal centre of cognitive-cultural economic production and consumption. The Greater Toronto Area (GTA) is a large and growing city-region with considerable economic weight. Its population of 5.6 million increases by approximately 100,000 people every year, and its Gross Domestic Product of $327 billion gives the city-region the economic size of Argentina, South Africa, Venezuela, Ireland, or Finland. The Toronto population is highly educated and very culturally diverse. In 2001, 52 per cent of the GTA's population of those twenty years and over held a post-secondary degree, diploma, or certificate (Gertler, Tesolin, and Weinstock 2006). As the headquarters of nearly 40 per cent of all business operations in Canada, Toronto is Canada's largest regional economic engine. Of the total labour force of 3.7 million in the Toronto region, 14.3 per cent of workers are employed in the manufacturing sector; 8.6 per cent in the professional, scientific, and technology services sector; 3.3 per cent in the information and cultural industries; and 1.9 per cent in arts, entertainment, and recreation (TRRA 2010).

Toronto enjoys a position as first among equals in the domestic cognitive-culture economy and has emerged as a major regional second-tier city on the basis of its strengths in a range of cognitive-cultural industries and activities, including information and communication technologies (ICT), R&D-based industries, financial services, and creative and entertainment industries. Many among Toronto's business, political, and social elite would like to see Toronto rise further in the world urban hierarchy to become a top-level global city, as articulated by the Toronto Board of Trade in its recent action plan for the Toronto Region titled *From world-class to world leader* (Toronto Board of Trade 2009). Toronto faces stiff competition internationally, however, as a supplier of cognitive and cultural products and services. Many organizations produce scorecards or benchmarks that compare top cities on a range of dimensions that are believed to matter when cities compete with each other to attract talent and investments. A persistent finding of these endeavours is that to get to the top, Toronto's performance needs to improve. Although Toronto consistently ranks among the top two dozen cities in the world according to many indicators, rarely does it rank as number one in something. Observes a recent report analysing Toronto's "3 Ts" of technology, talent, and tolerance: "Toronto is not currently a leader within its group of competitive peers" (Martin Prosperity Institute 2009: 22).[1] In core industries of the CCE (biotechnology, screen arts, finance, and ICT), Toronto consistently ranks third among North American cities.

This sort of diagnosis strikes a nerve among those who believe that Toronto must excel globally in some area in order to differentiate itself from other world cities. As Toronto's recent economic development vision statement *Agenda for prosperity* (2008) puts it, "Toronto is a city on the cusp" of success but other jurisdictions are investing more effectively to develop capabilities needed for global prominence. The Innovation Systems Research Network (ISRN) project "Social Dynamics of Economic Performance in City-Regions" postulates that urban economic performance critically depends on successful *innovation* as a core capability. This chapter critically addresses Scott's concept of the cognitive-cultural economy from the perspective of *innovation* – the creation of value through new Schumpeterian combinations. The CCE is a vast generator of innovation, creating value through "scientific knowledge inputs, continuous innovation, product multiplicity and differentiation, the provision of customized services, symbolic elaboration and so on" (Scott 2008c: 64). It is not straightforward, however, to apprehend innovation in the cognitive-cultural economy. Much cognitive-cultural innovation is not solely or principally technological. Instead, it is a combination of organizational, technological, and "soft" (aesthetic) innovation which is often enabled and amplified by ICT adoption. Accumulated conventional knowledge, conceptual frameworks, and stylized facts about technological innovation dynamics are of limited utility as reliable guides to innovation in the CCE.

This chapter has two parts. In the first, we address a pair of key conceptual and methodological issues: how to operationalize the concept of the CCE, and how to conceptualize, observe, measure, and assess innovation in the cognitive and cultural branches of the CCE. The second part of the chapter applies the CCE concept to the case of Toronto, a major regional city but not a central hub in the network of global cities. We provide a high-level review of innovation dynamics in Toronto's cognitive, cultural, and ICT sectors and briefly discuss three principal kinds of value-creating spillovers (knowledge, product, and network) typical of the cognitive-cultural sector, illustrated with examples from Toronto's screen-based media industry. Our analysis draws on the scholarly research literature, a survey of innovation among hundreds of firms in Toronto's creative and entertainment industries, and well over one hundred interviews undertaken in conjunction with the ISRN's "Innovation and Creativity in City-Regions" project.

The Cognitive-Cultural Economy

Following Boyer, Scott uses four parameters to describe the capitalist cognitive-cultural economic and social order: leading sectors, the technological foundation, characteristic forms of labour relations, and typical competitive practices (Scott 2007). The *leading sectors* of the CCE are "technology-intensive manufacturing, services of all varieties (business, financial, personal), fashion-oriented neo-artisanal production, and cultural-products industries (including media)" (2007: 1467). The common *technological foundation* of the CCE is provided by ICTs, which enable critically important capabilities in adopter industries. The characteristic forms of *labour relations* in CCE industries, according to Scott, are substantially flexibilized working conditions, deroutinized labour processes, and destandardized outputs. *Competition* in CCE industries is intense because of globalization and very high rates of innovation.

The concept of the CCE captures and foregrounds the two major waves of primarily non-technological innovation that animate contemporary capitalism: industrial proficiency in creating and commodifying culture, and analogous proficiency in the rationalization of economic production. Cultural production and rationalization of production are enabled and accelerated through embodied technological progress provided by ICTs, which have become a necessary but not sufficient condition of successful cognitive-cultural innovation.

Scott notes that in the cognitive-cultural economy "the realm of human culture as a whole is increasingly subject to commodification" (1997: 323). Proficiency in creating and commodifying culture stimulates and caters to a very wide range of consumers' subjective and affective needs and desires. The commodification of culture on an industrial scale is a historically novel feature of contemporary capitalism. It has induced a widespread aestheticization of daily life (Featherstone 1987), leading to what various authors call the "society of the spectacle," the "dream society," the "emotion economy," the "attention economy," or, as expressed in the title of Pine and Gilmore's 1999 classic, *The experience economy*, in which "work is theater and every business a stage."

Cognitivization, the other major wave of innovation in the contemporary post-industrial economy, provides capability for strong information handling, storage, and analysis in support of highly rationalized control of information flows, organizational processes, coordination, and feedback loops (Beniger 1986). This capability is directly enabled by the adoption of ICTs. Cognitive functions and tasks are major beneficiaries of embodied technical progress realized through the adoption of capital equipment and the use of external technical services. Strong information-handling ability and high levels of rationalized control constitute the backbone of "regimes of computation" (Hayles 2005) which arise from adoption of ICTs in industry and government for the purposes of communication, coordination, administration, analysis, risk assessment, optimization, and decision making. The information intensification and rationalization of the organizational world requires increasing degrees of "cognitivization of work" (Kallinikos 1999). Information intensity, rationalized control, and cognitivization of work imply an economic and social order variously called the "information economy," "knowledge society," or "learning economy."

Clearly, the concept of a cognitive-cultural economy faces major challenges of operationalization before accurate comparative observation and benchmarking can take place. Nevertheless, the broad outlines of the theoretical framework are clear. The cognitive-cultural economy is represented by certain *industries or sectors* that disproportionately rely on *work tasks, functions*, and *occupations* with high *cognitive or cultural intensity*. The three major *institutional components* of the cognitive-cultural economy are the *ICT supplier industry*, the *cognitive sector*, and the *cultural sector*.

- The ICT supplier sector, a major R&D-intensive cognitive industry in its own right, plays a uniquely important role in the CCE economy as supplier of foundation technologies to adopter industries, enabling cognitivization of work tasks and functions and commodification of cultural products and services on an unprecedented scale. The cognitive and cultural industries are major customers of the ICT supplier industry. While cultural commodities originate and are consumed in the "content layer" of the digital economy, cultural commodification is enabled and amplified by advances in the software and transport layers (Yoo, Henfridsson, and Lyytinen 2010).

- The cognitive sector is marked by occupations with highly analytical, systemizing, and synthesizing capabilities, including analysts, managers, engineers, researchers, technicians, and accountants. It encompasses all R&D-intensive industries, all information-processing industries, most of the knowledge-intensive service sector, and the higher-level management, technical, and administrative functions of all industries, notably the control, coordination, technical support, analytical, and technology-based novelty-producing functions of industries and government. Paradigmatic cognitive industries are the knowledge-intensive business services, R&D-based industries, and industries that are based primarily on information handling and analysis, such as financial services.
- The culture-producing sector and associated work functions, tasks, and occupations deliver experiences, affect, meaning, social positioning, sensation, and subjective gratification as final outputs. The cultural sector is marked by occupations with highly specialized and distinctive creative, symbolic, expressive, and communicative capabilities, notably designers, writers, singers, actors, and interpreters of culture. The set of cultural industries included in the CCE is very broad, comprising the design, advertising, fashion, art, music, architecture, crafts, live and mediated entertainment, and tourism, as well as trans-sectoral cultural work tasks, functions, or occupations such as design, public relations, marketing, and advertising.

Combined, these three components of the CCE make possible flexible specialization, mass customization, commodification of culture, possessive individualism, financialization, and ubiquitous surveillance (Harvey 2005).

Understanding Innovation in the Cognitive-Cultural Economy

Schumpeter's view of entrepreneurial innovation encompasses the broad range of innovation dynamics visible in the contemporary CCE. Most national and regional innovation surveys, however, in line with standards established by the OECD over the past three decades, focus on the much narrower "Technological Product and Process" (TPP) innovation, which defines innovation almost exclusively in terms of R&D-based improvement of technical efficiency. Economic theory underlying TPP models of innovation conceptualizes value creation and economic growth as consequences of functional improvement in technology, the principal driver of production efficiency. Surveys that measure TPP innovation consider only technologically new or improved products or production methods and their related inputs, especially R&D spending, scientific publications, highly qualified manpower, patents, and product or process innovations.[2]

Innovation research has focused largely on the determinants and effects of R&D-based technical progress (Baregheh, Rowley, and Sambrook 2009). But it is well established that TPP-oriented innovation surveys do not capture the full range of value-creating non-technological innovation activities of firms. This issue has arisen regarding measurement of innovation in the service sector, e-business innovation, innovation in industrial clusters, and innovation in non-market environments such as the social economy or the public sector, prompting researchers to argue for broadening the innovation concept (Arthurs et al. 2009; Bloch 2007; Hauknes 2003; Salazar and Holbrook 2004). More recently, the limitations of the TPP-oriented innovation framework have been emphasized with respect to the value-creation processes in the cultural sector (Eltham 2012; Jaaniste 2009; Stoneman 2010).

Since 2005, OECD-inspired innovation surveys include certain aspects of marketing and organizational innovation, but only when they are deployed in support of TPP innovation (OECD 2010a, 2010b). Left aside is quite a broad range of value-creating Schumpeterian combinations which are not directly related to productivity-enhancing technological improvements (Hawkins and Davis 2012). For example, the Sawhney-Wolcott-Arroniz (2006) model of innovation, which transcends the TPP model, identifies twelve dimensions of business innovation (offerings, brand, platform, networking, presence, supply chain, organization, processes, value capture, solutions, customers, and customer experience).

Innovation along most of these dimensions is enabled by ICT adoption, relying on cognitive work tasks, functions, and occupations within the firm. Research and policy statements about Canada's productivity gap with the United States always point to Canada's lower levels of investment in ICTs and slower ICT adoption rates as key causal factors (Sharpe 2006). Investment in ICTs improves business performance in two ways: by increasing production efficiencies and through qualitative enhancement of brands, positioning, relationships with customers, and relationships with suppliers.[3] The latter is noticeably the case when small and medium-sized enterprises adopt ICTs; firms with fewer than fifty employees represent nearly 98 per cent of all firms in Canada. Thought leaders on value creation with ICTs advocate first determining the value the firm wishes to create and only then considering how ICT can help support the firm's value-creating strategy (Hopkins 2010).

Roles for cultural occupations, functions, and tasks are mostly present in the product innovation, service, human resources, and marketing activities of the firm, especially in the branding and customer experience dimensions of the model. From the perspective of the cultural sector, the weakest part of this model is the *customer experience* dimension, which remains largely an unknown in the model. Much past research conceptualizes customer experience in terms of satisfaction, loyalty, or intent to purchase. These outcomes do not do justice to firms in the cultural sector, for whom the principal value proposition is to produce experience through "soft innovation," which is defined by Stoneman as "changes in goods and services that primarily impact upon sensory or intellectual perception and aesthetic appeal rather than functional performance" (2010: 329). Since firms in the cultural sector create value by providing experiences, research on innovation in cultural industries needs to focus on how commodified experiences create value for customers.[4]

Analysis of TPP innovation usually assumes that the rate and magnitude of firm-level product innovation are in general positively related to economic performance. But since product innovation is ubiquitous in the cultural sector, no straightforward relationship exists between rate or magnitude of product innovation and economic performance among cultural firms. Typically, cultural products are horizontally differentiated. Horizontal product innovations "entail a new product that is more highly valued by consumers who place a lower value on the existing product" leading "to products with different characteristics that are preferred by consumers with different tastes" (Soberman and Gatignon 2005: 168). Horizontal product differentiation results in the proliferation of products and an increasingly fine segmentation of taste markets, a frequently observed phenomenon in cultural industries (Caves 2000). Because in creative industries product innovation is pervasive, the principal indicators of the importance of a new cultural product are *extent of adoption* (Stoneman 2010) or *critical acclaim* (Simonton 2009).

In TPP innovation, formal R&D is the key source of new technical knowledge, complemented by suppliers, sales teams, customers, and service providers. But in the cultural sector, innovation in the content layer rarely calls on formal R&D, so levels of R&D spending, patents, or knowledge transfers from formal R&D institutions are inappropriate ways to measure knowledge flows. Instead, to capture the key dynamics of innovation in the cultural economy requires a conceptual framework that emphasizes knowledge flows among cultural producers, their intermediaries, and their customers. For example, Cohendet, Grand-adam, and Simon (2010) theorize the interactions, flows, and transactions in an urban cultural economy in terms of three institutional layers: an upper ground constituted of firms and other institutions (including universities) that finance and launch products, an underground constituted of creative practitioners and their communities of practice, and a middle ground constituted of intermediaries, hang-outs, associations, and activities. Increasingly, frameworks must also take into account co-evolution among the content, software, and transport layers of a highly digitized cultural economy (Yoo et al. 2010).

Firm-level product innovation capability is a necessary but not sufficient condition for economically successful value creation. Experience goods face substantial uncertainty regarding market acceptance, as expressed in the famous "nobody knows anything" syndrome in cultural product innovation (Caves 2000; De Vany 2004). A recent survey of innovation in the Ontario Media and Entertainment Cluster, which is highly concentrated in the GTA, shows that television production, book publishing, and musical recording exhibit very high levels of new product development capabilities, and at the same time very low levels of profitability (HAL 2009). Business capabilities that permit the commercial exploitation of creative product innovation capabilities require deliberate development in firms in the cultural sector (Davis, Vladica, and Berkowitz 2008). The following sections situate the current dimensions of Toronto's cognitive-cultural economy and explore the nature of the innovation process within the CCE using this more comprehensive definition of innovation.

Toronto's Cognitive-Cultural Economy: Innovation and the ICT and Media Industries

Most recent assessments of Toronto's competitiveness include an evaluation of the city-region's innovation capability, which is generally interpreted as science- or technology-based innovation (TPP innovation as discussed above). For example, the OECD's recent *Territorial review* of Toronto (2010c) provides a diagnosis of Toronto's R&D-intensive industries and assesses Toronto's innovation performance in terms of indicators of TPP innovation, largely leaving aside innovation in the cultural sector. Another recent assessment of Toronto's innovation status is provided by the Toronto Region Research Alliance (TRRA), a regional economic development organization that promotes increased investment in research and innovation in the Toronto Region Innovation Zone. Like the OECD *Territorial report*, the TRRA finds that the Toronto region has a strong foundation in science-based industry but without a commensurate impact for its research, relatively low propensity to commercialize intellectual property, and a comparatively low level of investment in innovation (TRRA 2010). Neither of these reports pays much attention to the broader conception of innovation in the CCE that we have outlined above.

The ICT Industry

The Toronto region has the third-largest ICT cluster in North America. Around 29 per cent of Canada's 40,000 ICT firms are located in the Toronto region. About half of Toronto's ICT firms are located within the City of Toronto, while the others, especially the largest firms, are located in Toronto's proximate cities of Markham and Mississauga. In 2004 the Toronto regional ICT industry employed 212,000 persons in services and manufacturing, making the Toronto cluster three times larger than that of Vancouver or Ottawa-Gatineau (Lucas, Sands, and Wolfe 2009). In 2009 Toronto-region ICT companies earned $52 billion in revenue, of which $22B were in manufacturing and $30B in services (City of Toronto 2010). Most of Toronto's ICT firms are very small; 95 per cent are software developers, communications providers, or consulting service providers with fewer than fifty employees. The region is well represented among fast-growing or award-winning small ICT firms. Nearly 40 per cent of the top 250 Canadian ICT companies are located in Toronto (ibid.), while nine of the twenty-five largest Canadian-owned ICT firms are located in the Toronto region.[5] Toronto also plays host to many of the largest transnational ICT firms with business operations in Canada, being home to more than three hundred subsidiaries of transnational ICT firms. Of the top twenty-five foreign-owned ICT firms in Canada, twenty-one are located in the Toronto region (Branham Group 2010).

The Toronto-based ICT industry is unlike other Canadian ICT clusters, not only in terms of size and variety of specialties and segments, but also in terms of market orientation. While other Canadian ICT clusters are highly extroverted, the Toronto ICT cluster primarily serves the domestic market (Lucas, Sands, and Wolfe 2009). In 2004 the cluster earned about $6 billion from export sales, representing around 15–20 per cent of the cluster's estimated $30–$35 billion in sales (E&B Data 2004). ICT firms in the Toronto region target four distinct customer markets: other firms operating in the ICT sector (24% of sales), consumers of digital communications and media products and services (24%), financial and business services (16%), and the industrial sector (34%) (ibid.). The strong local market (in size and rate of growth) is regarded by the Toronto ICT industry as the most valuable of its competitive advantages, eclipsing infrastructure and the availability and talent of the Toronto workforce (ibid.).

Strong domestic demand for ICT products and services from Toronto-based firms is evidence that the Toronto ICT industry plays an important role in diffusing cognitive occupations, work tasks, and functions into other domestic industries. Further evidence is provided by labour-force data: of the approximately 173,000 ICT workers in the GTA in 2001, about 39,000 were employed in non-ICT sectors, notably finance and insurance and non-ICT manufacturing (E&B Data 2004). Furthermore, the emergence of specialty ICT sub-clusters related to end markets in mobile media, digital media, e-learning, financial services, social networking media, and e-health is attributed to strong demand from end-user industries in the local market. The emergence of early-stage IT specialties in the GTA is evidence of the importance of the local cognitive-cultural economic and social environment in innovation path creation. But Toronto's growth potential in telecommunications, a core ICT sub-sector of the CCE, is affected by the geographic decentralization of the telecommunications industry in Canada across five or six headquarter cities.[6]

In addition to its thick labour market, the Toronto region enjoys a dense ICT research environment. In 2004, there were nearly one hundred ICT R&D centres within GTA universities, community colleges, and other public institutions. Toronto's three universities each house a variety of specialized communications and advanced technology research labs. The Toronto ICT industry also has a complex support network for promotion of ICT business activity and professional development. There is evidence, however, of an important trade-off between scale and effective governance: the "size, diversity and geographic spread of the ICT sector in the Toronto Region make it difficult to bring companies and organizations in the region together in a coordinated effort to improve the business environment and to establish the same level of brand awareness that their counterparts have achieved in other cities" (Impact Group 2006: 20).

The Media Industry

Toronto hosts the principal media agglomeration in English-speaking Canada and the third-largest in North America after New York and Los Angeles (Davis 2011). In this agglomeration are found most of anglophone Canada's major screen production houses, public broadcasters, and many of its private broadcasters. Many Canadian book, magazine, music, and newspaper publishing headquarters are located in Toronto, as are four of the eight principal Canadian media conglomerates. The agglomeration includes the country's largest concentration of independent screen-content producers, specialty broadcasters, supporting institutions, and many suppliers of specialized services and inputs: sound recording studios, law firms, post-production services, media marketing and publicity agencies, financial and insurance services, theatrical exhibitors, Internet publishing firms, technical service suppliers, advertising agencies, below-the-line crews and their craft unions, and public and private post-secondary educational programs. Tens of thousands of media micro-enterprises are present in the GTA (Davis 2010). All three levels of media policy and program agencies are strongly represented in the city. Altogether the content layer of the Toronto media cluster (including film and television production, books, magazines, music, and interactive media) employed around 40,000 people and generated about $4.5B in revenues in 2007 (Davis 2011).

The six sub-sectors mentioned above are the designated components of the Ontario Entertainment and Creative Cluster, one of the few areas of the Toronto CCE to have been mapped in some detail using innovation survey methods and concepts adapted from technology-based innovation cluster analysis (HAL 2009). In comparison with Canadian technology clusters, the Ontario media cluster reports a more difficult business climate, better government support, far greater importance of domestic competitors and customers, a larger critical mass, a higher level of self-awareness as a cluster, less-dense linkages with local support institutions and associations, analogous business development and product innovation capabilities, and (with the exception of digital media) far lower expectations of growth and profitability (HAL 2009).

If it were not for public policies and programs that provide production support and (in some cases) impose national-origin requirements, many Ontario producers of books, television programs, films, music, and magazines would find it difficult to survive (Davis 2011). It is a challenge for smaller English-language countries to develop competitive export industries for their cultural products unless these products are accepted among US consumers, who serve as de facto taste makers due to the size of the US market (Grant 2008). At the same time, imported English-language cultural products take market share away from local cultural producers. High levels of imported English-language cultural products in the local market are a fact of life for Ontario cultural producers.

Innovation and Spillovers in the Cognitive-Cultural Economy

According to Jacobs's hypothesis on the urbanization externalities that arise in larger and more diverse urban economies (Beaudry and Schiffauerova 2009; Ejermo 2005; Van der Panne 2004), a strong ICT sector in a diversified metropolitan economy with cognitive and cultural sectors should yield two main benefits of co-location with these other sectors: (1) greater ICT product and service variety, more rapid innovation cycles, greater market success for new products and services, and faster growth for ICT firms in the metropolis than for firms in smaller urban areas, and (2) more rapid uptake and effective use of ICT across the range of industries in the metropolis than in outlying regions. Cognitive-cultural innovation, which we emphasize has major non-technological dimensions, is even more susceptible than technological innovation to such urban externalities. Cultural industries are most likely to flourish in an economic and social environment with high local variety and heterogeneous demand. Large metropolitan environments offer a much greater variety of potential adopters and a broader and deeper range of potential Schumpeterian combinations than socioeconomic environments with lower diversity, giving metropolitan regions with great social and economic variety an important advantage in the cognitive-cultural economy. In such regions, innovation is driven by the many flows of knowledge via linkages, interactions, and spill-overs within the cognitive-cultural economy proper and between the cognitive-cultural sectors and other sectors.

Thus, in principle, co-location of the ICT sector and user industries is beneficial to both, because knowledge and technology spillovers from the ICT sector into adopter industries provide abundant opportunities for exchanges of tacit knowledge and interactive learning between ICT suppliers and users with novel needs. The home market may serve as a driver of development for new cognitive products and services which later may be exported. On the other hand, the economic role of cultural industries is currently a matter of debate (Potts and Cunningham 2008). Some authors argue that cultural industries may play a previously unsuspected strategic role as catalysts of variety creation across many industrial sectors (Bakhshi and McVittie 2009; Frontier Economics 2007; Potts and Cunningham 2008). Further, the cultural sector may play a "general purpose" role in culturization, analogous to that played by ICTs in cognitivization.

The cultural sector affects the broader economy in two principal ways (Frontier Economics 2007). The first is by direct commercialization of cultural products and services. Here, business-to-business relationships are unexpectedly important. Bakhshi and McVittie (2009) and Bakhshi, McVittie, and Simmie (2008) find that, in the United Kingdom, in the aggregate nearly 60 per cent of the outputs of the fashion, software, architecture, publishing, advertising, arts, radio and television, and film industries are intermediate inputs for other businesses.[7]

The second way the cultural sector affects the broader economy is by localized spillovers of cultural knowledge into adopter or user sectors in ways that induce and enable innovation. Co-location is one indication of likely knowledge spillovers. Chapain et al. (2010) report substantial co-location among firms in the cultural industries in the United Kingdom: between advertising and software firms, among music-film-publishing-radio-TV firms, and among advertising, software, advanced manufacturing, and knowledge-intensive business services. Toronto's cultural sector displays patterns of co-location among book and magazine publishers, electronic equipment repair firms, broadcasters, recording studios, software publishers, technical and trade schools, specialized design services, independent artists-writers-performers, performing arts organizations, advertising, grant-making services, event promoters, artists' agents, and spectator-sports companies (Vinodrai and Gertler 2007) suggesting the existence of largely undocumented knowledge transfers and spillovers among these industries.

Three kinds of spillovers are proposed for the creative sector: knowledge spillovers, product spillovers, and network spillovers (Frontier Economics 2007). Below we describe and briefly discuss these three kinds of spillovers, each illustrated with a mini-case study from Toronto's media industry.

Knowledge Spillovers: Digital Visual Effects and Computer Animation Firms

In the cognitive-cultural economy, knowledge spillovers occur when technologies developed in the ICT sector are adopted in other sectors in a cognitivization process, or when aesthetic attributes developed by one industry are transferred to another (or, more properly, are created for customers) in a culturization process. Cognitivization takes place in the cultural sector, for example, when cultural firms analyse customers' data trail to develop new product or service offerings (Zwick and Knott 2009). Analogously, culturalization takes place in the cognitive sector when the value proposition of a functional product or service is enhanced with aesthetic or symbolic attributes provided by a cultural discipline, as in the case of design, for example (Vinodrai 2009).

The distinction between the cognitive and the cultural attributes becomes blurred when technology in the form of powerful animation and digital effects software contributes to culturization by enabling the creation of photorealistic moving images. Complex cognitive and cultural spillovers occur when a user industry borrows liberally from the worlds of technology and visual culture, and also contributes to these and other sectors.

This is the case in the computer animation and visual effects industry, which in Toronto numbers around a hundred firms, has estimated revenues of $170–$200 million, and employs around two thousand people (Nordicity 2008). In Canada, 3D computer animation has largely supplanted 2D hand-drawn animation, a television staple that has now largely migrated to countries with low labour costs. Digital visual effects (digital VFX) are typically used as photorealistic extensions of live-action cinematography to create new scenes, characters, or effects, although today entirely computer-generated feature films are familiar to everyone. The computer animation/digital VFX industry services several segments of the screen industry: feature films, television shows and movies, commercials, games, mobile content, broadband and Internet content, music videos, and scientific visualization. Some firms in the post-production sector also provide digital VFX services.

Most of the computer animation/digital VFX companies in Toronto are small or medium-sized firms, hiring staff on a project-by-project basis, although the city also has several larger firms with in-house studios that produce programming for the TV market and outsource some of their production tasks. Computer animation/digital VFX firms in Toronto have provided special effects and scenes for many very well-known films and television shows. An illustrative list of firms and shows to which Toronto firms have made contributions includes CORE Digital Pictures (*Resident Evil: Apocalypse, Blade 2, X-Men, The Tudors, The Wild*), Mr X Inc. (*Tron: Legacy, Resident Evil: Afterlife, Scott Pilgrim vs. the World, Amelia, Whiteout, Eastern Promises*), Copperheart Entertainment (*Ryan, Splice*), Rocket Science (*Chloe, Saw 3D, Everest*), Intelligent Creatures (*Day of the Triffids, Piranha 3D, Babel*), Coptor (*The Cassandra Syndrome*), and Starz Animation (*9, Gnomeo and Juliet*). The latter firm is an affiliate of a US entertainment conglomerate that recently expanded into digital VFX in Toronto. As is clear from the titles, digital effects are now used very widely outside the science fiction and children's genres where they originated, and have become so extraordinarily realistic that viewers frequently are unable to distinguish artifice from reality.

The production of a fully animated digital feature film represents the highest degree of firm-level capability in the computer animation/digital VFX industry. It is a very complex and expensive process requiring substantial infrastructure and expertise. The production process requires specialized proprietary software tools for management of workflow, pipelines, projects, and tens of thousands of digital content "assets." The industry-standard software animation tool Houdini is produced by Toronto-based Side Effects Software.

Talented digital effects workers provide the key technical and artistic capabilities. The production process involves story designers, computer model builders, software developers, animators, texture painters, and technical specialists in computer networking, storage, and communications. Many VFX/animation workers in Toronto received their training at such leading animation schools as Sheridan College. Toronto's reputation as a digital animation hotspot was enhanced when the computer-animated documentary *Ryan*, produced at Seneca College, won an Academy Award for best animated short in 2004.

A typical digital feature film production lasts two years, employs 300 people, utilizes 1000 or more computer processors, and upwards of 100 terabytes of online storage. It takes between 1500 and 2000 shots, each of which is from two to five seconds in duration, to complete a feature film. Every frame and every element in it must be created from scratch, and every element must be consistent in colour, texture, shape, and opacity. A typical feature film is made of 115,000 individual pictures, often made up of huge files, and data processing, storage, network, and communications requirements are substantial and critical.

A key firm in Toronto with the capability to produce all-digital feature films was CORE Digital Pictures, the city's largest independent animation studio, which suddenly closed its doors in March 2010, unable to raise cash from private or public sources. The reasons for the failure of this firm have not been disclosed, but three notable California-based computer animation/digital VFX firms also recently closed their doors (CafeFX, Asylum VFX, and ImageMovers Digital), citing poor economic conditions. The computer animation/digital VFX sector has become volatile because capabilities are coming on stream around the world, and fast broadband connections allow animation and VFX work to be parsed into tasks and distributed widely among subcontractors. Since labour represents about half the cost of a digital feature production, outsourcing animation/VFX production to low-cost labour locations is economically attractive. This is exacerbating price-based competition among providers of digital animation/VFX services around the world, and is spurring incentive-based competition among host jurisdictions. For example, the Ontario government advanced $23 million to Starz Animation Toronto in 2009 to help the firm create and retain over 250 jobs in the province. Part of this advance went towards the production of the animated feature film *Gnomeo and Juliet* (2011), which was released in 3D and distributed by Toronto-based E1 Entertainment.[8]

The volatility of the animation/visual effects industry, which is structured mainly as a service industry, shows that it is not enough for a city like Toronto to have a co-located leading training institution, a leading software provider, leading digital VFX firms, and a leading distributor. A sustainable and competitive media cluster also requires a cohort of indigenous firms that can successfully create and commercialize digital products for which they retain intellectual property rights: that is, they need to develop and successfully exploit a rights-management business model.

Product Spillovers: Innovating and Commercializing
"Convergent" Transmedia Properties

Product spillovers occur when goods and services "increase demand for complementary goods in other sectors" (Chapain et al. 2010: 25).

Product spillovers are common in media domains. Often they involve a firm and its partners extending intellectual property such as a brand, characters, or a narrative into the merchandising of toys, games, food, clothing, or live events. The latter is the preferred business model in the children's screen entertainment industry, although it is regarded as too risky and complex for small firms to engage in alone (Davis, Vladica, and Berkowitz 2008). Said the executive of one Toronto-based children's production firm, "Merchandising is a very, very inexact science, part of rights management; it is a combination of inexact science and lottery." Nevertheless, this company "will not produce something that does not have merchandising potential." In Toronto a network of firms and organizations specializing in entertainment for kids has emerged, encompassing firms in digital media, film and television production, games, toys, education, and live events.

In an example of path creation in transmedia, approximately two dozen Toronto firms are developing highly specialized capability to create cross-platform content for film, television, and mobile screens. Movement of content from one medium to another is not new, of course – books become movies, songs are recorded and broadcast, theatrical plays can end up on television, and TV shows can circulate on websites and mobile devices – but the emergence of digital production and distribution platforms makes transmedia (cross-platform) strategies compelling.

The advent of transmedia properties and transmedia audiences obviously creates dilemmas for small and medium-sized production companies. The ability to develop and implement business models that work across platforms provides economies of scope among transnational and national media conglomerates, and is a source of competitive advantage (Aris and Bughin 2009; Vukanovic 2009). Also, as Jenkins (2006) points out, convergence is very much about participatory media culture. But user-generated content tends to drown out small content producers and undermine small service providers such as audio recording studios.

Small media producers engaging in transmedia strategies must create complementary media goods and services across platforms, devices, and channels, but what kinds of complementarity, and who will pay for it? In transmedia business ventures, frequently one profitable media platform subsidizes distribution on other platforms. This is currently the case with digital content, which in Canada is generally subsidized via the broadcasting system. In Toronto, profitable digital firms can be found among Web 2.0 applications developers, games developers, and advertisers, but the core of the Toronto screen-content development industry still gains most of its income from television, not from products distributed on digital platforms such as the Internet or mobile media. In 2009, only around 2 per cent of the Ontario film, television, and cross-platform interactive media industry's revenues of nearly $900 million were generated from interactive media production (Nordicity 2009). Of 189 firms, 20 firms were active in interactive media (generally designed in conjunction with film or television properties) and five had true transmedia capabilities – the capacity to work in all three areas of film, television, and cross-platform interactive media.

Small interactive media firms behave strategically regarding their transmedia activities, which are expected to increase in economic significance in coming years. Said the executive of one highly regarded firm that produces television shows, games, animation, and interactive websites, "Most of our work comes from doing interactive properties that are linked to television shows. We are dubious that interactive projects have revenue potential, but we view it as a way to grow and get a name for ourselves." At the same time, servicing the broadcasting industry provides a competitive advantage to this firm because, they said, "we have a really good knowledge of the TV industry and how TV people think. Other companies that do what we do often come from the online gaming world, but they don't know how TV functions." The firm's perceived advantage over competitors that focus strictly on digital gaming is that, while game creators try to create a complex and technically sound puzzle, an interactive media producer tries to create an interactive experience that works as part of a film or television story.

We noted earlier that transmedia innovation involves many kinds of extensions across media, not just interactive media. But the current technological state of the media industry has given interactive media enormous strategic significance for the Toronto media industry as a whole, since successful transition to a converged media environment is widely regarded as a critically important step in the evolution of the Toronto media cluster. The issue, therefore, is how a screen-media innovation system that is path-dependent on the established broadcast industry can create new innovation pathways yielding profitable transmedia properties. We conjecture that firms that develop internal transmedia capabilities are best positioned to discover or invent viable business models for transmedia innovation. However, it must be noted that while the development of transmedia capabilities in the indigenous screen production sector in Toronto is highly dependent on co-located pools of film, television, and interactive-media technical, creative, and business labour, certain non-local factors, notably national broadcast policy, also play a determining role (Davis 2011).

Network Spillovers: Creative Districts and the Conversion of Cultural to Economic Capital

Network spillovers occur "where the mere presence of creative businesses in a given place benefits other local firms" (Frontier Economics 2007). Most ambitious creative-city initiatives favour investments in signature cultural buildings, high-profile infrastructure for cultural production and consumption, and major cultural events in the downtown core, seeking network spillovers in the form of attraction of domestic and international talent, investment, high-value-added business activities, high-income local and transient cultural consumers, and domestic gentrifiers.[9] The best-documented case of a spillover-intensive cultural district in Toronto is Queen West–Liberty Village, which encompasses art, music, screen media, digital media, and cultural consumption services, and displays the expected gentrification effects (Catungal, Leslie, and Hii 2009; Matheson 2004; Sharpe et al. 2004).

In many respects, creative-city and creative-class initiatives are "symbolic policies" (Rousseau 2009) designed to persuade owners of financial, symbolic, or relational capital to invest in the host city. For local policymakers, the ultimate expected outcome of investments in "creative clusters and cultural scenes" is job creation and economic growth. As illustrated by the case of the Toronto International Film Festival (TIFF) and the Bell Lightbox centre, TIFF's home since September 2010 in Toronto's Entertainment District, a major cultural event can function as a catalyst in the urban growth machine, effectively converting cultural capital into economic capital, while responding to a broader range of cultural, policy, and economic interests.

Thousands of film festivals take place annually around the world. TIFF, which launched in 1976 as the Festival of Festivals, has become one of the top five film festivals in the world, screening over 300 features and shorts during its 11-day run and attracting more than 250,000 visitors, including key industry players and film celebrities. Associated with the TIFF Group are ongoing activities: a cinémathèque, a student film showcase, a film reference library, a distribution system for Canadian and international independent films that serves 163 Canadian communities, a lecture series, special exhibits, and an annual event honouring Canadian cinema. TIFF also organizes the Sprockets Toronto International Film Festival for Children and Youth. In addition to its success within the very competitive field of film festivals, TIFF has become a well-known Toronto icon, a key component of the Toronto brand.

TIFF's stated goals can be characterized as the promotion of cultural betterment through cinema by engaging audiences in ways that local cinemas, even multiplexes with their bars and game galleries, have been unable to do:

> Our mission is to transform the way we see the world through the film, and TIFF Bell Lightbox is an essential means towards achieving this end. Our new building will allow us to reach many more people, young and old, from every walk of life and from every ethnicity. Film is the most accessible art form, its reach ubiquitous and its influence as a cultural force pervasive ... Understanding this powerful medium is TIFF's role. (Piers Handling, TIFF director and CEO, in TIFF 2009: 3)

A film festival and its associated activities must satisfy up to three separate constituencies whose interests and objectives do not necessarily coincide – the public, industry professionals and celebrities, and public-sector partners. Many festivals cater to one or two of these constituencies, but TIFF is unusual among film festivals in that it successfully creates value for all three. As an open (curated) film festival, TIFF must attract the general public, consisting of local and tourist film enthusiasts who are looking for interesting, high-quality, and novel films, as well as the "'get-together' ambiance of festivals, which are convivial, party-like meeting places for movie lovers, and opportunities to encounter filmmakers" (SECOR Consulting 2004: 4). Film industry professionals, a group composed of directors, producers, distributors, buyers, and writers, seek three things: access to quality films, the opportunity to launch a film, and a gathering of key professionals for deal-making purposes. Public-sector partners seek to promote national films and build audiences for them, demonstrate national culture and its components, and rally the local stakeholder community (SECOR 2004), and they require corresponding evidence of economic, social, and cultural benefits. In 2008–9, TIFF generated economic benefits for Ontario estimated at $162 million, including $61 million in tax revenues (TCI 2010).

The Bell Lightbox, located on the corner of John and King streets in downtown Toronto, is a signature architectural accomplishment in glass designed by Bruce Kuwabara of KPMB Architects, taking its place with the other downtown architectural achievements of Toronto's cultural renaissance: the Royal Ontario Museum, the Art Gallery of Ontario, the National Ballet School, the Royal Conservatory of Music, and OCAD University. The Bell Lightbox contains five cinemas, two galleries, three studios, administrative offices, a rooftop terrace, an open-air amphitheatre, and retail space, allowing the TIFF Group to expand its portfolio of festival and non-festival activities. The opening of the Bell Lightbox in September 2010 was heralded in the blogosphere as the coming-of-age of film festivals, "an enormous evolutionary step in the history of film festivals ... a step that reflects the increasingly expansive roles festivals play in world cinema culture" (Dargis 2010). Specifically, film festivals, with TIFF among the leaders, are "growing up," "even as screens grow small" (ibid.), becoming permanent events that offer special kinds of access to the film world unavailable to spectators in theatres, much less to mere home DVD watchers. This entails investment in real estate and year-round operational infrastructure. An observer at the Lightbox's opening could not refrain from commenting on the "complex ballet of competing interests" that produced the Lightbox, wondering "what exactly this new era represents. At the risk of exploiting too obvious an analogy, this new structure has become the vessel into which all manner of interested parties – festival organizers, city planners, industry participants, cineastes in Toronto and the world over – have poured their sometimes-aligned, sometimes-competing interests" (Kredell 2010).

The process by which the Bell Lightbox emerged illustrates the competing expectations and interests regarding promotion of cinema in Toronto's cognitive-cultural economy, which does not have an economically viable indigenous motion picture industry of its own. The Lightbox was financed by contributions from government, various corporate sponsors, and many private citizens during a ten-year campaign. The most striking aspect of the Lightbox is the way that it allows real-estate developers and film producers to directly convert cultural capital into economic capital. The land on which the Lightbox now sits was donated by the Reitman family, whose parents emigrated to Toronto after the Second World War and operated a car wash on the site in the days when the current Entertainment District was a depressed and unattractive section of the city, and whose progeny include celebrated expatriate

Canadian filmmakers Ivan Reitman (*Ghostbusters*, *Animal House*) and his son Jason (*Juno*, *Up in the Air*). The management of the building's construction was donated by the notable Toronto builder-developer Daniels Corporation. The quid pro quo to induce network spillovers was the forty-four-storey luxury condominium built by the Reitman-Daniels interests atop the Lightbox. Advertised as "one part condo, one part film festival," Festival Tower suites are named after Hollywood stars, and residents receive three-year membership to the Lightbox, preferred pricing to Lightbox events, invitations to parties with film VIPs, exclusive screenings accompanied by special film guests, a special film program for residents developed by TIFF Group experts for screening in the private Festival Tower cinema, special passes, and exclusive direct indoor access to the Bell Lightbox to bypass ticket lines. The value proposition to Tower residents and investors is access to film culture's aura. The websites of the Festival Tower and its forthcoming sister property, the Cinema Tower, are cinema-themed, featuring trailers, flickering images, and ticket metaphors, communicating the invitation to "wrap yourself in the cinematic lifestyle." It is not much of a stretch to suppose that this might result in "looking out from your living room windows and seeing George Clooney on the terrace below – or passing director Ivan Reitman in your condo's front lobby" (Hanes 2010).

This formula of combining luxury downtown real estate with the promise of physical and cultural proximity to cinema celebrities in the heart of the Entertainment District, with easy access to high-end cultural consumption in the form of theatres, musical venues, sports facilities, nightclubs, and restaurants, proved irresistible. Festival Tower became "the hottest address in town" in Toronto, which at the end of 2010 had 286 active condominium construction projects, the largest number of any North American city. Most suites in Festival Tower were reserved long before the building was completed, and the developers are currently constructing a sister building, the 43-storey, 440-suite Cinema Tower, nearby.

Conclusions: Toronto, Innovation, and the Cognitive-Cultural Economy

Toronto's game plan for global competitiveness requires the city to excel as a centre of cognitive-cultural production and consumption. Cognitive-cultural economic growth offers a route towards global-city status that does not depend on sheer size: "size and density are not in themselves essential to new rankings of cities in, for example, global city theory ... Rather what matters is the continuing influence which the city as a socially structured space exerts in the conduct of human life" (Lloyd and Clark 2001). Promoting the urban cognitive-cultural economy signals competitiveness and attractiveness (because the CCE is considered, among targeted mobile investors and talented knowledge workers, to represent the leading edge of production and consumption), and at the same time strengthens the performance of the domestic CCE sector.

The realization of Toronto's growth ambitions will be determined by the city's ability to increase the rate of public and private investment in innovation in its cognitive-cultural economy. The city-region has maintained its status as first among equals on the Canadian domestic scene, but this status translates into runner-up status globally. In this sense, Toronto's growth ambitions and potential are hampered by the political difficulty in Canada of channelling strategic public investments into the designated CCE growth areas of R&D and cultural industries in the Toronto metropolis, by the shrinking latitude for public investment in innovation due to the recent economic crisis, by voter discontent with global cultural ambitions when necessary infrastructure and services such as transportation appear neglected, and by the growing economic and social distance between players in the cognitive-cultural economy and Toronto's urban underclass.

A particularly important challenge for Toronto will be to maintain effective investments in indigenous cultural industries, most of which are facing high levels of foreign competition in the domestic market. As we have seen, much CCE innovation revolves around ICT adoption and non-technological value creation, activities that are largely unrecognized by innovation policies and programs at the federal, provincial, and metropolitan levels. The Canadian TPP-inspired, R&D-based innovation policy framework has little to say about how to improve the performance of indigenous cultural industries.

To fully deploy Scott's concept of the cognitive-cultural economy for purposes of analysis, policy intervention, and improvement of management capabilities would require development of an economic and social knowledge base substantially different from that which is currently available. It would require, notably, a reliable way of characterizing, measuring, and comparing the cognitive and cultural intensity of occupations, firms, industries, and specific regional economies. Our discussion of CCE innovation identifies a further methodological challenge concerning the accurate apprehension of CCE innovation dynamics through theorization and observation of the value-creation processes in CCE innovation. Much greater attention is called for to map and analyse local and extra-local spillovers, interactions, linkages, and knowledge flows, including deeper analysis of the ways that knowledge, product, and network spillovers occur. Insofar as cognitive-cultural innovation is at the core of processes of metropolitan economic and social transformation, this is an area that calls for further comprehensive investigation.

ACKNOWLEDGMENTS

Research reported here was conducted with the support of a grant from the Social Sciences and Humanities Research Council of Canada on "Innovation and Creativity in City Regions." This support is gratefully acknowledged. Also, we thank one of the anonymous manuscript reviewers and the editor of this volume, David Wolfe, for their helpful feedback.

NOTES

1 The study compares Toronto with North American city-regions of similar population size, including Detroit, Montreal, Boston, New York, Chicago, Atlanta, Dallas, Seattle, Vancouver, and Los Angeles.
2 Guidelines for measuring innovation are set out in the *Oslo Manual*, currently in its third edition (OECD, 2005).
3 For a recent review of the literature on firm-level effects of ICT adoption see Brynjolfsson and Saunders (2010).
4 In his influential theory of consumer value, Holbrook (1999) identifies eight kinds of perceived hedonic and utilitarian value: efficiency, excellence, status, esteem, play, aesthetics, ethics, and spirituality. See also Smith and Colgate (2007) for a more recent formulation of types of value. For examples of research on value creation through production of experience in specific segments of cultural industries see Chan (2009) on museums, Davis and Vladica (2010) on animated film, Fiore and Kim (2007) on retailing, Gallarza and Gil (2008) on tourism, Hume and Mort (2008) on performing arts, Park (2004) on restaurants, and Zomerdijk and Voss (2009) on services.

5 The comparison with Waterloo, which has grown two of the largest and most dynamic Canadian-owned ICT firms, Research in Motion and Open Text, is unfavourable to Toronto.
6 Unlike in the United States and many other countries, no single Canadian city predominates as a telecommunications headquarters (Rice 2006).
7 Research on Toronto's media and entertainment industries finds comparable levels of orientation towards the local and regional markets in the book, magazine, film, television, and music industries (HAL 2009), but further research is required to determine how much of this production serves as an intermediate input.
8 Starz Animation Toronto was recently rebranded as Arc Productions after being acquired by a private consortium of Canadian investors in April 2011.
9 On these points see Bontje and Musterd (2009), Currid (2009), Evans (2009), Kipfer and Keil (2002), Mathews (2010), Oakley (2009), Ponzini and Rossi (2010), and Rousseau (2009).

REFERENCES

Aris, Annet, and Jacques Bughin. 2009. *Managing media companies: Harnessing creative value*. Chichester, UK: Wiley.

Arthurs, David, Erin Cassidy, Charles H. Davis, and David A. Wolfe. 2009. "Indicators to support innovation cluster policy." *International Journal of Technology Management* 46 (3/4): 263–79. http://dx.doi.org/10.1504/IJTM. 2009.023376.

Bakhshi, Hasan, and Eric McVittie. 2009. "Creative supply-chain linkages and innovation: Do the creative industries stimulate business innovation in the wider economy?" *Innovation: Management, Policy & Practice* 11 (2): 169–89. http://dx.doi.org/10.5172/impp.11.2.169.

Bakhshi, Hasan, Eric McVittie, and James Simmie. 2008. *Creating innovation: Do the creative industries support innovation in the wider economy?* London: National Endowment for Science, Technology and the Arts.

Baregheh, Anahita, Jennifer Rowley, and Sally Sambrook. 2009. "Towards a multidisciplinary definition of innovation." *Management Decision* 47 (8): 1323–39. http://dx.doi.org/10.1108/00251740910984578.

Beaudry, Catherine, and Andrea Schiffauerova. 2009. "Who's right, Marshall or Jacobs? The localization versus urbanization debate." *Research Policy* 38 (2): 318–37. http://dx.doi.org/10.1016/j.respol.2008.11.010.

Beniger, James R. 1986. *The control revolution: Technological and economic origins of the information society*. Cambridge, MA: Harvard University Press.

Berube, Alan, A. Friedhoff, C. Nadeau, P. Rode, A. Paccoud, J. Kandt, T. Just, and T. Schemm-Gregory. 2010. *Global MetroMonitor: The path to economic recovery. A preliminary overview of 150 global metropolitan economies in the wake of the great recession*. Washington, DC, and London: Metropolitan Policy Program, the Brookings Institution, and LSE Cities, London School of Economics.

Bloch, Carter. 2007. "Assessing recent developments in innovation measurement: The third edition of the Oslo Manual." *Science & Public Policy* 34 (1): 23–34. http://dx.doi.org/10.3152/030234207X190487.

Bontje, Marco, and Sako Musterd. 2009. "Creative industries, creative class and competitiveness: Expert opinions critically appraised." *Geoforum* 40 (5): 843–52. http://dx.doi.org/10.1016/j.geoforum.2009.07.001.

Branham Group. 2010. *Canada's ICT industry: A national perspective*. Ottawa: Branham.

Britton, John N.H. 2007. "Path dependence and cluster adaptation: A case study of Toronto's new media industry." *International Journal of Entrepreneurship and Innovation Management* 7 (2/3/4/5): 272–97. http://dx.doi.org/10.1504/IJEIM.2007.012885.

Brynjolfsson, Erik, and Adam Saunders. 2010. *Wired for innovation: How information technology is reshaping the economy*. Cambridge, MA: MIT Press.

Catungal, John Paul, Deborah Leslie, and Yvonne Hii. 2009. "Geographies of displacement in the creative city: The case of Liberty Village, Toronto." *Urban Studies* (Edinburgh, Scotland) 46 (5–6): 1095–114. http://dx.doi.org/10.1177/0042098009103856.

Caves, Richard E. 2000. *Creative industries: Contracts between art and commerce*. Cambridge, MA: Harvard University Press.

Chan, J.K.L. 2009. "The consumption of museum service experiences." *Journal of Hospitality Marketing and Management* 18 (2–3): 173–96. http://dx.doi.org/10.1080/19368620802590209.

Chapain, Caroline, Phil Cooke, Lisa de Propis, Steward MacNeill, and Juan Mateos-Garcia. 2010. *Creative clusters and innovation: Putting creativity on the map*. London: NESTA.

City of Toronto. 2010. *Toronto: Canada's high-tech hub*. Toronto: City of Toronto and others.

Cohendet, Patrick, David Grandadam, and Laurent Simon. 2010. "The anatomy of the creative city." *Industry and Innovation* 17 (1): 91–111. http://dx.doi.org/10.1080/13662710903573869.

Creative Metropoles. 2010. *Creative metropoles: Situation analysis of 11 cities. Final report*. http://www.creativemetropoles.eu/uploads/files/report_cm_final_formatted_02.2010.pdf.

Currid, Elizabeth. 2009. "Bohemia as subculture, 'Bohemia' as industry: Art, culture, and economic development." *Journal of Planning Literature* 23 (4): 368–82. http://dx.doi.org/10.1177/0885412209335727.

Dargis, Manohla. 2010. "Festivals grow up, even as screens grow small." *New York Times*, 24 September.

Davis, Charles H. 2010. "New firms in the screen-based media industry: Start-ups, self-employment, and standing reserve." In *Managing media work*, ed. M. Deuze, 165–78. Thousand Oaks, CA: Sage.

Davis, Charles H. 2011. "The Toronto media cluster: Between culture and commerce." In Charlie Karlsson and Robert Picard, eds, *Media clusters across the globe: Developing, expanding, and reinvigorating content capabilities*, 223–50. Cheltenham: Edward Elgar.

Davis, Charles H., and Florin Vladica. 2010. "Consumer value and modes of media reception: Audience responses to the computer-animated psycho-realist documentary *Ryan* and its own documentation in *Alter Egos*." *Palabra Clave* 13 (1): 13–30.

Davis, Charles H., Florin Vladica, and Irene Berkowitz. 2008. "Business capabilities of small entrepreneurial media firms: Independent production of children's television in Canada." *Journal of Media Business Studies* 5 (1): 9–40.

De Vany, Arthur. 2004. *Hollywood economics: How extreme uncertainty shapes the film industry*. New York: Routledge.

Donegan, Mary, Joshua Drucker, Harvey Goldstein, Nichola Lowe, and Emil Malizia. 2008. "Which indicators explain metropolitan economic performance best? Traditional or creative class." *Journal of the American Planning Association* 74 (2): 180–95. http://dx.doi.org/10.1080/01944360801944948.

D'Ovidio, Marianna. 2010. "Network locali nelle economia cognitiva-culturale: Il case di Milano." *Rassegna Italiana di Sociologia* 51 (3): 459–84.

E&B Data. 2004. *Greater Toronto Information and Communication Technologies (ICT) industry profile 2004*. Toronto: E&B Data for Greater Toronto Marketing Alliance and others.

Ejermo, Olof. 2005. "Technological diversity and Jacobs' externality hypothesis revisited." *Growth and Change* 36 (2): 167–95. http://dx.doi.org/10.1111/j.1468-2257.2005.00273.x.

Eltham, Ben. 2012. "Three arguments against 'soft innovation': Towards a richer understanding of cultural innovation." *International Journal of Cultural Policy* 1–20. http://dx.doi.org/10.1080/10286632.2012.658044.

Evans, Graeme. 2009. "Creative Cities, creative spaces and urban policy." *Urban Studies* (Edinburgh, Scotland) 46 (5–6): 1003–40. http://dx.doi.org/10.1177/0042098009103853.

Featherstone, Mike. 1987. "Lifestyle and consumer culture." *Theory, Culture & Society* 4 (1): 55–70. http://dx.doi.org/10.1177/026327687004001003.

Fiore, Ann Marie, and Jihyun Kim. 2007. "An integrative framework capturing experiential and utilitarian shopping experiences." *International Journal of Retail and Distribution Management* 35 (6): 421–42. http://dx.doi.org/10.1108/09590550710750313.

Frontier Economics. 2007. *Creative industry spillovers: Understanding their impact on the wider economy*. London: Frontier Economics Ltd.

Gallarza, Martina G., and Irene Gil. 2008. "The concept of value and its dimensions: A tool for analysing tourism experiences." *Tourism Review* 63 (3): 4–20. http://dx.doi.org/10.1108/16605370810901553.

Gertler, Meric S., Lori Tesolin, and Sarah Weinstock. 2006. *Toronto case study*. Munk Centre for International Studies, University of Toronto.

Grant, Peter S. 2008. *Stories under stress: The challenge for indigenous television drama in English-language broadcast markets*. Report prepared for the International Affiliation of Writers Guilds, December.

HAL. 2009. *Ontario's creative cluster study: Music – film and TV – magazines – books – interactive digital media*. Ottawa: Hickling Arthurs Low for the Ontario Ministry of Culture.

Hanes, Tracey. 2010. "Living in the heart of the festival." Toronto.com, 15 September.

Harvey, David. 2005. *A brief history of neoliberalism*. Oxford: Oxford University Press.

Hauknes, Johan. 2003. *Innovation and economic behaviour: Need for a new approach?* Oslo: STEP - Centre for Innovation Research. STEP report 20-2003.

Hawkins, Richard, and Charles H. Davis. 2012. "Innovation and experience goods: A critical appraisal of a missing dimension in innovation theory." *Prometheus* 30 (1): 235–59.

Hayles, Katherine. 2005. *My mother was a computer*. Chicago: University of Chicago Press.

Holbrook, Morris B. 1999. "Introduction to consumer value." In *Consumer value: A framework for analysis and research*, ed. M. Holbrook, 1–28. London, New York: Routledge. http://dx.doi.org/10.4324/9780203010679.ch0.

Hopkins, Michael S. 2010. "Value creation, experiments and why IT does matter: Interview with Michael Schrage." *MIT Sloan Management Review* 51 (3): 57–61.

Hume, Margee, and Gillian Sullivan Mort. 2008. "Satisfaction in performing arts: The role of value?" *European Journal of Marketing* 42 (3–4): 311–26. http://dx.doi.org/10.1108/03090560810852959.

Impact Group. 2006. *ICT Toronto: An Information and Communication Technology (ICT) cluster development strategy for the Toronto region*. Toronto: Impact Group for City of Toronto Economic Development and others.

Indergaard, Michael. 2009. "What to make of New York's new economy? The politics of the creative field." *Urban Studies* (Edinburgh, Scotland) 46 (5–6): 1063–93. http://dx.doi.org/10.1177/0042098009103855.

Jaaniste, Luke. 2009. "Placing the creative sector within innovation: The full gamut." *Innovation: Management, Policy & Practice* 11 (2): 215–29. http://dx.doi.org/10.5172/impp.11.2.215.

Jenkins, Henry. 2006. *Convergence culture: Where old and new media collide*. New York: New York University Press.

Kallinikos, Jannis. 1999. "Computer-based technology and the constitution of work: A study on the cognitive foundations of work." *Accounting, Management and Information Technology* 9 (4): 261–91. http://dx.doi.org/10.1016/S0959-8022(99)00011-9.

Kipfer, Stefan, and Roger Keil. 2002. "Toronto Inc? Planning the competitive city in the New Toronto." *Antipode* 34 (2): 227–64. http://dx.doi.org/10.1111/1467-8330.00237.

Kloosterman, Robert C. 2010. "Embedding the cognitive-cultural urban economy." *Geografiska Annaler. Series B, Human Geography* 92 (2): 131–43. http://dx.doi.org/10.1111/j.1468-0467.2010.00338.x.

Kredell, Brendan. 2010. "The different hats of Toronto's Bell Lightbox." *In Media Res*, 17 September. http://mediacommons.futureofthebook.org/imr/2010/09/17/different-hats-torontos-bell-lightbox.

Lloyd, Richard, and Terry Nichols Clark. 2001. "The city as an entertainment machine." *Research in Urban Sociology: Critical Perspectives on Urban Redevelopment* 6:357–78. http://dx.doi.org/10.1016/S1047-0042(01)80014-3.

Lorenzen, Mark, Alan J. Scott, and Jan Vang. 2008. "Editorial: Geography and the cultural economy." *Journal of Economic Geography* 8 (5): 589–92. http://dx.doi.org/10.1093/jeg/lbn026.

Lucas, Matthew, Anita Sands, and David A. Wolfe. 2009. "Regional clusters in a global industry: ICT clusters in Canada." *European Planning Studies* 17 (2): 189–209. http://dx.doi.org/10.1080/09654310802553415.

Matheson, Sarah A. 2004. "Projecting placelessness: Industrial television and the "authentic" Canadian city." In *Contracting out Hollywood: Runaway productions and foreign location shooting*, ed. Greg Elmer and Michael Gasher, 117–39. Lanham, MD: Rowman and Littlefield.

Mathews, Vanessa. 2010. "Aestheticizing space: Art, gentrification and the city." *Geography Compass* 4 (6): 660–75. http://dx.doi.org/10.1111/j.1749-8198.2010.00331.x.

Nordicity. 2008. *Economic profile of the Ontario computer animation and visual effects industry*. Toronto: Nordicity for Computer Animation Studios of Ontario.

Nordicity. 2009. *Ontario profile 2009: An economic profile of domestic film, television, and cross-platform interactive media production*. Toronto: Nordicity for Ontario Producers Panel and Canadian Film and Television Production Association.

Oakley, Kate. 2009. "Getting out of place: The mobile creative class takes on the local." In *Creative economies, creative cities: Asian-European perspectives*, ed. L. Kong and J. O'Connor, 121–34. Berlin: Springer. http://dx.doi.org/10.1007/978-1-4020-9949-6_8.

OECD. 2005. *Oslo Manual: Guidelines for collecting and interpreting innovation data*. 3rd ed. Paris: Organisation for Economic Co-operation and Development.

OECD. 2010a. *Measuring innovation: A new perspective*. Paris: Organisation for Economic Co-operation and Development.

OECD. 2010b. *The OECD innovation strategy: Getting a head start on tomorrow*. Paris: Organisation for Economic Co-operation and Development.

OECD. 2010c. *OECD territorial reviews: Toronto, Canada*. Paris: Organisation for Economic Co-operation and Development.

Park, C. 2004. "Efficient or enjoyable? Consumer values of eating-out and fast food restaurant consumption in Korea." *International Journal of*

Hospitality Management 23 (1): 87–94. http://dx.doi.org/10.1016/j.ijhm.
2003.08.001.

Pine, B. Joseph, and James H. Gilmore. 1999. *The experience economy: Work is theater and every business a stage.* Cambridge, MA: Harvard Business School Press.

Ponzini, Davide, and Ugo Rossi. 2010. "Becoming a creative city: The entrepreneurial mayor, network politics and the promise of an urban renaissance." *Urban Studies* (Edinburgh, Scotland) 47 (5): 1037–57. http://dx.doi.org/10.1177/0042098009353073.

Potts, Jason, and Stuart Cunningham. 2008. "Four models of the creative industries." *International Journal of Cultural Policy* 14 (3): 233–47. http://dx.doi.org/10.1080/10286630802281780.

Rice, Murray D. 2006. "Dominant national centres: A comparative analysis of the headquarters communities of New York and Toronto." *Industrial Geographer* 4 (1): 29–51.

Rousseau, Max. 2009. "Re-imagining the city centre for the middle classes: Regeneration, gentrification and symbolic politics in 'loser cities.'" *International Journal of Urban and Regional Research* 33 (3): 770–88. http://dx.doi.org/10.1111/j.1468-2427.2009.00889.x.

Roy, Ananya. 2009. "The 21st-century metropolis: New geographies of theory." *Regional Studies* 43 (6): 819–30. http://dx.doi.org/10.1080/00343400701809665.

Salazar, Monica, and Adam Holbrook. 2004. "A debate on innovation surveys." *Science & Public Policy* 31 (4): 254–66. http://dx.doi.org/10.3152/147154304781779976.

Sawhney, Mohanbir, Wolcott, Robert C., and Inigo Arroniz. 2006. "The 12 different ways for companies to innovate." *MIT Sloan Management Review* 47 (3): 75–81.

Scott, Allen J. 1997. "The cultural economy of cities." *International Journal of Urban and Regional Research* 21 (2): 323–39. http://dx.doi.org/10.1111/1468-2427.00075.

Scott, Allen J. 2004. "Cultural-products industries and urban economic development: Prospects for growth and market contestation in global context." *Urban Affairs Review* 39 (4): 461–90. http://dx.doi.org/10.1177/1078087403261256.

Scott, Allen J. 2007. "Capitalism and urbanization in a new key? The cognitive-cultural dimension." *Social Forces* 85 (4): 1465–82. http://dx.doi.org/10.1353/sof.2007.0078.

Scott, Allen J. 2008a. "Inside the city: On urbanisation, public policy and planning." *Urban Studies* (Edinburgh, Scotland) 45 (4): 755–72. http://dx.doi.org/10.1177/0042098007088466.

Scott, Allen J. 2008b. "Resurgent metropolis: Economy, society and urbanization in an interconnected world." *International Journal of Urban and Regional Research* 32 (3): 548–64. http://dx.doi.org/10.1111/j.1468-2427.2008.00795.x.

Scott, Allen J. 2008c. *Social economy of the metropolis: Cognitive-cultural capitalism and the global resurgence of cities.* Oxford: Oxford University Press. http://dx.doi.org/10.1093/acprof:oso/9780199549306.001.0001.

Scott, Allen J. 2010. "Cultural economy and the creative field of the city." *Geografiska Annaler. Series B, Human Geography* 92 (2): 115–30. http://dx.doi.org/10.1111/j.1468-0467.2010.00337.x.

Scott, Allen J. 2011. "A world in emergence: Notes toward a resynthesis of urban-economic geography for the 21st century." *Urban Geography* 32 (6): 845–70. http://dx.doi.org/10.2747/0272-3638.32.6.845.

Scott, Allen J., J. Agnew, E. Soja, and M. Storper. 2001. "Global city-regions." In *Global city-regions: Trends, theory, policy,* ed. A. Scott, 11–30. Oxford: Oxford University Press.

SECOR Consulting. 2004. *Analysis of Canada's major film festivals.* Montreal: SECOR.

Sharpe, Andrew. 2006. *The relationship between ICT investment and productivity in the Canadian economy: A review of the evidence.* Ottawa: Centre for the Study of Living Standards.

Sharpe, Dean, Tim Jones, Tony Lea, Ken Jones, and Sue Harvey. 2004. *Critique and consolidation of research on the spillover effects of investments in cultural facilities.* Toronto: Artscape and Ryerson University. Report submitted to Canadian Heritage.

SICA. 2010. *The economic crisis and the prospects for arts and culture in Europe.* Amsterdam: Dutch Centre for International Cultural Activities. http://www.sica.nl/sites/default/files/EN_Crisis_en_bezuinigingen_in_Europa_juni2010.pdf.

Simonton, Dean Keith. 2009. "Cinematic success criteria and their predictors: The art and business of the film industry." *Psychology and Marketing* 26 (5): 400–20. http://dx.doi.org/10.1002/mar.20280.

Smith, J. Brock, and Mark Colgate. 2007. "Customer value creation: A practical framework." *Journal of Marketing Theory and Practice* 15 (1): 7–23. http://dx.doi.org/10.2753/MTP1069-6679150101.

Soberman, David, and Hubert Gatignon. 2005. "Research issues at the boundary of competitive dynamics and market evolution." *Marketing Science* 24 (1): 165–74. http://dx.doi.org/10.1287/mksc.1040.0065.

Stoneman, Paul. 2010. *Soft innovation: Economics, product aesthetics, and the creative industries.* Oxford: Oxford University Press.

Storper, Michael, and Allen J. Scott. 2009. "Rethinking human capital, creativity and urban growth." *Journal of Economic Geography* 9 (2): 147–67. http://dx.doi.org/10.1093/jeg/lbn052.

TCI. 2010. *Economic activity associated with the 2008–2009 operations of TIFF.* Toronto: TCI Management Consultants for Toronto International Film Festival.

TIFF. 2009. *TIFF. 2009 annual report.* Toronto: Toronto International Film Festival.

Toronto Board of Trade. 2009. *From world-class to world leader: An action plan for the Toronto region*. Toronto: Toronto Board of Trade.

TRRA. 2010. *2008 annual Toronto Region innovation gauge*. Toronto: Toronto Region Research Alliance.

Van der Panne, Gerben. 2004. "Agglomeration externalities: Marshall versus Jacobs." *Journal of Evolutionary Economics* 14 (5): 593–604. http://dx.doi.org/10.1007/s00191-004-0232-x.

Vinodrai, Tara. 2009. *The place of design: Exploring Ontario's design economy*. Toronto: Martin Prosperity Institute working paper 2009-WPONT-014.

Vinodrai, Tara, and Meric S. Gertler. 2007. "Measuring the creative economy: The structure and economic performance of Ontario's creative, cultural and new media clusters." Toronto: University of Toronto, unpublished report prepared for the Ontario Ministry of Culture.

Vukanovic, Zvezdan. 2009. "Global paradigm shift: Strategic management of new and digital media in new and digital economics." *International Journal on Media Management* 11: 81–91.

Wolfe, David A. 2009. *21st century cities in Canada: The geography of innovation*. Ottawa: Conference Board of Canada.

Wolfe, David A., and Allison Bramwell. 2008. "Innovation, creativity and governance: social dynamics of economic performance in city-regions." *Innovation: Management, Policy & Practice* 10 (2–3): 170–82. http://dx.doi.org/10.5172/impp.453.10.2-3.170.

Yoo, Youngjin, Ola Henfridsson, and Kalle Lyytinen. 2010. "The new organizing logic of digital innovation: An agenda for information systems research." *Information Systems Research* 21 (4): 724–35. http://dx.doi.org/10.1287/isre.1100.0322.

Zomerdijk, Leonieke G., and Christopher A. Voss. 2009. "Service design for experience-centric services." *Journal of Service Research* 13 (1): 67–82. http://dx.doi.org/10.1177/1094670509351960.

Zwick, Detlev, and Janice Denegri Knott. 2009. "Manufacturing customers: The database as a new means of production." *Journal of Consumer Culture* 9 (2): 221–47. http://dx.doi.org/10.1177/1469540509104375.

4 Living on the Edge: Knowledge Interdependencies of Human Capital Intensive Clusters in Vancouver

BRIAN WIXTED AND J. ADAM HOLBROOK

What place is there in the global structure of innovation for specific locations? We know that innovation is spiky (Florida 2005) – high peaks standing above the deserts that essentially correspond to population geography. However, many of these important places are co-located among large mega-regions (Florida et al. 2008). Taylor (2004) expresses the idea that the global economy is structured around a limited set of global city-regions (in fact mega-regions). Not surprisingly, then, recent trends in the literature on innovation systems has focused on cities and, specifically, on the role and the relative importance of knowledge specialization versus diversity in stimulating *urban* innovation and growth (Glaeser et al. 1992; Beaudry and Schiffauerova 2009).[1] The implementation of this research question has focused on the internal dynamics of the city rather than the external relations and setting – essentially treating each city as an island.[2] Typically, a key variable for the urban system context (see, for example, Filion 2010) has been city size (see, e.g., Duranton and Puga 2000, 2005 – bigger cities innovate more).

But what of cases, such as Vancouver, on the west coast of Canada, an urban island cut off from other population centres by water, mountains, and an international political boundary. In this chapter we examine the role of knowledge specialization and diversity in industry innovation within the context of the influence of the larger physical geographic setting. To do this, we focus on the issue overlooked by many researchers:[3] the extent of the knowledge specialization and diversity dynamic in an inter-city (multi-spatial) framework.

The Metro Vancouver region, while having a high degree of industrial diversity (Beckstead and Brown 2004), is not a locus for manufacturing

(Spencer and Vinodrai 2009). To explore the nature of knowledge dynamics in Vancouver's urban economy, we investigated five human-capital-based segments of the economy: fuel cells, wireless (mobile phones and related technologies), new media, bio-pharma, and Hollywood North (feature movies). These sectors have been labelled human-capital-intensive because it is not clear that the label of high technology is useful in this situation, as Hollywood North, bio-pharma, fuel cells, and new media are focused on intellectual property (IP) production. However, each of them is highly reliant on skilled technicians, scientific researchers, and creative "talent" (Florida 2002).

The second key aspect of our analysis is to layer the analysis of specific economic activities and their patterns of knowledge flows with concepts from the core-periphery and agglomeration literature in the new economic geography (see, for example, Krugman 1998) to investigate inter-city *innovation* system structures. To undertake this analysis we utilized the idea that the life cycle of the Vancouver cluster relative to the global life cycle of the sectors helps explain the nature of its dependence on indigenous knowledge flows or knowledge spillovers from elsewhere.

Taken together, these three perspectives, knowledge specialization and diversity in specific economic sectors in the context of their life-cycle history, allows us to learn much about the dynamics of place and the social characteristics of innovation for Vancouver. The entrepreneurship and creativity is locally embedded, but exhibits the unusual characteristic that economic activity sits within a value chain where intellectual property is the principal output. Mostly, knowledge is assembled in Vancouver based on localized specializations and augmented by strategic interactions across local and distant clusters. Apart from fuel cells, all the clusters relied upon significant knowledge exchanges with organizations in cities in the eastern North American mega-region or in southern California (sometimes even further afield). Some inter-city relations were based in specialization (bio-pharma), while others (new media) were based in diversity locally and inter-regionally. From the evidence, a few fundamentals for Metro Vancouver policy development can be formulated, mainly related to lifestyle and maintenance of the "stickiness" of the labour market.

The current level of innovative activity in Vancouver is probably due to its ability to attract a degree of talent well beyond its local buying power.[4] Our analysis reveals that in many instances, the clusters in Vancouver emerged early relative to the emergence of their global counterparts. However, while the clusters have been successful in survival terms, their success in economic terms is not straightforward.

It appears certain that the broader setting of Vancouver's urban economy has been crucial in this development. The favourable climate and geo-physical position as a necessary airline and shipping gateway contributes to the local innovation culture and activity (Barnes et al. 2011). We therefore conclude by postulating that cities on the Pacific Rim have developed important innovation-system properties on a trajectory that differs from the mega-regions of Europe and eastern North America.

Innovation in City Systems

Specialization versus Diversity: Innovation in City Regions

During the 1990s and early 2000s the innovation system construct primarily focused on national and sub-national (regions) jurisdictions, including clustering over various scales of aggregation (see, for example, Edquist 1997; Carlsson 2006; Wixted 2009).

There is still active discussion of innovation systems at the regional and national scales, but recently there has been substantial growth of interest in cities and the clustering of economic activities of particular sectors in cities (metropolitan regions). The research question at the heart of much of this work is whether innovation in cities benefits from so-called MAR (Marshall-Arrow-Romer) spillovers or Jacobs externalities (Glaeser et al. 1992; Beaudry and Schiffauerova 2009; Wolfe this volume).

A basic finding of this literature is that larger cities tend to be more conducive to Jacobs-type externalities and knowledge flows. However, Beaudry and Schiffauerova (2009) also argue that the MAR and Jacobian approaches are industry- rather than city-specific. Clusters that are highly IP-dependent are more likely to follow the Jacobs model, while clusters that include manufacturing activities are more likely to follow the MAR model. This finding is highly relevant to Vancouver's history. Given the low level of manufacturing in Vancouver, it was logical to study activities that produce mainly IP to test this hypothesis. Nevertheless, the evidence presented here reveals that we also need to go beyond the simple MAR-Jacobs dichotomy of intra-city activity and include multi-spatial knowledge spillovers.

As many of Vancouver's rising clusters are human-capital-driven, rather than resource-sector-based (see Barnes and Hutton 2009), Jacobs's arguments on the attractiveness of place are of special interest here. Sticky labour markets are important. In Canada there are often local cultural traits: for example, the entrepreneurial traditions of western Canada and the francophone culture of Quebec, which are major factors in determining the growth and nature of specific industrial clusters. It has been argued by Florida (2002) that highly skilled personnel move to locales where they want to live and then seek employment. One might add that, just as importantly, highly skilled people might also stay where they like it rather than move away. But this still leaves the question of what are the main critical factors that attract and hold these human-capital-based sectors in Vancouver?

Agglomeration, Centrifugal and Centripetal Forces

It is surprising that, while there is much written on agglomeration (particularly with regard to clusters), within innovation theory there is no body of research which examines agglomeration as a property of the macro-geographic environment.

In terms of the MAR-Jacobs debate, patterns of agglomeration are driven by specialization and economic cost structures (transport, land, and labour mobility), with diversity essentially irrelevant. However, Florida's analysis of the movement of creative people may find its place in the system with the idea of labour mobility. The key implication of this perspective is the evolution of activities inwards to larger centres and away from smaller "centres." In the Canadian context, with its relatively free labour markets and moderate transport-system cost structure, this might support the concentration of activity in the mega-region from Windsor through to Montreal (Canada's "main street" – Filion 2010), which itself is geographically close to the eastern US mega-region. However, Vancouver, the third largest city in the country, is 3300 kilometres away from the Canadian industrial heartland.

Taking this further, what does local mean, and how far is close enough? If we examine the relations of location/distance of industries and activities from one another, we find first that over the long term city agglomeration patterns have not been altered by innovations such as IT (Polèse and Shearmur 2004). Using data from a Canadian innovation survey, Shearmur (2008) examined the relationship between urban and rural geographies in innovation. He comments:

> For certain types of innovation it is not local context, but access to resources, that is important. Access is a spatial concept: except for information available by electronic means, most other inputs to innovation have a geographic dimension: face-to-face contacts, access to knowledge

intensive business services, entry points to global pipelines, delivery of products – all these are easier from some locations and more difficult from others. (25)

Going further he continues: "In order to maintain the necessary proximities, physical infrastructure and basic services (such as air transport, highway maintenance, hotels etc. ...) are required. It is no coincidence that many – if not most – functioning innovation systems are in or near major metropolitan areas" (25).

Such modelling is highly informative and enables us to attribute a certain number of economic geographic phenomena to macro developments. Nevertheless, one must bear in mind Martin's (1999) warnings against the shortcomings of a mathematically based economics of space; instead, Martin lauds economic geography's turn towards the particularities of place. His emphasis on the richness of understanding history, culture, and the local determinants of growth and decline is highly relevant to our case study of Vancouver.

As a broad generalization, many cluster studies are so particular they give no indication as to their macro-geographic setting, a criticism made by Markusen (1999), although this does not appear to have altered the trajectory of research.

Knowledge-intensive Industries in Vancouver

In this chapter we develop an understanding both of Vancouver's place in the larger Canadian urban system, as well as of its unique history on western edge of the North American continent. Our analysis is based on approximately one hundred interviews conducted across nearly a decade, supplemented with archival material and other data, where available.

The study of Vancouver focuses on an examination of knowledge-based (or human-capital-intensive) economic activities as the basis for both the interviews and historical pattern of development. The five activities chosen were:

- fuel cells (predominantly based on hydrogen technologies), using interviews and published papers based on in-depth interviews;
- bio-pharma (a range of firms creating human-health-oriented biotechnologies or traditional pharmaceuticals), based on interviews;
- motion pictures based on extant literature;

- new media (mainly the electronic games sector), based on interviews; and
- wireless technologies, based on interviews and extant literature.

The central question of the Innovation Systems Research Network (ISRN) project relates to the importance of industrial and knowledge specialization relative to industrial and knowledge diversity for innovation. Within this broader context, we were particularly interested in the "why Vancouver" question. This led us to look at the work on the forces of agglomeration and de-agglomeration, with an implied framework of specialization and economic cost structures across a wide geographic system (more than one city). An important component of the study was to use the interview and historical material to understand agglomeration (centripetal/centrifugal) forces by examining the life-cycle stage of the Vancouver clusters in relation to global industry life-cycle patterns.

The logic here is that the start-up time of local clusters can show whether there are centrifugal pressures (labour/land costs, knowledge specialization, etc.) at work driving the activity away from initial centres and finding niches in new cities or agglomeration forces sucking activity away from small centres to mega-centres. If the first case is true, we could expect that start-up times of clusters will be delayed relative to global start-up. If the alternative is true, then we could expect that, regardless of start-up time, the cluster will have experienced difficulty in maturing.

For example, a technology industry may have started up somewhere else in the 1980s, but only emerged in Vancouver in the late 1990s. This would be a primary indicator that industry evolution was expanding the geographic frontiers and creating a space for new players. *This would imply that command and control, as well as knowledge, spillovers are based in distant centres and not in internal accumulation.* This would be prima facie evidence of a form of outsourcing, although as we shall . see, that term is not an accurate way of thinking about Vancouver even in late emergent cases. Alternatively, if the cluster start-up times were relatively close together, then it would indicate that enough research and technical expertise resided within the city itself to generate innovation and new industries.

In current analyses of cluster lifecycles (figure 4.1), Immarino and McCann (2006), Audretsch and Feldman 1996 and Klepper (2001) maintain that what is important is the life cycle of a cluster/industry

Figure 4.1   Cluster life cycles

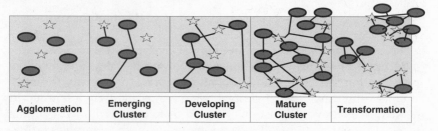

| Agglomeration | Emerging Cluster | Developing Cluster | Mature Cluster | Transformation |

Source: Redrawn from Andersson et al. 2004: 29

in a particular geography. Developing a correspondence between the local and the global has two dimensions: first, the start-up dates of local clusters vis-à-vis the global start-up date and, second, the current conditions of the clusters vis-à-vis the global industry. A typology of life cycle that is attractive because it is not couched in the usual S-curve language, instead using terminology representative of network formation, was presented in Andersson et al. (2004).

In adopting this language we decided that we could use the same concepts for the Vancouver city region as we could for the global industry and adopted a matrix frame of life cycles locally and internationally.

Vancouver in the Canadian Urban System

Vancouver is a mid-sized city (approx. 2.1 million) on the western fringe of the North American continent (figure 4.2). It is a population island bounded by water, mountains, and the US border. Vancouver's sister city to the south, Seattle, has very different dynamics, hosting both a large aerospace cluster with Boeing's assembly base and an IT cluster that includes Microsoft. A recent OECD report (2006), for example, reported that Vancouver had a US purchasing-power-parity per capita income of $32,000 compared to Seattle's $54,000.

Vancouver has particular underlying dynamics that include the attraction of talent (see below) and offshore capital (particularly Chinese investment in real estate; Penner (2010). The formal economy of Vancouver is diversified across a number of sectors and has important infrastructure, including the fourth largest shipping port in North America and an important international airport.

Figure 4.2 Vancouver in the North American city system context

Vancouver as the Quintessential Talent Attractor

Vancouver, and British Columbia, has had huge growth in micro businesses during the late 1990s and 2000, and this is noticeable in the business demography statistics. Vancouver is more weighted to very small businesses and as a share of the business population lacks the middle-sized and larger business.[5] This is important because larger businesses tend to provide a more stable labour market and higher income sources (Harrison 1994). An interesting insight into the dynamics of the BC economy was provided in the course of 2008, which started in the boom and ended with the beginnings of the economic crisis. BC Stats (2009) reported:

> In 2008, self-employment reached a low early in the year (approximately 403,300 self-employed, province-wide), while its highest level was recorded in September (454,100) ... At the outset of 2008, British Columbia was still feeling the effect of a labour crunch, which is likely to have had at least some influence on self-employment levels in the province. In particular, a tight economy and labour shortages can potentially draw self-employed workers into the employee workforce. Conversely, the economic downturn in the latter part of 2008 is likely to have had its own impact on people's tendencies to choose self-employment.

For all its industrial diversity and rich history of technological innovation, Vancouver does not profit from these activities as much as one might expect. If we compare Vancouver's pay scales for bachelor degree and above, it is below the national average. For certificates and diploma degrees Vancouver is marginally above the national average (figure 4.3). If we were to use occupational data, a similar picture could be presented, with the higher skill categories receiving less pay than the average and the lowest occupational categories receiving par or more than the national average.

Barnes et al. (2011) show that, except for a few of the lower-level occupational groups, more people in Vancouver work less than a full working week than their counterparts in Toronto, Montreal, or indeed the Canadian average. They suggest this is a sign of a "more leisure oriented labour force" (306). It is equally plausible that the small entrepreneurial businesses are unable to embed sufficient value into the local system of innovation.

Despite the pay scales, Vancouver keeps its talent and can attract new talent. For example, Helliwell and Helliwell (2001) reveal that between the early 1950s and late 1990s there has been a trend for greater numbers of University of British Columbia graduates across a number of degree levels and types (notably with the exception of MBAs and medicine) to reside in British Columbia. This is an indication of talent retention. While analysis of the ISRN city-region profiles suggests that between 2001 and 2006 there was a net outmigration intra-provincially, there was net in-migration across provinces, and each year British Columbia attracts about 44,000 immigrants. Possibly driven by the

Figure 4.3 Average annual income by level of qualification, 2006

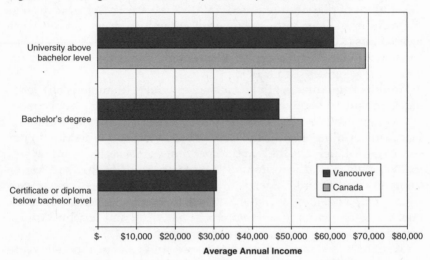

Source: ISRN city-region profiles 2006 (http://www.utoronto.ca/isrn/city%20profiles/
index.html)

high rates of immigration, Vancouver is rated as the most expensive
city to live in Canada.[6]

Case Histories

It is not sufficient to examine the local knowledge specialization and
diversity patterns; we also examine questions of "spill-ins," in other
words, the benefits flowing into particular activities from outside the
region, and also the labour markets for skilled people in these knowledge-intensive clusters.

Fuel Cells

Geoffrey Ballard, a Canadian scientist who moved to Vancouver in the
1980s, took an abandoned fuel cell design whose patents had expired
and searched for ways to improve its power and build it out of cheaper
materials. In doing so, he founded Ballard Power, which is today a
major player in the global fuel-cells industry. Ballard Power developed

fuel cells with a significant increase in power density while reducing the amount of platinum required. By 1992 Ballard Power Systems had produced a fuel cell system powerful enough to drive a small commercial bus (Koppel 1999). By 1998 Ballard was partnering with Daimler-Benz to pursue the commercialization of fuel cell technologies for automotive engines (Panik 1998).

From its beginnings with just a single, though technologically leading, company, Vancouver today is the base for a leading global cluster of fuel cell expertise. Price Waterhouse Coopers (2004) has estimated that British Columbia (mostly Metro Vancouver) accounted for 43 per cent of global fuel cells R&D and 38 per cent of that sector's employees. In 2008 Ballard sold its automotive research assets to the Ford Motor Company and Daimler-Benz.[7] In addition, Vancouver is the home of research organizations such the National Research Council Institute for Fuel Cells Innovation (see Holbrook et al. 2010), and nearby Victoria has an important university research program.

Of all the knowledge-intensive activities studied here, fuel cells is the smallest and most specialized sector. Although it has benefited from R&D capital from outside the region, it is significantly an undiversified cluster, as it relies on local R&D facilities and accessing near-local talent from the University of Victoria (Holbrook et al. 2010). It does not appear to require extensive supply chains,[8] as most of the materials for fuel cells are experimental and commercial materials appear to be machined in Vancouver.

Although Vancouver has been a first mover in this industry, we make no claims as to what the future for the activity is. It may be that research and design will remain and manufacturing start up elsewhere. Alternatively, it could be that the activity will move to larger research centres (see discussions in Molot 2008 and Molot and Mytelka 2009). Indeed, there is recent evidence that through the earlier partnership with Daimler-Benz, some of the key knowledge and technology developed in the Vancouver fuel cell cluster was subsequently transferred back to Germany and has become a key factor in contributing to the growth of the fuel cell industry in that country (Tanner 2012). However, Daimler AG is not cutting ties with Vancouver completely and operates a facility in the same building as Ballard Power Systems.The entrepreneurs are local and the technological developments have been local, with the global firms coming to the Vancouver cluster to play in the field, but there is only small-scale manufacturing occurring in the region and it lags well behind the industrial capacity found in the major European countries.

Bio-Pharma

According to Rothaermel, the founding of Genentech in 1976 is often referred to as the start of the commercialized biotechnology industry (2000: 155). The genesis of the bio-pharma cluster in Vancouver started with major investments in bio-pharma research at the University of British Columbia in the 1970s. Significant breakthroughs were made in these labs, not the least of which led to a Nobel Prize for Dr Michael Smith. One of the early successful spin-offs was QLT (founded 1981: LifeSciences BC 2010), a company that successfully developed and marketed a treatment for macular degeneration. This research effort and a series of successful spin-offs (see Clayman and Holbrook 2004) have led to a vibrant industry, albeit one that develops, but does not manufacture, its products. The bio-pharma group in Vancouver specializes in developing new IP-based products and then selling the IP rights to other, larger firms to bring to market. Today there are about one hundred firms in the bio-pharmaceuticals cluster, but the majority remain very small (LifeSciences BC 2010).

The prime characteristic of the bio-pharma cluster in Vancouver is one largely of IP rent seeking. For example, it does not have the critical mass of other important centres east and south, but companies do have strategies for managing the construction of value in Vancouver. So, for example, as one manager explained, key inputs, such as strategic pieces of knowledge or people, are sourced from across North America (or the world):

> Research materials and R&D contractors are mostly sourced from other parts of North America, very commonly in the States. In our case, materials and labour are mostly sourced from locations on the west coast of the United States. The contract organizations generally follow the distribution of the major universities and the major technical centres in North America. So they're exactly as you'd expect. There's an east coast concentration, there's a west coast concentration, and there's a very big concentration in North Carolina because North Carolina was extremely effective a number of years ago at attracting the pharmaceutical industry and it gathered around it all the suppliers and other experts.

Why are they in Vancouver at all? Quite possibly it is due to an excellent university, which over the years generated enough spin-out and sticky human capital that there is now a respectable cluster of expertise.

> [The] University of British Columbia is a main advantage. Legal and financial services [are] sourced from Vancouver ... To a limited extent the company does benefit from ideas generated in the local milieu and the skilled people that reside here. But all these things are comparative. Compared to being in Saskatchewan, in Saskatoon, I would think probably yes. Compared to being in the Bay Area, absolutely no. We're here because some groundbreaking science was done here, and the scientists who created those innovations decided it was worthwhile trying to commercialize them, hence we're here. We get enormous benefit from continuing to work with them. (Confidential interview)

The knowledge and human capital base is somewhat specialized in this cluster, but it is highly reliant on maintaining a multi-spatial structure to its relations. Individual firms have networks that spread east, south, or west – depending upon their structure and strategy. Authorship networks based in Vancouver appear to have similar dynamics, linking academics in Vancouver and firms outside of Vancouver.[9] The strategy of local entrepreneurs has been to position themselves within the IP value chain as there is very little in the way of bio or pharma production in Vancouver, as there is in Toronto or Montreal.

Wireless

The wireless cluster started in British Columbia before the Second World War with the work of Donald Hings, one of several separate inventors of the class of portable communications devices known commonly as the "walkie-talkie" (Hayter et al. 2005; Hanson 2001). Since then the cluster has experienced many ups and downs, with expansionary periods leading to the creation of major firms (MPR Technologies, Glenayre), their subsequent takeover by multinational firms such as Motorola, followed by their absorption and ultimate disappearance (see also Langford and Wood 2005). Today, Nokia maintains a significant R&D presence in the region, although this facility was down-sized during the global financial crisis.

In many ways, the Vancouver cluster is representative of the emerging inheritance model of cluster formation (Klepper 2001). Two early wireless organizations, Glenayre Electronics Ltd and Mobile Data International (MDI) Inc., played a role in schooling engineers and entrepreneurs. Glenayre sold radio-telephone equipment for vehicles (Globe and Mail 1980a) as well as for use as an early form of GPS, deployed by BC Rail (Globe and Mail 1980b).[10]

One exemplar story might be useful. MDI was a spin-off from another BC company, MacDonald Dettwiler and Associates, an aerospace communications company, which has received significant government funds for technology programs. MDI won a significant contract with Federal Express in competition with Motorola,[11] and shortly afterwards was taken over by Motorola. Although Motorola did not continue its investment in the Metro Vancouver region, the people and money stayed.

As figure 4.4 shows, over the past decade, both employment and exports have been relatively static in Vancouver. However, R&D shows strong growth, reflecting confidence in the future. In 2009 there was significant economic turbulence in the ICT industry globally (OECD 2009), which was also reflected in the local Vancouver cluster. In 2009 Nokia reduced staff at its R&D centre which crosses over wireless and new media.[12] Despite this there seems to be optimism that Vancouver will retain its talent base, as it has done so in previous global crises.

Today's cluster has close ties to the new-media cluster. This closeness is highlighted by the very recent amalgamation of the two industry associations WinBC (for wireless) and New Media BC into DigiBC, an organization representing approximately 22,000 employees and 1300 companies. Nevertheless, there is also anecdotal evidence that there is considerable labour market turbulence in the Vancouver area over a long period of time. The local innovation system appears to be unstable, which for a cluster in the transforming phase of its existence could lead alternatively to greater success or a major decline.

Knowledge flows appear to have been intra-cluster and cumulative for many years, with organizations being created and dying, but with the serial entrepreneurs staying in the region. In recent years, the wireless cluster seems to have also developed significant knowledge spillovers with the new-media cluster. Capital inflows from outside the region which have become vested in the region have also been an important factor for many years.

Within the global distribution of cluster activities this figure highlights the point made with respect to other economic sectors examined in this chapter; the key area of specialization of Vancouver is in

Figure 4.4 Growth in the ICT sector in Vancouver (normalized to 2007)

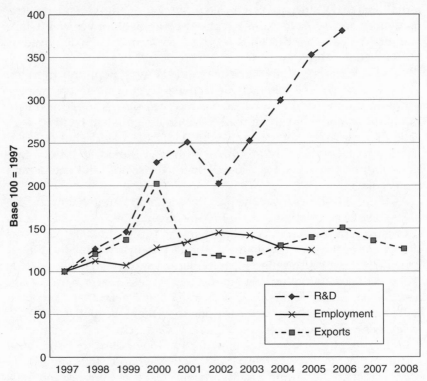

Source: Industry Canada 2007 and Industry Canada 2009a and 2009b.
Note: Base year 1997

knowledge creation, not production. The static nature of employment and production could not make the point any more clearly.

New Media

Cooke defines the new-media segment of an economy as follows:

Making their appearance from the mid 1980s in San Francisco, especially Silicon Valley, and Hollywood, Los Angeles, they intersected high technology and cultural industries. For Scott, the industry was organized into four hierarchical levels. At the base was machinery and hardware, notably computing, communication and associated peripherals and components. Resting on this base was programme interfacing software and systems design. Utilizing the software platform were the visual, audio and print industries at whose intersection were found multimedia. (2002: 289)

So new media is not just about electronic gaming in its various formats, but covers a range of activities that relates to electronic mediated communications (see Britton et al. 2009). This also distinguishes it from personal/business software, which is designed to achieve particular purposes (word processing, etc.).

The new-media cluster in Vancouver has grown from the cultural base of artists, independent filmmakers, and software developers who have been attracted to Vancouver over the past three decades by the climate, lifestyle, and cultural "buzz." Britton et al. (2009) argue that each of Canada's new-media clusters (Toronto, Montreal, and Vancouver) was preceded by TV and motion picture production, although this connection was weakest in Vancouver. Today there are many firms in the business, with entries and exits occurring regularly. The Vancouver cluster has been valued at approximately $2.2 billion, with 850 firms having more than 12,000 employees.[13] The most successful entry into new media has been Electronic Arts (EA), a California-based company which purchased the highly successful Vancouver start-up Distinctive Software Inc. in 1991 for $10 million. Vancouver has been built up by EA to be one of its flagship centres as EA has gone on to become a multinational giant in this field. At one stage it employed nearly 1800 people in its Vancouver operations,[14] which has particular global product mandates. Much of the work done in Vancouver across the cluster is contract-based, providing material for big games developers (such as Electronic Arts) or computer animation for film studios in Hollywood and "Hollywood North" (Metro Vancouver). A recent example of this increasing association is that both *Tron: Legacy* the movie and the game tie-in *Tron Evolution* were developed in Vancouver.[15]

The new-media specialization and diversity pattern is summed up excellently by the following quote from a new-media-producer manager in response to a question regarding the major benefits of Vancouver.

Vancouver is a beautiful city – a draw for recruits. Sea to sky [a reference to the seaside highway between Vancouver and the ski resort of Whistler]. Has a lot of amenities. Great place to make video games because there are lots of companies that make video games. Very incestuous here ...

Hollywood North helps. Strong film and television. Cross fertilization, we use those professionals. Great educational institutions that produce high caliber potential hires. On the Pacific rim gateway including to the east … It is cheaper to develop here than in the US. It was the case, anyway for many years.[16]

This one quote underscores that the knowledge advantages for Vancouver are simultaneously local (amenities), diverse (interactions with the motion picture industry), and trans-locational. Interestingly, new media is also a highly local activity. For example, within EA each flagship centre such as Vancouver has a mandate for particular games and there is apparently little cooperation across centres. Other, small firms also develop their work mainly in-house. However, while there is a high reliance on access to the milieu of talented workers that exist in Vancouver, experience is valued more than mere talent and there is an emphasis on job training (Britton et al. 2009: 221).

In 2009 this industry in Vancouver experienced significance economic turbulence during the global financial crisis, with EA sacking nearly 400 employees (and other smaller companies downsizing or closing). Although the industry did not collapse through 2010 and 2011, along with its global competitor centres the sector in Vancouver continued to experience economic hardship. Several further layoffs from key gaming companies, as well as Microsoft, in the summer of 2012 raised further questions about the long-term competitiveness of the Vancouver cluster, especially with respect to those in Montreal and Toronto. However, what we expect to see is a rebound in activity in Vancouver once the international market stabilizes due to the strong local talent base.

Motion Pictures

Of the selected clusters, this is the only example of one that has been greatly influenced by external investment and outsourced activities. After all, Vancouver is not called "Hollywood North" (Gasher 2002) without reason. As a very important destination for outsourced Hollywood productions, Vancouver has been of interest to researchers for some time (see, e.g., Coe 2000, Coe 2001; Scott 2002; Scott and Pope 2007; Christopherson and Storper 1986).

There were surprisingly early attempts (in the 1930s and 1940s) to establish Vancouver as a filming location for Hollywood films (Spanner 2003). But nothing really happened until the early in the 1980s (Gasher 2002). Strangely enough, what aided this process was the vertical disintegration of the industry away from the production houses and into independents simultaneously with the agglomeration of specialist activities in Hollywood (Christopherson and Storper 1986). The Vancouver industry has grown from virtually nothing in the late 1970s (Gasher 2002) to $1.8 billion in 2008 (BC Film Commission 2009), most of which ($1.4b) was financed externally.

What is interesting about the story of Vancouver and motion pictures is the traditional split that exists between foreign film development in Vancouver and Canadian film industry activities in Toronto (Coe 2001). This trend has been weakening in recent years. The dependency of Vancouver on external interaction is seen by the need to continually reassess the tax credit[17] status of foreign-financed films (primarily out of the United States). Coe, for example, highlights the closeness of the Vancouver and Hollywood products:

> As a location, Vancouver was extremely well placed to benefit from the opportunities being created by the on-going vertical disintegration of the Hollywood studios. The city is close to Los Angeles (2.5 hours' flying time), provides an enviable "west coast" quality of life, is in the same time-zone thereby allowing easy co-ordination of activities between the two centres, has a mild climate which allows all-year-round filming and offers a large range of different scenic locations within 1–2 hours' drive of central Vancouver. (2001: 1760)

In 2009, despite the economic turmoil, the province benefited from the second highest production activity ever (BC Film Commission 2010).

Analysis

There is something very West Coast about the economic sectors analysed in this chapter. Business services and manufacturing in Canada are based in Calgary (see Langford et al. this volume), Ontario, and Quebec, while in America they are based in Detroit and eastwards. New media, wireless, and motion pictures, in particular, are West Coast–based in the United States. However, Vancouver definitely has Canadian, rather than US-styled, institutional and governance frameworks.

We can summarize our main findings as they relate to the MAR-Jacobs debates as follows:

• The fuel cells cluster relies on a deep local labour market and, although it was largely unacknowledged in the interviews,

Vancouver has benefited from flows (particularly of finances) from major auto companies. Yet, the cluster does not rely to any significant extent on input–output relations as it engineers many of its own parts in small-scale manufacturing.

- The emerging bio-pharma cluster has benefited over an extended period from research findings from the University of British Columbia as well as from inputs in the form of contracted research results from other major centres of pharmaceutical companies as a strategy of IP building.
- The new-media cluster relies upon a diverse range of inputs, particularly people with experience and contacts with the major motion picture cluster (Hollywood North – see Coe 2001).
- Wireless has also relied on a deepening of the local talent pool, connections with the new-media cluster, and international contacts.
- Hollywood North is reliant on continued spin-out activity from Hollywood driving the local cluster.

Apart from fuel cells, all the other cases rely on a diversity of inputs, either across geographies or from other industries in our cases. This reliance on diversity is at the very least local, but in many cases stretches across the North American continent both east and south. The evidence we have on outputs[18] is that the "products" of these clusters – mostly IP or intangibles – are marketed globally.

Turning our attention back to the question of whether the cases analysed above are built on indigenous sticky talent or are the result of outsourced work from elsewhere, we can compare the local and global trajectories.

- While the local fuel-cells cluster did not invent hydrogen fuel cells, Ballard Power rejuvenated interest in the technology, which had been largely thought to be a dead end (Koppel 1999). Vancouver is now a major centre for the technology, with a number of entrepreneurial start-ups.
- The bio-pharma cluster emerged early in the life cycle of the industry, but remains a complicated case due to the combination of a high and increasing level of activity, with a high number of entry and exit of start-up companies.
- The new-media cluster took root in Vancouver early, and though not escaping the ramifications of a worldwide drop in sales during

Table 4.1 Industry start-up timeline

Cluster	Vancouver / BC origins	Global origins
New media	1980s	1980s (California)
Bio-pharma	1981	1976 (approx.)
Fuel cells	1980s (Ballard Power)	1950s (space program) / re-emerged 1980s
Wireless communication	Late 1930s	1930s
Motion pictures	1980s	Los Angeles, 1900s (1920s – beginning of the boom)

the global financial crisis, appears to be continuing towards consolidation and strength.
- The wireless cluster, which has long been embedded in the city with a steady stream of local start-ups, continues to develop, albeit in a constant state of flux and transformation.
- Hollywood North took time to take root in Vancouver and relies on government support particularly with the Canadian dollar's rise in comparison to the US dollar, but the sector came through 2009 and 2010 surprisingly well. The cluster plays an anchor role in Vancouver for wireless and new media.

A number of aspects stand out from this analysis (see table 4.1). First, apart from the example of motion pictures, the clusters started early in British Columbia relative to the global start-up. In the case of fuel cells, BC led the world in the re-examination of the technology (Koppel 1999). This would suggest that mostly these are not spin-out clusters, but initially developed out of local technological capabilities and entrepreneurship. Naturally, their trajectory since has been a combination of local and international factors. As has been shown, as firms decline in Vancouver, they are replaced by new firms started by employees of the previous firms.

Clearly, being an early mover in a new economic activity does not determine that a locale will become a major world centre in that activity in future years; there are other drivers behind that dynamic. These cases show that a small to medium-sized city located on the periphery of a national economy can become an early mover in new activities, and if the conditions of talent attraction are good, it can retain a *presence* in those new industries. Global shifts in economic geography driven by political change, transport costs, or industry dynamics all influence the local trajectories of these sectors.

Moving beyond this historical picture, our case histories indicate that there is significant alignment between the current local and international life-cycle positions. Against the pull of larger centres in their respective technologies, particularly in the case of bio-pharma, Vancouver has – despite some turbulence and difficulty in increasing the number of firms in these sectors – nonetheless maintained viable clusters.

The fuel-cells sector is latent, while bio-pharma is still developing, with numerous companies but without a strong profit profile. While both new media and motion pictures are established and increasingly co-dependent, motion pictures is the more significant of the two, and wireless, as it has been for ten years, is continuing to be shaped by global structural transformation.

The sum of the evidence can be summarized as follows. With respect to specialization versus diversity: while there is evidence of specialized knowledge spillovers in fuel cells and bio-pharma (with the caveat that it operates multi-spatially), there is strong evidence of diversity and cross-cluster spillovers in new media, wireless, and motion pictures in particular.

With respect to agglomeration: while relying on a diverse range of inputs within and from outside the city, the life-cycle evidence suggests that all the clusters are mostly of long standing in the Vancouver region. With the exception of the major motion picture activity, all clusters have indigenous roots, suggesting that they are not spin-outs from more innovation-intensive regions. But there are two possible counter-arguments: (1) it is possible that transport costs within Canada (maybe due to the Rocky Mountains) are not cheap enough to induce specialization externalities or, alternatively, the labour market has peculiar characteristics; or (2), Vancouver as a talent attractor pulls people in, but generates an immobile labour force that is unwilling to move away despite the evidence of lower wages.

These two styles of analysis enable us to capture in a single image both the life-cycle aspects and the knowledge spillover dimensions of our clusters (figure 4.5). The only cluster that seems ill aligned with the global scene at present is bio-pharma, which although well established, consists of mostly small developing businesses (compared to other clusters, with a few significant anchor firms).

Figure 4.5 Specialization, diversity, and multi-spatial knowledge spillovers and case trajectories

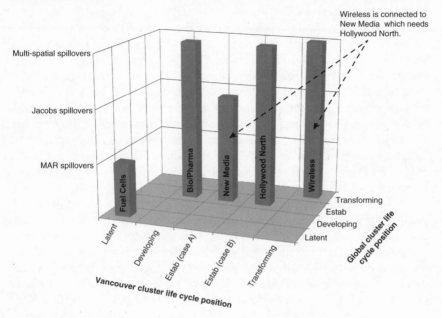

Conclusions

The evidence presented in this chapter may appear contradictory in a number of ways.[19] Vancouver is particularly reliant on knowledge spill-in from other regions, yet its agglomeration pattern indicates that it has a high capacity for the self-generation of innovations. The Vancouver case histories provide strong support for an inheritance model of developments as suggested by Klepper (2001), with the exception that Vancouver's firms rarely become large.[20] Apart from the motion picture industry, the other cases reveal that economic and technological activity started early relative to the emergence of other clusters. This provides evidence that local universities and local talent has been quite significant. Further, despite considerable apparent churn of names, businesses, and ownership structures, Vancouver's technology-oriented activities continue to develop world leading technologies and applications.

This process appears to be driven in this location by the desirability of Vancouver as a place to live. While in general it is possible to question Florida's hypothesis about people choosing cities (see, e.g., Niedomysl and Hansen 2010), for Vancouver there is evidence to support the hypothesis. The data reveals that Vancouver continues to attract people even though on average workers should expect to earn less and pay more for living expenses. On the surface this is prima facie a consumer city (Glaeser and Gottlieb 2006), but the evidence points to the city having a creative history of innovation and entrepreneurship.

In terms of our initial research question regarding knowledge patterns, we found that only one of our five case studies (fuel cells) could be understood as relying primarily on local knowledge spillovers. All the other cases should be understood as instances of Jacobian spillovers (within a multi-cluster or multi-spatial setting). All our cases specialize in human capital and IP, which provides support for the conclusions of Beaudry and Schiffauerova (2009), who argue that specialization is a factor for manufacturing, and diversity spillovers were found to be important for IP-based activities. However, here we find that only wireless could possibly be understood as being based in Jacobian knowledge externalities within the region; but as external relations are also important for the diversity of knowledge sources, our preference is to give that priority. More important, for this region on the semi-periphery, the evidence indicates that multi-spatial spillovers play a very important role.

To summarize, early and sticky talent has been important in building the presence of various economic activities in Vancouver, but those same industries often rely on creative strategies for obtaining new talent or knowledge within an inter-regional (North American or global) setting. The factors that make Vancouver a vibrant entrepreneurial economy (its position on the Pacific as a gateway away from population mega-centres) are also part of its vulnerability.

What lessons can policymakers derive from this evidence? It is clear that conventional innovation policy tools, such as support for R&D, are necessary, but not sufficient. The bio-pharma cluster in Vancouver certainly started through major bio-pharma R&D investments, but that by itself is not sufficient to explain its continued existence. Clearly, conditions must also exist to attract and retain highly skilled workers (Florida 2002).

In Vancouver these conditions exist: the natural setting is spectacular and the various levels of government have combined to provide a strong infrastructure and amenities base. The Economist Intelligence Unit (2011) regularly rates Vancouver as one of the most desirable places to live in the world. The infrastructure includes significant investment in mass transit, airports, hospitals, schools, and universities. Moreover, government policies have also favoured the development of significant cultural and recreational amenities: theatres, galleries, sports venues,[21] and green spaces. The key question for the future is whether there is a break point where talent can no longer be attracted or retained due to a growing gap between income and costs (particularly housing). Further, there are very practical urban policies to enhance innovation. Specific points that could be addressed to local policymakers are:

1. Liveability. The earn less–pay more phenomenon imposes real limits on the city as a destination for mobile talent. Policy must remain focused on the liveability of Metro Vancouver. Without such standards, it is quite foreseeable that talent will decide it can reside elsewhere.
2. It is abundantly apparent from this and every study of Vancouver's high technology sectors that the universities have been crucial as a source of knowledge and new research. Policy must stay focused on ensuring they remain at the cutting edge.
3. The histories of these cases (particularly wireless) reveal that experience in the relevant industry has been important to entrepreneurial activity. There is an opportunity to develop policies that assist students gain not just a university "education," but experience in the relevant industry – much like professional degrees that have developed over the years.

Yet, such conclusions do not totally solve the contradiction with which we opened this section. We need to determine the degree to which Vancouver's role as an international gateway plays an important part in its innovation successes. The dominant feature of the Pacific Rim is the presence of a series of major cities such as Vancouver, San Francisco, Sydney, Singapore, Hong Kong, and Tokyo. There are other strong local innovation systems on the Pacific rim, including "city-states" such as Singapore, Hong Kong, Sydney, and some large Chinese port cities, but only a very few are real gateways to larger manufacturing-based systems of innovation. Is it possible that, because of the distances involved, the Pacific system of innovation can be thought of as being a chain of local systems of innovation, with pivot points where continental lines of communication intersect the Pacific Rim?[22]

While much of the Canadian economy still trades north–south, the BC economy and the city of Vancouver, in particular, have shifted their

orientation towards Asia. This despite the links of many of the industries discussed here to the west coast of America. So although Vancouver
is a Canadian city shaped by factors internal to Canada's constitutional
constructions of policy, its institutional arrangements, its immigration
patterns, and its development are influenced by the super states and
city-states of the Pacific Rim. Vancouver's high-human-capital industries remain on the leading edge – both continually transforming and
attempting to find solid niches from which to grow sustainably – but
also on the edge of a continent. The latter has advantages in orient-
ing firms towards the growth markets of the Pacific while simultaneously disconnecting it from the institutional and political power bases
nationally.

ACKNOWLEDGMENTS

An earlier version of this paper was presented at the DRUID conference at
Imperial College, London, UK, in June 2010, and we acknowledge a number
of helpful comments.

NOTES

1 The debate is has been referred to as MAR- Jacobs. MAR: (Marshall-
   Arrow-Romer = specialization is important) versus (Jane) Jacobs (known for
   her views on a variety of diversity-related themes in city development).
2 See Wixted (2009).
3 This focus on intra-city dynamics is not unprecedented, as Wixted (2009)
   argues this is largely the overarching narrative within which neo-
   Schumpeterian studies has been framed – even though that analysis only
   dealt with national and regional innovation systems and clusters.
4 Vancouver is a high-cost city with lower than average pay rates – see
   discussion below.
5 98 per cent of all businesses in BC employ fewer than 50 people. 80 per cent
   employ between 0 and 4 people (BC Stats 2009).
6 See Cox and Pavletich 2011 – Vancouver is the 3rd most unaffordable city of
   325 cities across Australia, Canada, China SAR (Hong Kong), Ireland, New
   Zealand, the UK, and the USA – behind Sydney and Hong Kong.
7 See http://phx.corporate-ir.net/phoenix.zhtml?c=76046&p=irol-news
   Article&ID=1100245&highlight=.

8 The details are still sketchy at the time of publication, but a late development confirms many of the positions argued for here and places a new
   spin on others. Dateline: 17 March 2011. Mercedes-Benz has announced
   that it would spend $50 million converting excess space in Ballard Power
   Systems' facilities into a *fuel cell stacks production centre*. The fuels cells
   would be destined for cars and buses rather than the existing commercial
   market for fuel cells technology – light industrial vehicles such as forklifts.
   See http://www.bivinteractive.com/index.php?option=com_content&task=
   view&id=3953&Itemid=32 or http://blog.gov.bc.ca/canadaspacificgate
   way/2011/03/17/fuel-cell-mercedes-benz/?WT.rss_a=Celebrating%20
   the%20next%20generation%20of%20automobiles%20with%20Mercedes-
   Benz%20in%20Vancouver,%20British%20Columbia&WT.rss_ev=a&WT.
   rss_f=Canada%5C. The inference can be drawn that the production facility
   is being located in Vancouver to be near skilled workers and the R&D
   rather than being located near assembly lines in the east of Canada or
   other countries. As such this example is the first to be identified in Vancouver of what has been called elsewhere a cluster complex – a group of
   clusters that are tied together in a production network (see Wixted 2009).
9 Correspondence with Bryn Lander.
10 Glenayre was eventually bought and sold by several companies, with its
   corporate history becoming hard to trace.
11 http://www.derekspratt.com/HTML/Business/Other/Motorola%20
   Overview.html.
12 Nokia in BC during the 2008–9 year laid off half its workforce; http://www.
   vancouversun.com/business/Nokia+cuts+jobs/1790820/story.html. See
   also http://www.techvibes.com/blog/july-17-weekly-vancouver-game-
   industry-news.
13 Price Waterhouse Coopers (2010) does not specify a breakdown of aggregate data by subsectors for these different categories, so the overall
   percentage breakdown has been used (26% wireless and 74% new media).
14 http://www.techvibes.com/blog/biggest-video-game-developers-in-bc.
15 See http://www.gamespot.com/news/6286589.html. Note: the game
   studio was closed after the release of the game.
16 The Canadian dollar was approximately 65 cents US in 2002, as of writing
   this paper it is now $1.03 US.
17 http://www.cbc.ca/canada/british-columbia/story/2010/02/03/bc-
   film-digital-tax-credits.html.
18 Motion pictures are distributed world-wide: some examples of pictures
   filmed in Vancouver include *X Men*, *I Robot*, and *Fantastic Four*, among
   many other well-known titles (as listed in Wikipedia in April 2011).

New-media products by their nature are also developed by single centres for global distribution.

19 For a macro picture of the divergent trajectories of the Vancouver economy, with a carefully nuanced analysis, see Barnes et al. 2011.

20 Cashing out after developing the intellectual property of a firm is a frequent exit strategy in Vancouver. Vancouver is not a location suited to large-scale manufacturing (less so for services) because of the high cost of both labour and land, as well as the need to bring into the city virtually all required inputs.

21 Vancouver hosted the 2010 Winter Olympics, with consequent major construction projects. While there is considerable debate as to whether this was a suitable use of public funds, there is no question that the resulting infrastructure, such as the new light rail system connecting the airport to downtown Vancouver, will be of lasting benefit.

22 Hutton (1998: 74–5) describes multiple gateway functions provided in Vancouver.

REFERENCES

Andersson, T., S.S. Serger, J. Sorvik, and E.W. Hansson. 2004. *The cluster policies whitebook*. Malmo: International Organization for Knowledge Economy and Enterprise Development.

Audretsch, D., and M. Feldman. 1996. "Innovative clusters and the industry life cycle." *Review of Industrial Organization* 11 (2): 253–73. http://dx.doi.org/10.1007/BF00157670.

Barnes, T., and T. Hutton. 2009. "Situating the new economy: Contingencies of regeneration and dislocation in Vancouver's inner city." *Urban Studies* (Edinburgh, Scotland) 46 (5–6): 1247–69. http://dx.doi.org/10.1177/0042098009103863.

Barnes, T., T. Hutton, D. Ley, and M. Moos. 2011. "Vancouver: Restructuring narratives in the transnational metropolis." In *Canadian urban regions: Trajectories of growth and change*, ed. L. Bourne, T. Hutton, R. Shearmur, and J. Simmons, 291–327. Toronto: Oxford University Press.

BC Film Commission. 2009. *British Columbia Film Commission production statistics 2008*.

BC Film Commission. 2010. *British Columbia Film Commission production statistics 2009*.

BC Stats. 2009. *Small business quarterly*. http://www.bcstats.gov.bc.ca/.

Beaudry, C., and A. Schiffauerova. 2009. "Who's right, Marshall or Jacobs? The localization versus urbanization debate." *Research Policy* 38 (2): 318–37. http://dx.doi.org/10.1016/j.respol.2008.11.010.

Beckstead, D., and M. Brown. 2004. "From Labrador City to Toronto: The industrial diversity of Canadian cities, 1992–2002." Statistics Canada catalogue no. 11-624-MIE – no. 003.

Berry, B. 1964. "Cities as systems within systems of cities." *Papers in Regional Science* 13 (1): 147–63. http://dx.doi.org/10.1111/j.1435-5597.1964.tb01283.x.

Brakman, S., H. Garretsen, and M. Schramm. 2006. "Putting new economic geography to the test: Free-ness of trade and agglomeration in the EU regions." *Regional Science and Urban Economics* 36 (5): 613–35. http://dx.doi.org/10.1016/j.regsciurbeco.2006.06.004.

Britton, J.N.H., D.-G. Tremblay, and R. Smith. 2009. "Contrasts in clustering: The example of Canadian new media." *European Planning Studies* 17 (2): 211–34. http://dx.doi.org/10.1080/09654310802553456.

Carlsson, B. 2006. "Internationalization of innovation systems: A survey of the literature." *Research Policy* 35 (1): 56–67. http://dx.doi.org/10.1016/j.respol.2005.08.003.

Christopherson, S., and M. Storper. 1986. "The city as studio; the world as back lot: The impact of vertical disintegration on the location of the motion picture industry." *Environment and Planning. D, Society & Space* 4 (3): 305–20. http://dx.doi.org/10.1068/d040305.

Clayman, B. P., and J.A.D. Holbrook. 2004. "Surviving spin-offs as a measure of research funding effectiveness." *CPROST report 04-03*.

Coe, N. 2000. "On location: American capital and the local labour market in the Vancouver film industry." *International Journal of Urban and Regional Research* 24 (1): 79–94. http://dx.doi.org/10.1111/1468-2427.00236.

Coe, N. 2001. "A hybrid agglomeration? The development of a satellite-Marshallian industrial district in Vancouver's film industry." *Urban Studies* (Edinburgh, Scotland) 38 (10): 1753–75. http://dx.doi.org/10.1080/00420980120084840.

Cooke, P. 2002. "New media and new economy cluster dynamics." In *Handbook of new media: Social shaping and consequences of ICTs*, ed. L. Lievrouw and S. Livingstone., chap. 17. London: Sage.

Cox, W., and H. Pavletich. 2011. 7th Annual Demographia International Housing Affordability Survey. Available at http://www.performanceurbanplanning.org/.

Duranton, G., and D. Puga. 2000. "Diversity and specialization in cities: Why, where and when does it matter?" *Urban Studies* (Edinburgh, Scotland) 37 (3): 533–55. http://dx.doi.org/10.1080/0042098002104.

Duranton, G., and D. Puga. 2005. "From sectoral to functional urban specialization." *Journal of Urban Economics* 57 (2): 343–70. http://dx.doi.org/10.1016/j.jue.2004.12.002.

Economist Intelligence Unit. 2011. *A Summary of the Liveability Ranking and Overview February 2011*. http://www.eiu.com/Handlers/WhitepaperHandler.ashx?fi=LINKED_Liveability_rankings_Promotional_PDF.pdf&mode=wp&campaignid=Liveability2011.

Edquist, C., ed. 1997. *Systems of innovation: Technologies, institutions and organisations*. London: Pinter.

Filion, P. 2010. "Growth and decline in the Canadian urban system: The impact of emerging economic, policy and demographic trends." *GeoJournal* 75 (6): 517–38. http://dx.doi.org/10.1007/s10708-009-9275-8.

Florida, R. 2002. "The economic geography of talent." *Annals of the Association of American Geographers* 92 (4): 743–55. http://dx.doi.org/10.1111/1467-8306.00314.

Florida, R. 2005. "The world is spiky." *Atlantic Monthly*, October: 48.

Florida, R., T. Gulden, and C. Mellander. 2008. "The rise of the mega-region." *Cambridge Journal of Regions, Economy and Society* 1 (3): 459–76. http://dx.doi.org/10.1093/cjres/rsn018.

Gasher, M. 2002. *Hollywood North: The feature film industry in British Columbia*. Vancouver: UBC Press.

Gertler, M. 2001. "Tacit knowledge and the economic geography of context or The undefinable tacitness of being (there)." Paper presented at the Nelson and Winter DRUID Conference, Aalborg, Denmark, June. www.business.auc.dk/druid.

Glaeser, E., H. Kallal, J. Scheinkman, and A. Schleifer. 1992. "Growth in cities." *Journal of Political Economy* 100 (6): 1126–52. http://dx.doi.org/10.1086/261856.

Glaeser, E.L., and J.D. Gottlieb. 2006. "Urban resurgence and the consumer city." *Urban Studies* (Edinburgh, Scotland) 43 (8): 1275–99. http://dx.doi.org/10.1080/00420980600775683.

Globe and Mail. 1980a. "Companies in the news: Glenayre." 19 August.

Globe and Mail. 1980b. "Companies in the news: Glenayre." 11 September.

Hanson, C. 2001 "Walkie-talkie inventor receives order of Canada." *Vancouver Sun*, 17 August: B1, B7.

Harrison, B. 1994. "The small firms myth." *California Management Review* 36 (3): 142–58. http://dx.doi.org/10.2307/41165759.

Hayter, R., K. Rees, and J. Patchell. 2005. "High tech 'large firms' in Greater Vancouver, British Columbia: Congregation without clustering?" In *New economic spaces: New economic geographies*, ed. R. Le Heron and J. Harrington, chap. 3. Aldershot, UK: Ashgate.

Helliwell, J., and D. Helliwell. 2001. "Where are they now? Migration patterns for graduates of the University of British Columbia." In *The state of economics in Canada: Festschrift in honour of David Slater*, ed. P. Grady and A. Sharpe, 291–324. Montreal: McGill-Queen's University Press.

Holbrook, J.A.D., D. Arthurs, and E. Cassidy. 2010. "Understanding the Vancouver hydrogen and fuel cells cluster: A case study of public laboratories and private research." *European Planning Studies* 18 (2): 317–28. http://dx.doi.org/10.1080/09654310903491648.

Holbrook, J.A.D., and D.A. Wolfe. 2005. "The Innovation Systems Research Network (ISRN): A Canadian experiment in knowledge management."

Science & Public Policy 32 (2): 109–18. http://dx.doi.org/10.3152/1471543057
81779623.

Hutton, T.A. 1998. *The transformation of Canada's Pacific metropolis: A study of Vancouver*. Montreal: Institute for Research in Public Policy.

Immarino, S., and P. McCann. 2006. "The structure and evolution of industrial clusters: Transactions, technology and knowledge spillovers." *Research Policy* 35 (7): 1018–36. http://dx.doi.org/10.1016/j.respol.2006.05.004.

Industry Canada. 2007. ICT sector regional analysis – British Columbia. http://www.ic.gc.ca/eic/site/ict-tic.nsf/eng/it07190.html.

Industry Canada. 2009a. Canadian ICT sector regional analysis, 2008 exports data. http://www.ic.gc.ca/eic/site/ict-tic.nsf/eng/it07914.html.

Industry Canada. 2009b. Canadian ICT sector regional analysis, 2006 ICT sector research and development data. http://www.ic.gc.ca/eic/site/ict-tic.nsf/eng/it07914.html.

Klepper, S. 2001. "Employee start-ups in high-tech industries." *Industrial and Corporate Change* 10 (3): 639–74. http://dx.doi.org/10.1093/icc/10.3.639.

Koppel, T. 1999. *Powering the future: The Ballard fuel cell and the race to change the world*. Etobicoke, ON: John Wiley and Sons Ltd Canada.

Krugman, P. 1998. "What's new about the new economic geography?" *Oxford Review of Economic Policy* 14 (2): 7–17. http://dx.doi.org/10.1093/oxrep/14.2.7.

Langford, C., and J. Wood. 2005. "The evolution and structure of the Vancouver wireless cluster: Growth and loss of core firms." In *Global networks and local linkages: The paradox of cluster development in an open economy*, ed. D. Wolfe and M. Lucas, 207–27. Kingston: McGill-Queen's University Press.

LifeSciences BC. 2010. LifeSciences British Columbia. http://www.lifesciences bc.ca/Publications/LSBCPublications/default.asp.

Malmberg, A., and P. Maskell. 2002. "The elusive concept of localization economies: Towards a knowledge-based theory of spatial clustering." *Environment & Planning A* 34 (3): 429–49. http://dx.doi.org/10.1068/a3457.

Markusen, A. 1999. "Fuzzy concepts, scanty evidence, policy distance: The case for rigour and policy relevance in critical regional studies." *Regional Studies* 33 (9): 869–84. http://dx.doi.org/10.1080/00343409950075506.

Martin, R. 1999. "The new 'geographical turn' in economics: Some critical reflections." *Cambridge Journal of Economics* 23 (1): 65–91. http://dx.doi.org/10.1093/cje/23.1.65.

Maskell, P., and M. Lorenzen. 2004. "The cluster as market organization." *Urban Studies* (Edinburgh, Scotland) 41 (5–6): 991–1009. http://dx.doi.org/10.1080/00420980410001675878.

Molot, M. 2008. "The race to develop fuel cells: Possible lessons of the Canadian experience for developing countries." In *Making choices about hydrogen: Transport issues for developing countries*, ed. L. Mytelka and G. Boyle, chap. 11. Tokyo: United Nations University Press.

Molot, M., and L. Mytelka. 2009. "First mover advantages in a costly and disruptive technology: The high-stakes fuel cell game." Presented to Canadian Political Science Association meetings, Carleton University, Ottawa, 27–9 May.

Niedomysl, T., and H.K. Hansen. 2010. "What matters more for the decision to move: Jobs versus amenities." *Environment & Planning A* 42 (7): 1636–49. http://dx.doi.org/10.1068/a42432.

OECD. 2006. *Competitive cities in the global economy*. Paris: OECD.

OECD. 2009. *The impact of the crisis on ICTs and their role in the recovery*. Directorate for Science, Technology and Industry, Committee for Information, Computer and Communications Policy. DSTI/ICCP/IE(2009)1/FINAL, 17 August.

Panik, F. 1998. "Fuel cells for vehicle applications in cars: Bringing the future closer." *Journal of Power Sources* 71 (1–2): 36–8. http://dx.doi.org/10.1016/S0378-7753(97)02805-X.

Penner, D. 2010. "Vancouver becoming tourist destination for house-hunting Chinese." *Financial Post*, 28 June.

Polèse, M., and R. Shearmur. 2004. "Is distance really dead? Comparing industrial location patterns over time in Canada." *International Regional Science Review* 27 (4): 431–57. http://dx.doi.org/10.1177/0160017604267637.

Price Waterhouse Coopers. 2004. "Worldwide fuel cell industry survey."

Price Waterhouse Coopers. 2010. BC digital media and wireless industry survey. http://www.kast.com/files/2010_bcdigitalmediawireless_survey.pdf.

Rothaermel, Frank T. 2000. "Technological discontinuities and the nature of competition." *Technology Analysis and Strategic Management* 12 (2): 149–60. http://dx.doi.org/10.1080/713698467.

Scott, A.J. 2002. "A new map of Hollywood: The production and distribution of American motion pictures." *Regional Studies* 36 (9): 957–75. http://dx.doi.org/10.1080/0034340022000022215.

Scott, A.J., and N. Pope. 2007. "Hollywood, Vancouver, and the world: Employment relocation and the emergence of satellite production centers in the motion-picture industry." *Environment & Planning A* 39 (6): 1364–81. http://dx.doi.org/10.1068/a38215.

Shearmur, R. 2008. *Neo-regionalism and spatial analysis: Complementary approaches to the geography of innovation?* Montreal: INRS-Urbanisation.

Spanner, D. 2003. *Dreaming in the rain: How Vancouver became Hollywood North by Northwest*. Vancouver: Arsenal Pulp Press.

Spencer, G., and T. Vinodrai. 2009. Innovation Systems Research Network city-region profile, 2006: Vancouver. ISRN internal document.

Tanner, A.N. 2012. "The mechanisms of regional branching: An investigation of the emerging fuel cell industry." Paper presented at the annual DRUID Summer Conference, Copenhagen, 19–21 June.

Taylor, P. 2004. *Global city network*. London: Routledge.

Wixted, B. 2009. *Innovation system frontiers: Cluster networks and global value*. Berlin: Springer. http://dx.doi.org/10.1007/978-3-540-92786-0.

5 Innovation and Social Actors in Montreal: Intersectoral Challenges of Place-Based Dynamics

DIANE-GABRIELLE TREMBLAY
AND JUAN-LUIS KLEIN

As mentioned in the introduction to this book, there are debates as to how the economic structure and characteristics of urban economies impact on innovation and knowledge circulation within cities. Indeed, some authors consider that the most important dynamics are created by the advantages that accrue to firms located in dense clusters of similar and related firms – that is, by a greater degree of specialization within city-regions. By contrast, other authors indicate that the potential for innovation is stronger when knowledge circulates through various sectors of activity within a city – that is, greater diversity in the economic structure of a city-region would be the source of more innovation. Here, the essence is that ideas that no longer appear novel in one sector can actually have novel value in another. Research on innovation has thus highlighted the fact that cross-fertilization arising from stronger variety in the economy can develop a higher potential for new ideas and innovation within the local economy. Thus, the issue of whether it is industrial specialization or diversity that has the greater potential for innovation and growth within city-regions has attracted much interest, but the issue still appears unresolved. It is clear, however, that it has critical implications for cities' capabilities to adapt in a knowledge-based and globalizing economy and this is why we chose to address this question in this chapter.

Montreal can be seen as a metropolitan arrangement of actors that in response to the crisis of Fordism resolutely adopted knowledge-oriented strategies and at the same time remained concerned with quality of life at the local level (Klein, Manzagol, Tremblay, and Rousseau 2005). Between the 1980s and the beginning of the 2000s, Montreal was clearly the scene of a broad array of innovations, from technological and productive innovation to cultural and artistic innovation (Rantisi and Leslie 2008; Pilati and Tremblay 2008) and finally to social and socio-territorial innovation (Fontan, Klein, and Tremblay 2005; Tremblay, Klein, and Fontan 2009). This chapter addresses the issues related to these various forms of innovation in Montreal. It starts out with a review of policy documents and economic strategies that lay out the base of Montreal's economic renewal and reorientation towards a knowledge-based economy. It then goes on to analyse innovation in three case studies which illustrate the beginning of some cross-sectoral fertilization patterns, while recognizing at the same time that sectoral specialization tends to be the more common way to innovation, even for bottom-up projects, possibly in relation to path dependency.

The points that we wish to develop in this chapter on the innovation process in Montreal address the following questions: (1) What are the main policies and economic strategies that have favoured innovation and the knowledge-based economy? (2) How can various cases highlight the interest of the diversity path and intersectoral knowledge crossovers?

The Twofold Reconversion Strategy in Montreal

Evidence shows that, after the industrial crisis at the end of the seventies and the beginning of the eighties, the Montreal metropolitan region has experienced a process of reconversion to the knowledge-based economy (Tremblay and Rolland 2003; Klein, Manzagol, Tremblay, and Rousseau 2005; Fontan, Klein, and Tremblay 2005). The crisis was provoked by the delocalization of the manufacturing sector, which had severe effects on manufacturing-based areas and on the city as a whole. The process of industrial delocalization provoked all the problems associated with social and urban restructuring (unemployment, low incomes, and population loss), but at the same time triggered new strategies for economic development.

On the one hand, there was the adoption of a strategy oriented towards high tech and innovation. A high point of this strategy was the task force created by the federal government in 1985 and chaired by Laurent Picard, a well-known academic figure. Based on the deliberations of its sixteen members, who were institutional leaders from Montreal, the task force produced what is known as the "Picard report," which became a point of reference on Montreal's economic reconversion for both public and private practitioners. The report contains a strategy that encourages private leadership, internationalization, and the development of high-technology sectors (telecommunications, aerospace, biopharmaceuticals, information technologies, and microelectronics). These objectives were to be implemented at the metropolitan level and were expected to result in planning operations at that level (Tremblay and Van Schendel 2004).

On the other hand, a specific strategy was proposed by grassroots organizations from devitalized neighbourhoods, especially those mainly affected by delocalizations. These neighbourhoods mobilized around the community leaders to defend their assets and undertake development strategies that were adapted to new economic conditions and respected the interests of the local population. The main results of this mobilization were the "Community Economic Development" strategy and the creation of the Community Economic Development Corporations (CEDCs).

As the result of this twofold strategy, aiming at high tech as well as at the community-based economy, combining both private business development and community economic development, the metropolitan region of Montreal has been engaged in a process of economic renewal over the last two decades. On one side, Montreal's economy has adopted the direction of knowledge and innovation, in the wake of various programs oriented towards developing innovative sectors and clusters. Therefore, clusters in Montreal's strongest business sectors (aeronautics, biopharmaceutics, IT, film, etc.) have been favoured, in conformity to the Picard report's recommendations, thanks to private businesses and both the Communauté métropolitaine de Montréal and the City of Montreal. On the other, local actors, community groups, and grass-roots organizations have become important economic stakeholders in the reconversion process of old manufacturing, or pericentral, districts, which have mobilized a large array of resources in order to ensure a diversified employment base and to improve their image.

These two movements have converged recently into a culture-based strategy. Culture-based leaders, the business milieu, and community-based actors have focused on the choice of culture as a major orientation in metropolitan development, strongly influenced by Culture Montréal (Darchen and Tremblay 2010; Klein and Tremblay 2010). This is demonstrated in prestige projects such as the Quartier des spectacles, the designation of Montreal as a city of design by UNESCO, as well as in initiatives of social and territorial reinsertion and diversification, such as TOHU in the Saint-Michel neighbourhood, stimulated by the Cirque du Soleil (Tavano Blessi et al. 2012; Pilati and Tremblay 2008), or the development of design and cultural activities in the Mile-End neighbourhood (Leslie and Rantisi 2006; Klein et al. 2010). Culture Montréal has been very vocal in the last few years in pushing for strong state support for these industries, putting forward the strong innovative capacity of sectors such as circus arts, design, and contemporary art, as well as creativity in a wider sector of artistic domains, like cooking or fashion, for instance, in order to create a new branding for Montreal as a "creative city" (Brault 2009). This has of course not eliminated the efforts in such industrial sectors as aeronautics or ITC, but has contributed to another source of innovative and creative activities geared towards economic development.

Conceptual Framework: Clustering and Innovation

Theories on agglomeration and clustering appeared some time ago, with the Marshallian district (Marshall 1889), and these theories resurfaced as a response to the Fordist crisis, as proposed by the Innovative Milieu approach. Both clusters (single sector) and the Innovative Milieu (multi-sector agglomerations) are conceived as geographical concentrations of firms and supporting organizations that "trust" one another and frequently exchange knowledge. The Innovative Milieu approach emphasizes the role of the milieu as a source of innovation and industry growth: in this theory, there is more emphasis on competencies, as it is considered that the proximity of competencies and cross-sectoral flows of such competencies favour the creation of new innovative firms (Camagni and Maillat 2006). This accent on competencies appears all the more important to us in the context of the knowledge economy, and in our view the development of competencies is one of the main conditions of success for a cluster.

The performance of a cluster is dependent on the performance of the individual firms that are part of it, but not only; indeed, this performance is moderated by cluster conditions and the wider socio-economic environment of the firms. Among the factors which are seen as having an impact on the firms, let us mention human and social capital, R&D capacity and infrastructure, information infrastructure, community resources and support, as well as government policies and programs. (Arthurs et al. 2009; Julien 2005; Holbrook and Wolfe 2002; National Research Council 1998). Arthurs et al. relate the model to the various factors included in the Porter Diamond of performance (Porter

1990, 2003, 2004), but highlight the fact that Porter's definition of re-lated and supporting industries has been enlarged to include public and non-profit organizations that support cluster development; these are known as supporting organizations or intermediate organizations, as we prefer to call them.

The main difference between this cluster view and other views of economic development or growth is the fact that it highlights the social and territorial nature of the innovation process, what we have elsewhere called "socio-territorial capital" (Fontan, Klein, and Tremblay 2005). This socio-territorial capital is seen as playing as important a role as economic or financial factors (price, financial support, and so on). The territory is seen here as being more than a simple repository for economic activity and the role of social relations of production and interactions is highlighted. This explains that our analysis of the birth of creative clusters is largely centred on the analysis of social relations and interactions between actors. Given the importance of social relations and interactions between actors, we could say that the issue of knowledge flows should be important, but while it should theoretically be important, we did not find cross-sectoral flows in all the cases studied. There appear to be more in some sectors, for example in fashion, film, and multimedia (cf. Tremblay and Cecilli 2009; Tremblay 2013; Darchen and Tremblay 2013a, b). It may be that there are more of these cross-sectoral flows precisely in the cognitive-cultural economy (Scott 2006), as the multimedia, film, and fashion sectors seem to maybe highlight this somewhat more clearly than the IT and aeronautics sector, at least from our study.

As mentioned in the introduction, we consider that innovation and creativity are not the exclusive territory of high-tech sectors, as is often thought. An innovative society requires more than just technological change; it requires a social and cultural change in the way wealth is generated and in the governance methods put forth by stakeholders. Such change can occur in the form of a multi-scalar process wherein a neighbourhood or area interacts with the metropolis, thereby becoming a node in a global network (Amin and Thrift 1992; Borja and Castells 1997). The more these processes are inclusive and integrated, that is, the more all segments of society participate in a converging process, with a variety of actors (community, business, government, etc.), the better the results in terms of diversity and the more those results are appropriated by communities at large (Hillier et al. 2004). We regard this characteristic of inclusiveness and integration as the basis of a truly innovative society.

From that perspective, resources mobilized by social economy actors are crucial for integrating sectors into the innovation process that would otherwise be excluded. These resources are the launching platform for socially as well as technologically innovative entrepreneurial initiatives (Moulaert and Ailenei 2005) and foster link-ups between economically depressed sectors and performing and dynamic milieus and networks (Amin et al. 2002). Some cases, such as the Lab Créatif (Klein et al. 2010), demonstrate the innovative potential of the actors and resources of the social economy.

However, our research in a few of Montreal's main sectors or clusters (aeronautics, IT and multimedia, garment industry, and film) has shown that there is a diverse level and type of innovation (product and process, as well as organizational or social as well) in the various sectors. It is difficult to summarize our observations since they are specific to each sector and difficult to generalize to the whole economy, so we have chosen to highlight three cases that put forward the idea of cross-sector fertilization and the contribution of diversity to innovation and creativity, by addressing the cases of the Lab Créatif and the garment industry, the TOHU and circus arts, and the aeronautics sector. But first, we situate these cases in the more general context of Montreal's clusters and economic strategies.

Recent Economic Policies and Strategies: The Clusters Strategy of Montreal

The policies applied by the city of Montreal since the 1990s to the 2000s supported the idea of a network-based local development (Rifkin 2004) or cluster strategy (Gertler and Wolfe 2005; Tremblay 2006). The current mayor of Montreal, Gerald Tremblay, introduced the concept of industrial clusters in 1991, when he was the Quebec minister of industry, commerce, science, and technology. He saw it as a model which was, in his own words, "designed to stimulate the creation of conditions within which new ideas and processes can pass from embryonic to commercialized stages and provide returns for stakeholders" (CMM 2008).

Almost twenty years after the first presentation of the concept in Quebec, and after some experience with the concept at the provincial industry department, this model brought the City of Montreal to adopt the cluster concept as a strategy for innovation. Alas, this strategy has been applied in a restricted way, oriented to the development of a product rather than based on developing competences that could support a regional system of intersectoral fertilization. But before presenting this strategy, let us present a short story to put this strategy in context (cf. Tremblay, forthcoming).

The Cluster Strategy and Sectoral Governance in Montreal

The CMM (Communauté métropolitaine de Montréal) economic development plan (2005) identified four types of clusters: the *competitive* clusters, that is, those that bring together internationally competitive segments (aerospace, ICT, life sciences, textiles and clothing); the *emerging technology* clusters, that is, cross-sectoral technologies with high, long-term growth potential (environmental technology, nanotech, and advanced materials); the *manufacturing* clusters, those with growth potential based on natural resources (energy, paper and wood products, bio-food, petrochemicals and plastics and metallurgy); and finally *visibility* clusters, defined as strategic sectors for a city region's socio-economic development and branding, which include film, culture, tourism, and related services. (G. Tremblay 2008: 12).

The CMM was responsible for the strategy and so served as the coordinator of the three fundamental steps in the creation of a formal cluster recognized by the CMM: pre-start-up, start-up, and operation. One of the eligibility conditions for the CMM and its governmental partners to lend financial support was that all the industry stakeholders had to be part of a non-profit organization related to the sector.

A Bottom-up Approach?

The approach to these clusters is considered bottom-up in the sense that initiatives must come from firms (demand-led), that the metropolitan cluster must indicate that it can organize itself and adopt a development plan. The CMM is responsible for the general planning and coordinating of activities, but does not substitute itself to local actors and decision makers.

In order to get financing the clusters must develop a business plan with growth objectives, value-added projects, and performance indicators and must function with a consensus-based approach. They should develop a three-year action plan and a ten-year development strategy. The financing is normally ensured by the three government levels (municipal, provincial, federal) and the private sector, at 25 per cent each.

Once potential is identified in a given sector, three phases follow. In the pre-start-up phase the cluster has to develop its business case and is self-financed. Then, at the start-up phase, there should be confirmation from the private sector for some $200,000 minimum per year over three years, which will make it possible for the public-sector actors to put up to $200,000 each as well. There is a possibility of receiving seed money for a provisional cluster committee in order to prepare the long-term plan, the triennial plan, and the private-sector solicitation plan. This would be shared by the Quebec government and the CMM, with up to $200,000 for eighteen months (CMM 2008).

Clusters obviously exist at different phases, and for the moment only four clusters are considered to be at the operational phase, as mentioned above: aerospace, ICT, life sciences, and film-audiovisual. Of course, one may consider that many other clusters actually exist or are in the developmental stages, but only a few are formally recognized by the CMM policy. This does not mean that no knowledge flows occur in these other sectors, but they are of course somewhat more difficult to detect. Also, it is possible that cross-sectoral knowledge flows might be more present in such cross-sectoral technology sectors as circus arts, the environment, or advanced materials, since these are cross-sectoral by definition. In the case of circus arts also, while there is no formal cluster policy support, it may be considered that this is a "natural" cluster, one which developed with little policy support. And indeed, some authors consider that these naturally occurring ones are "real" clusters, while those depending on public policy would be clusters created intentionally, but not always as successfully.

The CMM cluster development strategy is based upon the individualization of the strategic sectors and the mobilization of the principal stakeholders around a leader or champion willing to rally the community around a common goal. The objective is to allow Montreal to project abroad the image of a city of knowledge, as well as a creative and prosperous one. The European Union's interest in these practices, which led to the development of CLUNET (Cluster Network) on 1 September 2006, is considered by many to be proof of Montreal's influence in this field. The three-year international project aimed to create a network of the most innovative regions with clusters. CLUNET is composed of sixteen partners, including Montreal, the only non-European participant, and its main objective is to launch pilot projects fostered by international cooperation "to achieve a common agenda for Europe that will lead to the creation of world-class clusters delivering global competitive advantage in lead markets."[1]

The main critique which may be put forward in relation to this strategy is that it is not so much centred on cross-sectoral or cross-cluster knowledge flows, but rather tends to reinforce intra-sectoral flows, which can of course lead to innovation, but which tend to be seen as somewhat limited, especially in a competency-based approach. Indeed, our interviews indicate that while there could be cross-sectoral flows between the aeronautics sector and some others – some research in France has shown flows between aeronautics and the automobile industry (Jalabert and Zuliani 2010) – our interviewees indicated that flows tended to be more within the sector. The aeronautics cluster is very active in getting people from the industry together, but it may be somewhat too centred around the aeronautics industry, and not open enough to innovation or ideas from other sectors or clusters. The same can be said for the IT sector, where firms tend to look at developments within the sector, not so much outside of it. On the contrary, the more creative sectors (film, multimedia, and fashion) tend to think a bit more "out of the box," although this does not mean all actors do so. There were, however, examples of a designer organizing shows with painters, photographers, or other artists, film people on the lookout for developments in the animation sector, and people in animation on the lookout for cooperation with multimedia. Although this is clearly not happening across the board, there appears to be a more open vision in the creative sectors (Tremblay and Tremblay, eds, 2010) than in the more traditional sectors, where the view of innovation is still quite centred on the R&D linear view.

The Role of Civil Society Organizations

We have centred here on the cluster and industrial strategy elements, but it is important to note a characteristic that marked the Quebec situation from the 1990s to the 2000s, that is, the participation of civil society organizations in economic development. While the CMM developed sectoral cluster strategies, an important turn was made by a large array of civil society actors that include community organizations, unions, and local development corporations. As a consequence of this turn, those civil society organizations became stakeholders in economic development. This tendency was put forward over recent decades, with the first CEDCs appearing in the wake of the 1981–2 crisis in many deindustrialized zones of Montreal as well as the establishment of job-creation-oriented union funds (Fondaction and Fonds de solidarité). The contribution of social actors to innovation and to socio-economic development in Montreal cannot be underestimated, although they are of course not the main actors in such development. In our view, under the imperatives of the new global economy, the reconversion of local spaces relies on the innovating capacity of the productive sectors, of course, but also of social actors, including the associative sector and social economy, who are very dynamic in Quebec and in Montreal (Klein et al. 2010), and whose action is sometimes neglected in studies on innovation. From this point of view, innovation is not the exclusive domain of high-tech sectors, as is often claimed; while of course community organizations do not have the same type or level of contribution, we want to highlight their role precisely because it often goes unnoticed or is neglected. In our view, all sectors must contribute to the process of building an innovative society and this requires more than just technological change, which is of course determinant but also often needs the support of organizational or social innovation in order to provide its full benefits. Building such a society requires a social and cultural change in the way wealth is generated and in the governance methods put forth by stakeholders, and in Montreal, some of the social, economic, and associative actors contribute to this process in cooperation with the more traditional stakeholders.

Three Case Studies

A comparison of three completely different cases allows us to show how organizations of civil society have become important stakeholders of economic development in Montreal in a diversity of sectors. The first case is the Lab Créatif, a cluster of fashion small businesses and creators located in a peri-central neighbourhood with the help of a CEDC. In this case, there are intra-sectoral knowledge flows between the participants in the project and a few outside, cross-sectoral if we consider the various artistic disciplines as different sectors, but intra-sectoral if we consider them all as arts (painters, photographers, and jewellers have sometimes been associated with fashion projects, leading to knowledge flows). The second one is the TOHU, a gathering of circus activities located in a devitalized neighbourhood under the impetus of the Cirque du Soleil. In this case there are more knowledge flows between various sectors, but again, mostly within the various artistic disciplines needed to put a show together. The third is the cluster organized around Bombardier's activities in aeronautics with the help of many intermediary organizations. In this case, the flows seemed to clearly be intra-sectoral, and the presence of the cluster seems to reinforce the tendency to look within the cluster or sector when seeking ideas or cooperation. These three very different cases address distinctive aspects of the collaborative culture which was the main characteristic of the Montreal economic strategy triggered during the 1990s. The community sector was an asset to economic dynamism alongside more traditional forms of innovation (as in the Bombardier case).

The Lab Créatif Case: A Social Innovation Based on a Diversity of Resources, Favouring Various Forms of Productive and Organizational Innovation[2]

Created in 1986, the Centre-Sud/Plateau Mont-Royal CEDC was one of the first such organizations in Montreal. According to its administrators, the organization has since worked to actively improve the population's quality of life in the Centre-Sud, Plateau Mont-Royal, Saint-Louis, and Mile-End[3] neighbourhoods. In 1998, the Quebec government conferred on this CEDC the status of CLD (centre for local development). As such, the CEDC cum CLD offers direct services to businesses, entrepreneurs, and self-employed workers. It seeks to improve employability in the borough and supports local initiatives aiming to offer services to the community. In an effort to consolidate the existing creative features in the Mile-End neighbourhood, the CEDC has actively participated in the creation of the Lab Créatif (hereinafter LAB), a cluster of young fashion designers.

LAB is the result of joint efforts between young fashion designers and the CEDC. The project had two objectives: (2) the revitalization of an industrial zone in the Mile-End neighbourhood, one of Montreal's areas most seriously affected by shutdowns and industrial relocation in the wake of the city's industrial crisis;[4] and (2) the reconversion of the apparel industry, formerly a flourishing industry and now in serious decline locally. Created in 2004, LAB is a non-profit enterprise that brings together young designers (under thirty years old generally) willing to work side by side in order to increase the profitability of their companies by developing knowledge flows and transferring "tricks of the trade" between them. LAB's charter was obtained in 2006, and all creators who work in textiles and make use of the equipment required for garment making, either in Montreal or elsewhere in Quebec, can be members.

The LAB gathers over 150 designers,[5] all of them from the Montreal region, with a certain number sharing premises. The other members are located nearby. LAB is not a business incubator; although it has assumed a companion role in many start-up projects, it does not launch new enterprises. Member enterprises are not always expected to move out after starting; they can remain and pursue development and growth on the premises.

A CLUSTER PROJECT BASED ON SHARING

The idea behind the LAB project stems from an offer made by a fabric cutter who wanted to share unoccupied space in his workshop. The designers who accepted the offer soon understood how such grouping was promising and decided to broaden the opportunity, thus extending the possibility for sharing resources and knowledge. Indeed, it started with a sewing pool, needles, and fabric, then contacts, tips, and finally ideas, some coming from various other artistic or cultural activities (arts and crafts, photo, contemporary art, etc.).

LAB members are small-run producers and their output runs in the range of 500 to 800 units for each collection. Designers lay out the design, they are pattern-makers, and they cut and sew samples or prototypes (the first stages of production). Most of the time the production is outsourced to subcontractors located in Greater Montreal, which turns out to be a problem. Qualified subcontractors who are available for such small runs are hard to come by, as one interviewee told us: "Since these subcontractors are trained to operate on an assembly line ... they are absolutely not prepared for designer work." Designers must face other problems as well – money lenders are hesitant about investing in young designers' exclusive collections.

With the support of a CEDC which offers a diversity of resources, knowledge, and support, designers created LAB precisely to handle these issues. LAB's mission consists in supporting the development of fashion-design businesses in the Montreal region and the province of Quebec.[6] To this end, LAB offers its members various services pertaining to design and creation, production, management, distribution, promotion, and financing. LAB is something of a resource centre, helping designers to find staff assistants and trainees, subcontractors, and suppliers ready and able to help out with production, and this often leads to knowledge exchanges between members and access to new resources.

For promotion purposes, LAB launched a website where designer collections are showcased. In addition, LAB offers its members, free of charge, participation in the Montreal Fashion Week show.[7] Other projects are planned for the future, including the creation of a collective showroom, an online sales interface, a boutique, and the hiring of sales representatives. In partnership with the Centre de transfert technologique de la mode (CTTM), LAB also provides access to production equipment and facilities, from standard sewing machines to sophisticated programmable equipment. The equipment is rented to LAB members at competitive hourly (or annual) rates. In order to solve labour requirements, LAB is developing novel cooperative training solutions in partnership with existing educational institutions, namely, the École des métiers des Faubourgs-de-Montréal, a school partly dedicated to fashion crafts (pattern design, made-to-measure tailoring and alterations, and garment making).

So, to a certain extent, we can speak here of inter-sectoral knowledge flows, inasmuch as some institutions are actually transferring knowledge (use of new materials, production techniques) between various sectors (arts and crafts, photo, contemporary art), but this is still emergent. While some designers mention collaborative projects with some arts and crafts people, with photographers or contemporary artists for a joint exhibit or other project, it is clearly not the dominant approach in the fashion sector. On the contrary, the sector tends to be very individualistic. However, it is interesting to find that the Lab Créatif project has managed to change this view and bring many designers and artists to work together on projects, and this has sometimes led to innovations in fashion design, the introduction of cloth into contemporary art, and the like. So there does seem to be some cross-sectoral knowledge flows in this project, and although the approach is not dominant in the fashion design sector, it seems the younger generation of fashion designers are very open to the idea of putting together creative ideas, and generating products or ideas that are on the frontier of two artistic fields.

THE LAB, AN INNOVATIVE SOURCE FOR A VARIETY OF RESOURCES

As was mentioned earlier, LAB resulted from the collaboration between fashion designers and the Centre-Sud/Plateau Mont-Royal CEDC. In the early stages, CEDC representatives consulted with the designers to work out original ideas concerning the relevance, feasibility, and format of the planned grouping or cluster. Then the CEDC provided designers with resources, making it possible both to hire consultants who would prepare a business plan and to finance the start-up phase using the development funds entrusted to the intermediary. It also carried out the day-to-day-business of the group and operationalized the business plan. Moreover, the CEDC provided designers with access to financial and organizational resources (Développement Économique Canada, Ministère du développement économique, de l'Innovation et de l'Exportation – Québec, Emploi-Québec) while the LAB mobilized sectoral and educational networks, such as the CTTM, the École des Faubourgs, etc.

After acquiring valuable experience in the LAB project, the CEDC now intends to call upon other Mile-End creatives in several other fields or trades, something which would eventually increase and formalize the knowledge flows. The fashion sector is seen as the first step in consolidating the creative potential that exists in the neighbourhood.[8] The CEDC intends to work with representative groups in various fields in order to offer "affordable production and creation premises" and which are likely to yield a "potential for synergy." The idea is to aggregate visual art creatives, scenic and theatre arts performers, and editing, publishing, and technology specialties. The latter project is carried out in partnership with the city authorities in the borough of Plateau Mont-Royal who recognize the industrial features of the neighbourhood buildings and provide access to premises which would not otherwise be available to creatives. The CEDC is committed to rent spaces for a ten-year period. The premises are thereafter sublet out to LAB and then to designers. Thus, one critical threat to the viability of designer projects, that is, uncertainty regarding location and the tensions involved in negotiating with property owners, is sidestepped.[9]

The LAB development experiment is a good example of the role a social economy agency can play as an intermediary for innovation. The CEDC's participation in the non-profit clustering of fashion designers contributes an original solution to some collective promotion and production needs of designers. This clustering provides young entrepreneurs with representation tools and means of action specific to their trade. They know they have access to diversified resources.

In addition to the pooling of equipment, the CEDC provided a form of legitimacy and access to otherwise unavailable knowledge because young fashion designers/entrepreneurs are not endowed with the type of social capital that confers financial credibility, nor have they yet availed themselves of a strong knowledge-transfer network.

CEDCs have access to arrays of human, organizational, and financial resources. LAB has developed collaborative efforts with educational and research institutions (CTTM and the École des métiers des Faubourgs-de-Montréal), with specialized consultants (for the business plan), with government (provincial and federal), with manufacturing representatives, with community business-management professionals, with other creatives, and with borough authorities. These, and access to financial programs earmarked for local initiative start-ups, are but a few examples of the resources that the CEDC has summoned and deployed to make the LAB project viable.

Access to adequate resources results from the community characteristic of the CEDC and its status as a local development agency. With the CEDC's support, the LAB has thus implemented mechanisms that make it possible both to share knowledge and to learn collectively. As mentioned previously, the culture of the fashion and artistic sectors is a very individualistic one, as everyone entering the sector considers him- or herself an "artist." Not everyone is open to learning from others and developing joint projects, or to taking the time to learn new techniques that could be introduced into their production. The existence of the LAB has brought people together, including people from other artistic disciplines who came to shows or other events, and this has started to introduce a culture of openness and cooperation.

Even if there has been some criticism of, and some managerial problems with, this case, we can still see this cluster as an innovative experiment which should inspire other sectors of activity, and it is surely useful in the fashion design sector. Whatever the evolution of this specific case over time, the idea of sharing knowledge and resources is surely to be seen as an important key to innovation.

The TOHU Case

The TOHU (la Cité des Arts du cirque) is located in one of Montreal's poorest districts, an area of the city which is undergoing a complete socio-economic transformation. This abandoned industrial site in Saint-Michel, a peri-central district of Montreal, near a former quarry, has become a landfill site. This district has a population made up mostly of immigrants, young people, and people with low educational and income levels.

TOHU resulted from the convergence, during the 1990s, of various local actors from the public and private sectors, that is, the Saint-Michel local community, civil society, and the circus movement, which illustrate the development of crossover knowledge flows between the actors of various sectors. Actors in the local community got together as a representative governance organization and created the Vivre

Saint-Michel en Santé (VSMS) non-profit organization, which fostered cooperation between other territorial actors of the public and private sectors (Tremblay and Pilati 2008, 2013). Through meetings between key individuals and exchanges of ideas between the various territorial actors, in November 1999 three organizations, Cirque du Soleil, the École nationale du cirque, and the EnPiste circus arts network, designed a project aimed at gathering in one place a critical mass of infrastructure for the creation, training, production, and diffusion of circus arts, thus establishing the conditions to make Montreal the international capital of circus arts (TOHU, 2006). To a large extent, this seems to have succeeded, since in July 2010 Montreal was the host of an international event called "Montréal complètement cirque."

The name TOHU is inspired from "tohu-bohu," an expression which conjures up the bubbling of ideas and actions, the chaos which precedes renewal, or the hustle and bustle of the big city (TOHU, 2006), and evokes the innovative and evolving nature of the project. TOHU's main interventions are centred on culture, that is, education, diffusion, and the development of an audience in the circus arts field. This project also includes two other components: an environmental one involving the rehabilitation of one of the largest urban landfill sites in North America and a social component involving artistic programming aimed at strengthening the socio-community aspect of the district. Here again, we have some elements of cross-sectoral knowledge flows, since there are some interactions between the three dimensions, that is, circus, environment, and society. For example, during the summer, the TOHU organizes an activity with children around environmental issues. It also takes into account a social dimension, having deep relations with the social organizations or community organizations of the district. From this perspective, TOHU has pledged to favour the local community for job opportunities. It also has pledged to offer some free activities for the local community. The relations between TOHU and environmental or community organizations are the place for some cross-sectoral knowledge flows. Of course, this is not in relation to technological innovation, as would be expected in aeronautics or other sectors, but rather in relation to organizational or social forms of innovation.

The combination of the three elements that characterize the theoretical bases of the "proactive cultural district" (Pilati and Tremblay 2007) can be found in the case of TOHU. First, TOHU is a geographical agglomeration of various organizations (Cirque du Soleil, EnPiste, École nationale du cirque, TOHU NPO, and, also in the same district, a residential centre for artists).[10] Thus, organizations from the same activity sector are concentrated here and share the same need for infrastructure, talent, and technology. TOHU has evolved through an initial phase of development of the circus arts companies that settled in the area. The specialized human capital and know-how needed to create innovation have contributed to the development of the district and to the dynamism of its production structure.

Second, TOHU is a place for the diffusion of artistic and cultural practices. In this particular case, TOHU includes creators, artists, producers, and choreographers, people who are directly involved in the circus arts field. The presence of these people has stimulated the springing up of many collateral businesses that are revitalizing the area. Today, TOHU is one of the largest districts involved in circus arts training, creation, production, and diffusion. It is a cultural crossroads which promotes collaboration between professional artists, specialists, associations, and citizens in the creation of circus-related activities. It also encourages integrated projects in order to provide a space for dissemination of these activities to the national and international circus communities, and allows the cluster's local creators to come into contact with the ideas and influences of creators from all over the world, including from other sectors.

Pierre Durocher[11] presented the TOHU project as a "diversified cluster in which young designers, entrepreneurs, and young people work in harmony in order to create a unique ambiance in Montreal." Indeed, many creative and talented people from fields and disciplines other than those pertaining to circus arts have come to TOHU, drawn by the presence of Cirque du Soleil. These people bring with them new ideas and artistic forms which translate into new lifestyles and trends. For many, TOHU has become a *creative ecosystem* where the organic, cooperation-based, and self-organization components are considered to be essential elements. This area, which brings together infrastructures and specialized human capital and has influence over circus arts in Canada and around the globe, becomes a physical and virtual base where the diverse know-how of talented and creative people from various sectors and trades is exchanged.

TOHU also carries out events, such as the circus show, Winter festival, and la Falla, with social objectives, where various groups in society can participate and which seek to reduce the gaps in terms of cultural capability (Sen 1999, 2000), gaps which often make it difficult for individuals to participate in cultural events that sometimes seem elitist.

Indeed, when events are not accessible and fail to meet the expectations of people in the community, these individuals cannot relate to them. Again, we find evidence of cross-sectoral knowledge flows here, much more than we found in other cases centred on the high-tech sectors, such as aeronautics, which tend to have more intra-sectoral knowledge exchanges, as we will see in the next section.

The TOHU project as a whole seems to be well oriented towards the three major effects sought, that is, "exerting an attraction for individuals with a creative ethos, producing innovations for the local territory as well as the economic and cultural system, and redirecting individuals and society towards activities laden with experiential value, while responding creatively to the challenges posed by the local context" (Sacco, Ferilli, and Lavanga 2006).

This research shows, however, that it is not easy to ensure social participation of the unemployed and other excluded members of the community. Indeed, while there is a certain degree of participation from the local community, not all segments of society or groups of immigrants have participated actively in TOHU's activities. (Tavano Blessi et al. 2012). This increase in social and cultural participation takes time, as habits need to be changed (ibid.). Nevertheless, we confirmed in a recent interview (Tremblay 2012) that TOHU jobs are still very much oriented towards the local immigrant community, and that TOHU is a sort of "stepping stone," helping the young local population to have access to the employment market, and then to move on to other jobs outside TOHU.

The Aeronautics Cluster

The aeronautics industry, even if it is largely globalized and very oriented towards technological innovation, also highlights some features of the consensus-building culture that has characterized economic development in Montreal. This culture explains why aeronautics was the first cluster implemented with the help of the CMM cluster policy and why it is the most efficient one. Let us mention that Montreal is considered as one of the world's main aeronautics hubs. The sector brings together many types of stakeholders: public actors, private businesses, unions, and intermediary organizations. At the business level, it is characterized by a pyramid structure divided into three groups of businesses: prime contractors, equipment manufacturers, and subcontractors. There are four main prime contractors: Bombardier, Bell Helicopters, CAE, and

Pratt & Whitney. The equipment manufacturers produce larger components and assemblies such as engine, engine accessories, and communications equipment. At the bottom of the pyramid, there are the subcontractors, in all some 220 small and medium-sized enterprises, offering products such as machined parts, casting and smelting works, machinery, and other products (CMM 2004). Subcontractors frequently work with aeronautics prime firms, but also with others sectoral prime contractors (Ben Hassen, Klein, and Tremblay 2011).

The Montreal aeronautics sector is characterized by a strong social network that has evolved through relations of proximity between stakeholders over the years. Contrary to other economic sectors in Montreal (e.g., textiles and clothing at the sectoral level), relations between the aeronautics actors are characterized by fairly solid cooperation. Of course competition and conflict do exist, particularly at the level of small businesses that compete for contracts, but the sector remains characterized by a climate of collaboration. Some even go as far as to speak of a "big aeronautics family."

This cooperation becomes even more pronounced during common projects that concern the future of the sector, such as Bombardier's C-Series. In July 2004, Bombardier Aerospace announced its plan to launch a new line-up of aircraft, the 110- and 135-seat C-Series (C-S models of short-haul, two-engine jet aircraft). However, the project encountered many difficulties, financial and otherwise. After stating, in May 2005, that it might have to move out of Quebec if it did not get financial support from the government, Bombardier received confirmation of the financial support it requested from the federal and Quebec governments. An agreement was reached with prospective suppliers, who committed themselves to support one-third of the development costs, estimated at more than $2 billion (CAD). Aéroports de Montréal also made an effort to locate the project in one of its airport facilities, and Bombardier announced it would locate the final assembly line of its C-Series aircraft at Montréal-Mirabel International Airport. However, in 2006, it put the project on hold due to a lack of orders. Some two years later, in February 2008, sensing a rekindled interest for this type of aircraft on the market, Bombardier relaunched the project. The governments of Canada and of Quebec then reiterated their financial commitment, and increased their support due to cost increases. Bombardier's business plan for the C-Series was based on a forthcoming government contribution covering one-third of the development costs.

In line with the consensus-building approach and cooperation fostered by various intermediate organizations, Bombardier's unions agreed to major cutbacks in employee benefits and working conditions. CAMAQ (Comité sectoriel de main-d'oeuvre en aérospatiale) supported the project, declaring that it would encourage Montreal's schools to train future C-Series workers. Bombardier eventually confirmed Mirabel Airport as the location of its assembly plant. This decision was made following a four-party arrangement between the company, the FTQ (Fédération des travailleurs du Québec), government agencies, and the Municipality of Mirabel. The evolution of this decision-making process illustrates how local mobilization and joint action at the metropolitan scale come into play when a sense of urgency is involved (Côté 2007). However, with regard to human resources, the relations between the businesses are characterized by competition for the recruitment of labour.

The intermediate organizations have played an important role in networking the stakeholders from the manufacturing side, which contributed greatly to the creation and active development of the aeronautics cluster. An important role played by the intermediate organizations concerns the support for innovation, but we need to mention here that this important cluster activity tends to limit the boundaries of innovation to the sector and not favour cross-sectoral knowledge flows much. This does not mean there are not any, in any firm. Rather, it was observed that people within the aeronautics sector do not have a very open attitude to innovation. For them, basically, R&D is the source of innovation, whether it be technological process or product innovation. While there are accounts of cross-sectoral flows in the literature (Jalabert and Zuliani 2010), they did not appear to be dominant in the Montreal cluster, and cluster policy tends to bring actors to centre upon their sector, rather than imagine that other sectors could be a source of new ideas or innovations.

In a context of limited natural resources and rising R&D costs, the local human resources necessary for innovation play an important role in the territorial embeddedness of the industry. In Montreal, the sector enjoys a dense network of stakeholders and intermediate organizations that encourage the mobilization of resources required by the sector to innovate (e.g., research, qualified labour, contacts). Among those intermediate organizations, CRIAQ (Consortium de recherche et d'innovation en aérospatiale) appears to be the most important with regard to innovation, in particular for precompetitive research in university-business partnerships. This gives businesses the possibility of having access to a skilled pool of labour to meet their specific needs. CRIAQ also informs Quebec universities of the technological challenges of the industry and enters into agreements with national and international organizations. Since its establishment in 2002, CRIAQ has managed thirty-one research projects involving over two hundred researchers and specialists.

However, many small businesses find that the large firms, such as Bombardier and a few others, are dominant in the cluster, and it is not always easy for the smaller ones to get their views heard at the provincial level. Of course, the large firms offer contracts to small firms, and so it is difficult for them to present views that are different from those of the large ones, as this may prove difficult for future contracts. Nevertheless, the development of the cluster over the years has made it possible for small firms to participate in R&D projects with large firms. So while there may be conflicts of interest, and even open conflicts at times, the consensus-building approach of Aéro Montréal and other intermediate organizations is appreciated by all (Ben Hassen 2012).

Discussion

In general, it is difficult to identify a main source or a few main sources of innovation in a region; rather, a complex series of factors, a web of interactions and knowledge exchanges, contribute to innovation (social, product, process, or organizational), as is shown particularly with the cases of Lab Créatif, TOHU, and aeronautics. Our research has shown that intermediate organizations and business associations play an important role in the acceleration of knowledge flows, in increasing the quality and the intensity of these flows (Dossou-Yovo, Tremblay, and Klein 2008; Dossou-Yovo 2010, Ben Hassen, Klein, and Tremblay 2011), although research still needs to be pursued to conclude on this more definitely.

It seems that technology-based innovation pushed by business organizations in order to improve the capabilities of a productive sector, like aeronautics, or some specific firms is rather sector based, while socially oriented innovation aiming at combining businesses and community objectives in a revitalization-oriented strategy rests rather on inter-sectoral flows. However, in both cases, sectoral dynamics are crucial, as the fashion domain (Lab Créatif) or the circus-art field (TOHU) show. But the expansion of these strategies on a larger scale implies cross-sectoral flows. Intermediate organizations and local associations can contribute to developing these knowledge flows, although there is still much work to be done along these lines. Indeed, in sectors like the garment industry, where there are many small firms, not organized hierarchically as in the aeronautics sector, there are fewer relations between firms, and knowledge flows just seem to be emerging within the fashion sector; however, they may be developing in a cross-sectoral direction, as the Lab Créatif case shows, or in some particular firms (some designers, for example, are working with photographers, contemporary artists, and the like) (Tremblay 2013).

In terms of product innovation, there are a few crossover knowledge flows, but these are difficult to identify and seem under-exploited. In the aeronautics industry, they seem rather uncommon, while in the garment industry, designers tend to be rather individualistic, and not really searching for opportunities for collaboration. As well, in the film industry, it was brought to our attention that special effects workers in film animation and gaming should collaborate more closely, but since they are formally part of two different clusters, this is only beginning to happen (Tremblay 2010; Tremblay and Cecilli 2009).Unfortunately, it seems that the official cluster policy does not particularly favour cross-sectoral flows, being rather oriented towards intra-sectoral activity.

So while cluster and innovative-milieu approaches stress the importance of knowledge flows, these seem to be mainly concentrated *within* sectors. In our view, this may have somewhat negative impacts. Indeed, in relation to the competence-oriented approach, which favours cross-sectoral knowledge flows, we can conclude that such an orientation does not dominate in Montreal, even in community bottom-up projects. While in all three cases studied the sectoral cluster has an influence on other sectors, triggering comprehensive place-based revitalization dynamics, the cross-sectoral dynamics do not dominate to this day. It may be useful to encourage clusters to open up somewhat; without changing the sectoral orientation totally, it may help to give the actors some idea of other sectors where they might find original, interesting ideas that may lead them along more creative paths than they had envisaged. It may be interesting to envisage more open seminars or activities, in order to bring people from various sectors to meet, especially where it is thought there may be some interesting cross-sectoral knowledge flows to be developed.

This is clearly not the dominant orientation at present, nor is it the orientation instituted by the CMM's cluster policy, which tends to favour intra-sectoral knowledge flows and not cross-sectoral flows. We thus wonder if the official cluster policy is not reinforcing the sectoral trends, already very present. While these are surely important for developing a competitive sector, it appears that cross-sectoral knowledge flows may be as important, if not more important at times, in spatial conversion to the knowledge economy.

In our view, the cluster strategy may also appear to be somewhat closed to the contribution of other social actors, mainly community actors or organizations that are more social in nature, or socio-economic (such as the CEDC or CLD). From this perspective, we might hope for a view on clusters that is more open to a variety of socio-institutional actors, and not too concentrated on business. Cluster dynamics show how important intermediators are for triggering productive collaboration and consensus building inside sectors. They could also play a similar role within the application of an intersectoral strategy, inasmuch as an open innovation-oriented policy is applied. This more open view would favour more cross-sectoral or cross-cluster knowledge flows, thus contributing to a better use of local resources and possibly more creativity and innovation. For the moment, the organizational richness of Montreal may not be used to its full extent, and thus the innovative potential of the city and the CMM may not be fully exploited.

Conclusion

We can thus conclude that Montreal clusters are strongly place based and that their governance involves active and efficient intermediary stakeholders. But even if some of them are not confined to one specific sector, the sectoral logic dominates productive and other interactions. They apparently concentrate the potential of intra-sectoral knowledge flows, although these may be the source of less innovative or creative ideas than some from other sectors. While we did observe some acceleration in the circulation of knowledge between members of some clusters, and some development of relational proximity, as well as learning and innovation, it would be difficult to conclude that innovation is strongly accelerated by knowledge flows between clusters or sectors in Montreal. The intersectoral fertilization of ideas is not yet at issue in the Montreal industrial strategy. Clearly, there are still efforts to be made to ensure the cooperation of many actors who are not yet active even in the existing clusters and to favour cross-cluster knowledge flows, as well as to ensure the transformation of competitive relations into cooperative relations throughout the cluster and beyond. A broader economic development strategy favouring more intersectoral collaborations would probably help many sectors develop more creative and innovative activities, and this would surely help the small businesses in the aeronautics sector (who may find clients in the auto industry or others) as well as the small businesses of the garment industry, where the Lab Créatif appears to be quite unique in terms of inter- and intrasectoral cooperation. For that, a renewal of the social coalition established as a reaction to the crisis of the 1980s is required. Public agencies, productive businesses, and civil society–based actors should reassert a collaboration-oriented strategy in order to reconvert the metropolis of Montreal into a creative region and make open and intersectoral innovation achievable.

NOTES

1 CLUNET conference, "Europe meets America," Montreal, 20 September 2007.
2 This section is based on Klein, Tremblay, and Bussières 2010.
3 CDEC Centre-Sud/Plateau Mont-Royal press release, 22 March 2007. www.cdec-cspmr.org.
4 This CDEC developed a multisectorial table for the reconversion of Mile-End.
5 See http://www.labcreatif.ca/createurs/.
6 Handout from a press conference, 22 March 2007.
7 The Montreal Fashion Week (MFW) is an annual event. The Spring-Summer 2006 show (24–8 October), prepared by 34 renowned designers, was attended by more than 7000 persons according to organizers . See www.patwhite.com/node/688.
8 CEDC press release, 22 March 2007.
9 Tensions can be sharp in buildings where the apparel industry is concentrated. Property owners impose pressures on tenants in order to sustain prices at the market rates they would earn for services or offices.
10 This artists' residence was completed in 2003 to house the Cirque du Soleil artists, who were on a visit to Montreal for intensive training and practice.
11 Interview with Pierre Durocher, Vivre Saint-Michel en Santé leader of projects related to the working group on urban and social revitalization of the Saint-Michel district. 18 January 2006, our translation.

REFERENCES

Amin, A., A. Cameron, and R. Hudson. 2002. *Placing the social economy*. London: Routledge.

Amin, Ash, and Nigel Thrift. 1992. "Neo-Marshallian nodes in global networks." *International Journal of Urban and Regional Research* 16 (4): 571–87. http://dx.doi.org/10.1111/j.1468-2427.1992.tb00197.x.

Arthurs, David, Erin Cassidy, Charles Davis, and David A. Wolfe. 2009. "Indicators to support innovation cluster policy." *International Journal of Technology Management* 46 (3–4): 263–79. http://dx.doi.org/10.1504/IJTM.2009. 023376.

Ben Hassen, T. 2012. "Le système régional d'innovation de l'aéronautique à Montréal entre dynamiques territoriales et sectorielles." PhD dissertation in urban studies, Université du Québec à Montréal.

Ben Hassen, T., J.-L. Klein, and D.-G. Tremblay. 2011. "Building local nodes in a global sector: Clustering within the aeronautics industry in Montreal." *The Canadian Geographer/Le géographe canadien* 55(4): 439–56.

Borja, J., and M. Castells. 1997. *Local & global: Management of cities in the information age*. London: Earthscan Publications.

Brault, S. 2009. *Le facteur C*. Montreal: Les Éditions Voix parallèles.

Camagni, R., and D. Maillat, eds. 2006. *Milieux innovateurs: Théories et politiques*. Paris: Economica.

Communauté métropolitaine de Montréal (CMM). 2004. *La grappe aérospatiale*. Montreal: Communauté métropolitaine de Montréal.

Communauté métropolitaine de Montréal. 2005. *Plan de développement économique*. Montreal: Communauté métropolitaine de Montréal.

Communauté métropolitaine de Montréal. 2008. *Les grappes et l'innovation: Libérer le capital créatif*. Montreal: Communauté métropolitaine de Montréal.

Côté, G. 2007. "Dynamiques territoriales et stratégies d'action publique: La genèse des projets technopolitains à Montréal et à Toulouse." Unpublished PhD dissertation in urban studies, Université du Québec à Montréal and INRS-USC.

Darchen, S., and D.-G. Tremblay. 2010. "What attracts and retains knowledge workers/students: the quality of place or career opportunities? The cases of Montreal and Ottawa." *Cities* 27 (4): 225–33 (Ref. No. JCIT-D-09-00070R1).

Darchen, S., and D.-G. Tremblay. 2013a. "Policies for the video game industry and the creative cluster model: A comparison between Melbourne and Montreal." *European Planning Studies*. http://www.tandfonline.com/action/sho wAxaArticles?journalCode=ceps20#.Us3iZuKxT1E.

Darchen, S., and D.-G. Tremblay. 2013b. "The local governance of culture-led regeneration projects: A comparative analysis between Montreal and Toronto." *Urban Research & Practice* 6 (2): 140–57. http://dx.doi.org/10.1080/ 17535069.2013.808433.

Dossou-Yovo, A. 2010. "Capacité d'innovation des petites et moyennes entreprises et contribution des organisations intermédiaires dans l'industrie des logiciels d'application multimédia à Montréal." Unpublished PhD dissertation in urban studies, Université du Québec à Montréal.

Dossou-Yovo, Angélo, Diane-Gabrielle Tremblay, and Juan-Luis Klein. 2008. "Territoire, processus d'innovation dans les PMEs et acteurs intermédiaires: Le cas du secteur des technologies de l'information dans la région métropolitaine de Montréal." Rimouski, Association de science régionale de langue française, 25–8 August. http://asrdlf2008.uqar.qc.ca/Papiers%20en%20ligne/DOSSOU-TREMBLAY.pdf.

Fontan, J.-M., J.-L. Klein, and D.-G. Tremblay. 2005. *Innovation sociale et reconversion économique: Le cas de Montréal*. Paris: Harmattan.

Gertler, M., and D. Wolfe. 2005. "Spaces of knowledge flows: Clusters in a global context." Paper presented at the conference "Dynamics of industry and innovation." DRUID 10th Anniversary Summer Organizations, Networks and Systems, 27–9 June, Copenhagen.

Hillier, J., F. Moulaert, and J. Nussbaumer. 2004. "Trois essais sur le rôle de l'innovation sociale dans le développement territorial." *Géographie, Économie, Société* 6 (2): 129–52.

Holbrook, A., and D. Wolfe, eds. 2002. *Knowledge clusters and regional innovation: Economic development in Canada*. Kingston: Queen's School of Policy Studies and McGill-Queen's University Press.

Jalabert, Guy, and Jean-Marc Zuliani. 2010. *Toulouse, l'avion et la ville*. France: Editions Privat.

Julien, P.-A. 2005. *Entrepreneuriat régional et économie de la connaissance*. Quebec: Presses de l'Université du Québec.

Klein, J.-L., C. Manzagol, D.-G. Tremblay, and S. Rousseau. 2005. "Les interrelations université–industrie à Montréal dans la reconversion à l'économie du savoir." In *Les systèmes productifs au Québec et dans le Sud-Ouest français*, ed. R. Guillaume, 31–54. Paris: L'Harmattan.

Klein, J.-L., and D.-G. Tremblay. 2010. "Social actors and their role in metropolitan governance in Montreal: Towards an inclusive coalition?" *GeoJournal* 75 (6): 567–79. http://dx.doi.org/10.1007/s10708-009-9270-0.

Klein, J.-L., D.-G. Tremblay, and D. Bussières. 2010. "Social economy-based local initiatives and social innovation: A Montreal case study." *International Journal of Technology Management* 51 (1): 121–38. http://dx.doi.org/10.1504/IJTM.2010.033132.

Leslie, D., and N.M. Rantisi. 2006. "Governing the design economy in Montréal, Canada." *Urban Affairs Review* 41 (3): 309–37. http://dx.doi.org/10.1177/1078087405281107.

Marshall, Alfred. 1889. *Principles of economics*. London: Royal Economic Society.

Moulaert, F., and O. Ailenei. 2005. "Social economy, third sector and solidarity relations: A conceptual synthesis from history to present." *Urban Studies* (Edinburgh, Scotland) 42 (11): 2037–53. http://dx.doi.org/10.1080/00420980500279794.

National Research Council. 1998. "Supporting the knowledge-based economy in the 21st century." Policy paper. Ottawa: National Research Council.

Pilati, Thomas, and Diane-Gabrielle Tremblay. 2007. "Cité créative et district culturel: Une analyse des thèses en présence." *Géographie, économie et société* 9 (4): 381–401.

Pilati, Thomas, and Diane-Gabrielle Tremblay. 2008. "Le développement socio-économique de Montréal: La cité créative et la carrière artistique comme facteurs d'attraction?" *Canadian Journal of Regional Science* 30 (3): 475–95.

Pilon, S., and D.-G. Tremblay. 2013. "A cultural perspective of the geography of clusters: The case of the video games clusters in Montreal and in Los Angeles." *Urban Studies Research* (Article ID 957630), 9 pages. http://www.hindawi.com/journals/usr/contents/.

Porter, M.E. 1990. *The competitive advantage of nations.* New York: The Free Press.

Porter, M.E. 2003. "The economic performance of regions." *Regional Studies* 37 (6–7): 549–78.

Porter, M.E. 2004. "Local clusters in a global economy." In *Creatives industries,* ed. John Hartley, 259–279. Oxford: Blackwell Publishing Ltd.

Rantisi, N., and D. Leslie. 2008. "The social and material foundations of creativity for Montreal independent designers." Presented at the annual meeting of the Innovation Systems Research Network, Montreal. http://www.utoronto.ca/isrn/publications/NatMeeting/index.html#nat08.

Rifkin, J. 2004. "When markets give way to networks ... everything is a service." In *Creatives industries,* ed. J. Hartley, 361–74. Oxford: Blackwell Publishing Ltd.

Sacco, P.L., G. Ferilli, and M. Lavanga. 2006. *The cultural district organizational model: A theoretical and policy design approach.* Venice: DADI, Università IUAV.

Scott, A.J. 2006. "Creative cities: Conceptual issues and policy questions." *Journal of Urban Affairs* 28 (1): 1–17. http://dx.doi.org/10.1111/j.0735-2166.2006.00256.x.

Sen, Amartya. 1999. *Un nouveau modèle économique: Développement, justice, liberté.* Paris: Éditions Odile Jacob.

Sen, Amartya. 2000. *Repenser l'inégalité.* Paris: Éditions du Seuil.

Tavano Blessi, G., D.-G. Tremblay, M. Sandri, and T. Pilati. 2012. "New trajectories in urban regeneration processes: Cultural capital as source of human and social capital accumulation – Evidence from the case of Tohu in Montreal." *Cities* 29 (6): 397–407. http://dx.doi:10.1016/j.cities.2011.12.001.

TOHU. 2006. *Tohu. Un modèle de développement durable au coeur du quartier Saint-Michel.* Montreal: TOHU.

Tremblay, D.-G. 2006. *Networking, Clusters and Human Capital Development,* Research report. 40 p. http://www.teluq.uqam.ca/chaireecosavoir/pdf/NRC06-08A.pdf

Tremblay, Diane-Gabrielle. 2010. "Montreal's cluster strategy in culture and technology sectors." In *Economic strategies for mature industrial economies,* ed. P.K. Kresl. Northhampton, MA: Edward Elgar Press.

Tremblay, D.-G. 2012. Video interview for the Teluq course on the management of creative territories. 19 March.

Tremblay, D.-G. 2013. "Creative careers and territorial development: The role of networks and relational proximity in fashion design." *Urban Studies Research* (Article ID 932571), 9 pages. doi:10.1155/2012/932571.

Tremblay, D.-G., and E. Cecilli. 2009. "The film and audiovisual production in Montreal: Challenges of relational proximity for the development of a creative cluster." *Journal of Arts Management, Law, and Society* 39 (3): 156–87. http://dx.doi.org/10.1080/10632920903218539.

Tremblay, Diane-Gabrielle, Juan-Luis Klein, and Jean-Marc Fontan. 2009. *Initiatives locales et développement socioterritorial*. Quebec: Presses de l'université du Québec.

Tremblay, Diane-Gabrielle, and Thomas Pilati. 2008. "The Tohu and artist-run centers: Contributions to the creative city?" *Canadian Journal of Regional Science* 30 (2): 337–56.

Tremblay, D.-G., and Thomas Pilati. 2013. "Social innovation through arts and creativity." In *International Handbook on Social Innovation*, ed. Frank Moulaert, Diana MacCallum, Abid Mehmood, and Abdel Hamdouch. Northhampton, MA: Edward Elgar Press.

Tremblay, D.-G., and D. Rolland, eds. 2003. *La nouvelle économie: Où? Quoi? Comment?* Quebec: Presses de l'Université du Québec. Collection d'économie politique.

Tremblay, D.-G., and V. Van Schendel, eds. 2004. *Économie du Québec: Régions, acteurs, enjeux*. Montreal: Éditions Saint-Martin.

Tremblay, G. 2008. "Clusters and innovation: Boosting creative capital." Presentation at 21st annual Entretiens du Centre Jacques Cartier. October.

Tremblay, Rémy, and Diane-Gabrielle Tremblay, eds. 2010. *La classe créative selon Richard Florida: Un paradigme urbain plausible?* Québec: Presses de l'université du Québec et Presses universitaires de Rennes. Collection Géographie contemporaine.

6 Firms and Their Problems: Systemic Innovation and Related Diversity in Calgary

COOPER H. LANGFORD, BEN LI, AND CAMI RYAN[1]

Calgary's phenomenal growth after 1947 has been almost entirely due to its position at the forefront of Canada's burgeoning petroleum and natural gas industries ... Calgary found the prosperity that was denied by the cattle industry and railroad development ... [It] enabled Calgary to transcend the ... dependence on a limited agricultural ... hinterland.

Foran and MacEwan-Foran 1982

Discovery of oil and gas in the Turner Valley region south of Calgary early in the twentieth century, followed by construction of pipelines in the 1950s, transformed Calgary's economic, political, and social structures. Alberta went from being one of the poorest provinces in Canada to the richest over several decades. Exploration and development in oil and gas have largely spurred the rapid growth of the Calgary census metropolitan area (CMA). Through the 1970s, 1980s, and 1990s, Calgary's energy sector attained critical mass, attracting the sector's national head offices. Now advertised as "Canada's Global Energy Centre," Calgary is home to the head offices of 87 per cent of the country's oil and natural gas producers (Calgary Economic Development 2009).

Calgary's dominant oil and gas cluster has differentially affected firms' capacities to function and develop. The primary focus on the oil and gas cluster has provided opportunities and positive spin-off effects from the associated economic growth, while also impeding some aspects of firm establishment and expansion (Langford, Li, and Ryan 2009a). Calgary scores high on the Conference Board of Canada's diversity rating (0.77), indicating a relatively high degree of firm-based diversity in the CMA (2007). This appears to be misleading. The myriad firm specializations in Calgary may be better viewed as providing a special sort of related variety (Cooke, this volume; Wolfe, this volume), discussed in more detail below, that channels knowledge into oil and gas innovation, now seen as a knowledge-intensive activity. Firm-based knowledge-generating activity in Calgary serves, either directly or indirectly, the demand of the oil and gas sector. The resulting tension between the dichotomies of specialization and diversity are illuminated by this case study. Emerging out of that tension, we report below on a rich platform of related diversity occupying an intermediate position between two classic approaches (Cooke, this volume; Wolfe, this volume). This presents problems of its own.

Calgary: A Related-knowledge Platform

Despite the central role played by the oil and gas industry in Calgary's overall economy, a closer examination of the structure of its urban economy suggests a much more diversified basis, albeit one which depends largely on production of oil and gas. However, "Calgary's role in the industry largely derives from the [managerial, technical] functions of the oil and gas industry head offices and the producer service firms that support them" (Miller and Smart 2011: 271). To use the framework developed by Innovation Systems Research Network (ISRN) researchers: Calgary's industries are highly clustered (Spencer et al. 2010). By these metrics 42 per cent of Calgary's employment is in clustered industries, compared to a Canadian average of 22 per cent. The cluster definition used by Spencer et al. (2010), which relies on location quotients[2] (LQ) as a core criterion, identified seven clusters defined by the standard North American Industry Classification System (NAICS) codes. The dominant cluster is the oil and gas cluster, with an LQ of 5.02. The closely related mining industry is second in employment LQ at 1.81, but the structure of employment (with firms in only 38 per cent of subcategories) does not satisfy the ISRN cluster definition. The sum of these two (NAICS code) industries accounts for 6.5 per cent of the labour force. Clusters also exist in Information and Communication Technologies (ICT) manufacturing (1.06) and ICT services (1.15), as well as finance (1.00) and business services (1.44). The latter three of these have close links to the oil and gas industry. In fact, a conventional qualitative cluster study (Langford et al. 2003) would include large parts of these clusters within the oil and gas industry as suppliers (see below). This may be better seen as providing a special sort of related variety[3] that channels knowledge from a variety of specializations brought into the Calgary economy to support oil and gas innovation. It is not based on related knowledge strictly speaking, defined as the sharing of a common science base (shown to be innovation-positive for firms in different sectors by Feldman and Audretsch 1999), nor does it reflect inputs from subcategories of a single high-level statistical category (as defined by Frenken et al. 2007). Rather, the degree of relatedness found in Calgary's economy originates from the common goal of meeting the demand for oil and gas directly or indirectly. Other Calgary clusters are found in construction (1.64), which is related to rapid growth, as well as oil and gas, and in logistics (1.19). The overall structure of Calgary's economy is scored at 0.77 on the Conference Board of Canada's diversity index, where 1 is the maximum (Lefebvre et al. 2009: 6), indicating a relatively high degree of diversity. However, as we argue below, this conclusion is misleading.

The key issue in examining knowledge flows is the question of what counts as a "sector" and how to define flows as intra- versus intersectoral. Analysing this issue yields the most interesting perspective to emerge from the study. If sectors were identified on the basis of conventional industrial classifications, the resulting analysis would yield an extremely rich, but noisy, pattern of knowledge flows low in information content. The pattern would suggest that inter-sector communication is rich, complex, and extensive, a misleading conclusion. The dominant oil and gas cluster, which does not actually handle much oil or gas in Calgary, but deals in technical, managerial, and financial knowledge for oil and gas extraction, is not limited to activity in companies classified in the oil and gas NAICS codes. Strong linkages are found from the sector to professional scientific- and engineering-services firms. ICT services play a key role in oil and gas production. A less widely appreciated aspect is that creative financial arrangements have also been crucial to the growth of the oil and gas industry. As well, a significant share of the construction cluster is closely linked to oil and gas customers. An examination of the Canadian input-output tables (Statistics Canada 2009) indicates that inputs to the oil and gas industry are concentrated in a very few sectors – mining and inorganic materials on the goods side. On the services side, business services (including professional and scientific services) and financial services are important components. In a qualitative study of a cluster, a large share of firms in these sectors could be identified as closely linked suppliers to oil and gas firms. The literature on innovation confirms that customers and suppliers lead all other factors as the sources of ideas for industrial innovation (e.g., Cohen et al. 2002). The widely accepted view about the degree of integration of firms in other sectors with those in the oil and gas industry was captured by a successful and widely influential Calgary serial entrepreneur: "I've learned over the years never [to] start into a business in Alberta that doesn't have direct implications and impact on the oil and gas business. This is an oil and gas town."

The reach of oil and gas activity in Calgary belies the apparent diversity of the economy in the city-region. Oil and gas is also an industry where innovation commonly occurs in the context of large projects with participation by a number of companies. The vast majority of activity in the city is concentrated in the assembly of the managerial, technical, and financial knowledge to manage resource extraction in its Alberta hinterland, regionally across Western Canada, and globally. The focus of this activity is the *mobilization of knowledge*. In contrast to usual perceptions, the production of oil and gas in Calgary is a knowledge-intensive, multidisciplinary activity. It is best conceptualized as an integrated *related-knowledge platform* that draws upon a significantly differentiated knowledge base (Asheim and Hansen 2009). The knowledge base incorporates a wide variety of firms that spans many standard industry sectors and several types of science base, but converges on the common problem of oil and gas extraction. There is an oil and gas culture across many areas, with a shared core of cognitive styles.

The characteristics of this related-knowledge platform raise an interesting issue about the role of diversity in a region. There are two classic positions with respect to the debate over the relative merits of specialization versus diversity (Wolfe, this volume). The first is the assertion that the main motivation for co-location and main contributor to innovation is knowledge spillover within a sector (Marshall 1890; Arrow 1962; Romer 1986; Glaeser et al. 1992). Porter (1998) makes a similar argument with a slightly different emphasis (on the stimulus of competition). In contrast, Jacobs (1970) argues that unrelated diversity plays a major role in the vitality of cities, and that critical creative ideas come from outside a sector. This has been identified with knowledge spillover among different sectors with a common science base (Feldman and Audretsch 1999) or with knowledge spillover among different sub-sectors of a broad umbrella (e.g., 2-digit) statistical category (Frenken et al. 2007). The related-knowledge platform with the variety

of backgrounds of engineering, managerial, IT, and finance professionals suggests something of a broad Jacobsian degree of diversity except,
critically, that all the work of the professionals serving oil and gas projects is aimed at a single overall goal with a common cognitive core. The
typical examples cited by Jacobs, or the examples from Frenken et al.
and Feldman and Audretsch do not fit with the focus on a common
problem of the contributors to this related-knowledge platform. In the
Calgary economy, community is created around the outputs, not the
inputs. On the other hand, the very breadth of the knowledge bases
seems to require a Procrustean trick to fit it into the argument made
by the proponents of specialization. A focus on the importance of local
competition is not relevant in the case of the Calgary firms, or units of
multinationals, who are competing with oil and gas producers worldwide in a commodity market. Thus, we find that a rich platform of related diversity occupies an analytical position intermediate between
the two dominant approaches in the literature and, thus, presents problems of its own.

The Calgary Census Metropolitan Area (CMA) is just over one
million in population and oil and gas has proved the key driver of
the growth not yielded previously by cattle or railroads (Foran and
MacEwan-Foran 1982). It is tempting to think this middle ground
may be a characteristic of a city of upper-middle size that owes its
emergence to a dominant industry. Perhaps *related-knowledge diversity*
(RKD), where "related" focuses on outcomes and "diversity" on inputs, is a distinct growth and innovation factor with its own dynamic
of advantages and limitations. One negative characteristic is that all the
diverse sectors are vulnerable to changes in demand (or price) of the
output commodities. In the next section, we turn our attention to an
analysis of the nature of knowledge flows and innovation within this
related-knowledge platform.

Firm-based Innovation in Calgary

This section analyses firm-based innovative activity and behaviour connected with navigating in and around the dominance of the oil and gas
sector. Overall it examines the dynamics of regional innovation and attempts to identify systemic characteristics (Cooke 2001). Echoing Foray's (2004: 38) interpretation of "innovation" as implied by the concept
of a "knowledge economy," we argue that the process of innovation is
non-discreet, in continuous flux and is – in essence – similar at all scales,

suggestive (perhaps misleadingly) of a fractal process. This suggests a primary classification of Calgary firm innovation, not by scope, but rather by *type of problems* that are ultimately addressed by innovative activities. Our analysis of the interview results distinguished between four different ways in which innovations help to resolve firm-based issues or problems:[4]

- Innovations are designed to solve a problem or problems that are identified by a client or set of clients.
- Innovations are developed internally – within the firm – and are motivated by goals of organizational learning. Essentially, these latter innovations address issues around firm-based capacities or competencies.
- Innovations may tackle a market barrier problem wherein strategies for circumvention or management are required.
- Innovations – often composed of a suite of related products and services – are required for the creation of a new market.

These categories of innovations can be distinguished between those that support the competitive advantage of a particular firm and those that are of a collaborative nature among a group of firms entailing collective advantages. This distinction is significant when considering network phenomena and questions of specialization versus diversity.

The flow of knowledge is categorized according to location with respect to the Calgary city-region:

1 *Internal knowledge,* which circulates primarily within a particular firm or local innovative unit and is strongly conditioned by that organizational culture.
2 *Local knowledge,* which circulates primarily among innovative individuals and organizational units within the CMA-based system. The fluidity of knowledge under this category includes both intra-sectoral flows (within a given sector or industry) and cross-sectoral flows (that transcend sectoral boundaries).
3 *Non-local knowledge,* which arises from specific sources outside the CMA and is transmitted to the CMA via various channels such as publications, patents, and especially through industry specific pipelines or linkages such as "invisible colleges" of expertise, conferences, shows, and informal contacts (Bathelt et al. 2004; Maskell et al. 2005).

Table 6.1 Location-oriented knowledge typology

	Internal	Local	Non-local
Codified	Manuals, formal procedures,etc.	e.g., Local grey literature, etc.	Scientific literature, trade papers, etc.
Tacit	Embodied, mentoring, acquisition of appropriate staff with a skill set, etc.	Mentoring, workshops, networks, etc.	Training workshops, invisible colleges, etc.

In addition, we distinguish between tacit and codified knowledge, adopting the common definitions found in the literature. Polanyi (1966) suggests that tacit knowledge arises when one acquires a skill or attains a corresponding understanding (know-how) that is almost impossible or too cost-prohibitive to codify (Cowan, David, and Foray 2000). Gibbons et al. (1994: 68) suggest that tacit knowledge resides in the heads of those working on a particular process or is embodied in a particular organizational context. Codified knowledge, by contrast, is expressed in documents, equations, and logically analysable statements that are systematic, reproducible, and related to facts or information and the principles that explain them. Unlike its tacit counterpart, it is more easily shared (Lundvall 2004) among all with suitable receptor capacity.

Table 6.1 outlines and gives examples of a location-oriented knowledge typology that integrates tacit and codified knowledge with location factors. The distribution according to this typology of factors contributing to innovation is critical to identifying the internal processes of the CMA and whether and how knowledge flows within a sector or, in contrast, across sectors.

Our interviews reveal that problem solving is a distributed cognitive process dependent upon knowledge acquisition and sharing. First, we attempt to determine, from the evidence concerning both the nature of knowledge sources and the process of problem solving, whether an "innovation system" exists that can be analysed from a study of the Calgary city-region. The test is whether local practices demonstrate a systemic character in which local knowledge circulation is prominent. The problem is particularly acute in a centre like Calgary, where the largest cluster is in oil and gas and Calgary's contribution is primarily managerial, technical, and financial knowledge applied to a wide range of resource sites that are distributed across the globe. Other resource-based systems are most often similarly geographically relational.

Table 6.2 Number of innovations by type

	Client needs	Capacity building	Market barrier	New market creation
Firm advantage	28	23	5	18
Collective advantage	4	5	4	8

Figure 6.1 Problem types vs. knowledge factors

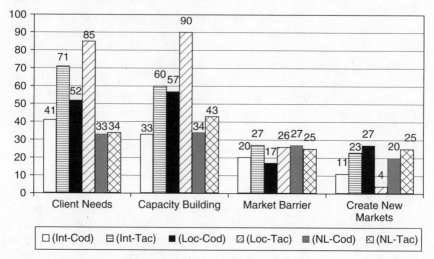

Note: Int-Cod = Internal Codified; Int-Tac = Internal Tacit; Loc-Cod = Local Codified;
Loc-Tac = Local Tacit; NL-Cod = Non-Local Codified; NL-Tac = Non-Local Tacit

Innovation and Knowledge Flow Patterns

The distribution of innovations by problem type is summarized in table 6.2. The distribution of types of knowledge flow supporting innovations classified by problem type is shown in figure 6.1. The importance of local inter-firm flow of knowledge is evident. As well, firm advantage is seen to outweigh collective advantage.

There is one central theme in the interviews that is not easily illustrated with one or two quotes or charts and tables. It is the pervasive description of innovations as complexes that might include products, but also entail custom consultation or knowledge-intensive services provided by the innovating firms (even for mass market items). Almost no innovations mentioned by interviewees were limited to individual products. This suggests a very close connection between knowledge flows and ongoing relationships among a network of firms, and emphasizes the importance of shared cognition in innovations that do not stop with product artefacts among the diverse professionals involved. Beyond this, much innovation was directed at assembling the competences required to provide a comprehensive suite of services. The clear lesson is that internal analyses of technological development are often skeletal at best and do not provide a sufficiently nuanced and detailed perspective on the nature of the innovation process. Innovation is seen here in its social guise as combining a variety of cognitive and cultural elements.

On average, the role of local tacit knowledge outweighed non-local tacit knowledge by a factor of 1.6, and local tacit knowledge similarly outweighed non-local codified knowledge as an input. Perhaps more surprisingly, tacit knowledge internal to the firm was less frequently cited than local tacit knowledge, and internal codified knowledge was an even weaker contributor compared to local codified knowledge. These aspects of our results and the fact that local tacit knowledge was the most heavily reported for three of the four problem types strongly supports the existence of a recognizable local system of rich knowledge flows. That tacit knowledge is more readily accessible locally than from afar is, of course, expected.

Knowledge flows predominately occurred in untraded form. The high level of local tacit knowledge associated with meeting client needs is connected to the commonly reported importance of clients as sources of ideas for innovation (e.g., Cohen et al. 2002). The fact that a large share of the firms interviewed provide service (engineering, IT, finance) to the dominant oil and gas activity, raised the above question of what is an intra-sectoral knowledge flow, as opposed to a cross-sectoral knowledge flow.

The extent of the outward linkage of this knowledge system is less clear. There is a degree of access to tacit non-local knowledge that establishes its use in all sectors, and with respect to all problems, and to professional or industry networks that are non-local. In some instances, these links are to a corporate structure, of which the Calgary establishment is a branch. Knowledge from afar seems to play a more prominent role when the problem concerned involves gaining a foothold in a new market or overcoming a market barrier. One local firm created both a new market in high-level safety skills for resource extraction, and simultaneously raised standards in the sector by making such safety skills a requirement for the type of work. Their outward networking was characterized as diverse: including international medical associations, specialty consumer product associations, and customer networking.

The fact that external codified knowledge also plays a role throughout similarly indicates that literature, patents, software, or trade publications are useful in all cases. For example, a junior exploration development firm develops new applications of resinous materials that depend on "intelligent applications" that must be protected. Bringing in knowledge to solve new problems (e.g., accessing knowledge in geoscience journals at the university) seems essential to both the business and talent retention functions of firms. This came up, for example, in the context of CO_2 management.

There is some evidence of a relationship between the four problem types that were identified and the relative importance in each of the geographic, tacit, and codified knowledge factors that illuminate the structure of the related-knowledge system. For example, the results suggest that local knowledge is more crucial for understanding problems that are locally centred to the firm and sector, including capacity building and meeting client needs. Recall that non-local knowledge emerges as relatively more important to innovations necessary for overcoming barriers or creating new markets. The way in which support firms serve *client needs* is illustrated by a quote from a geosience firm executive: "developing new methods, new concepts ... we don't have the resources ... we're far too small ... What we do is look at what might be new ... equipment or methodologies and bring them to a client's attention and then apply them [here]." In contrast, the following is a characteristic story of *capacity building*: "Probably one of our stronger connections ... is [the] LEAN consortium. We're part of the Canadian Manufacturers and Exporters Association. Within that, 40 different companies [are] broken into three consortiums that collaborate together on sharing best manufacturing practices ... A [major telecom manufacturer] is in [our consortium]."

Since innovations in the categories that are externally influenced include the more visible forms of innovation that are likely to produce patents or broad user recognition, we observed a larger role for the non-local codified knowledge of papers, reports, and patents. As well, the tacit knowledge from conferences, industry networks, professional networks ("invisible colleges"), and workshops enters the picture more prominently. "Pipelines" involving specific common interests on a global scale (Bathelt et al. 2004) exist. A medical technology firm executive describes a typical case: "The patents that the current products are based on are licensed from [major US university]. We're probably early in the stage of generating our own IP."

Inter-sectoral Knowledge Flows

For this analysis, we classify oil and gas, mining, a number of firms interviewed from construction, business, and engineering services, and ICT services that are closely integrated with and into oil and gas projects as the single related-knowledge platform. Data from our interviews that reveal close integration of the activities into oil and gas are therefore grouped into the oil and gas platform and knowledge flows among them and to oil and gas firms are considered "intra-sectoral." Among professional, scientific, and technical services which have a location quotient (LQ) of 1.7, many environmental services firms can be recognized as being at arm's-length from oil and gas to a significant degree, and these constitute a useful, second *environmental consulting* sector which is represented in the interview data. A sub-sector of both the business services (LQ = 1.57) and creative and cultural industries (LQ = 1.12) that was identified in the data demonstrates a cluster of firms that can be grouped as *a multimedia and advertising* sector. Interviewees suggested that Calgary is generating a certain degree of "buzz" in this area. Finally, scattered innovative firms drawn mainly from manufacturing are grouped here simply as "other." Thus, "inter-sectoral" knowledge flows are treated here as occurring among four groups of firms: fossil fuels, environmental consulting, advertising and multimedia, and other. Data from the interviews yield the pattern of knowledge flows between the four groups of firms shown in table 6.3.

The dominance of the diagonal in this matrix supports the distinction adopted on independent grounds as identifying knowledge communities.[5] The oil and gas platform is the only significant supplier of innovation support knowledge to other sectors. The central role of oil and gas is even more evident in analysis of knowledge flows into the not-for-profit sectors (Langford, Li, and Ryan 2009b).

Table 6.3 Intra- and inter-sectoral knowledge inputs (percentage of knowledge received from originating sectors)

Receptor sector	Originating sector				
	Oil & gas	Environment	MM/Adv.	Other	Not listed
Oil and gas	79	1	–	–	20
Environment	7	38	–	–	55
Multimedia/Advertising	4	–	61	–	35
Other	1	–	–	53	46
Originating sector's share of all inputs not identified	24	9	14	14	39

Diversification by "Spin-off" from the Platform

A major part of the diversification that has occurred in Calgary has depended upon "spin-off" from the oil and gas sector. For example, talent valued within the oil and gas platform has exploited the relevant knowledge to create industries with different customer profiles. One successful Calgary industry, Global Positioning Systems (GPS), received an early impetus from establishment of a GPS unit within Shell Oil, which was spun out as a non-core activity. Calgary GPS now serves a variety of user sectors, notably including agriculture (Langford et al. 2003). A pipeline company, Nova (privatized from Alberta Gas Trunk Lines), formed a partnership with Alberta Government Telephones (later Telus) to create NovAtel, which played a central role in development of wireless telecommunication in Calgary (ibid.). The serendipitous recognition of the usefulness of knowledge gained in one sector for the launching of an enterprise in a distinct sector was also observed in the interviews. In one example, imaging knowledge was transferred from the geophysical domain to the medical imaging industry. A geophysist was the key actor in development of a low-cost digital X-ray system, resulting in a company that has achieved international reach. This illustrates the breadth of the knowledge base of the related-knowledge diversity platform and its potential to spin off new industrial directions.

The relatively low incidence of inter-sector knowledge exchanges does not imply low value or impact for such "weak links" (Granovetter 1983). These links can be the source of fresh insights not in common circulation among firms in those sectors that are strongly linked. The origin of inter-sectoral knowledge flows may emerge from one of the features seen in the interviews with creative individuals (Langford, Li, and Ryan 2009a). Among the attractions of Calgary to individuals are activities and organizations from music or mountaineering to motorcycle clubs that do not recruit members from any particular professional activity. These provide rich opportunities for the development of strong personal links that accompany weak professional links. One interesting feature of the environmental-services firms not integrated into the oil and gas platform is that they report having gained significant advantages as a result of their expertise honed in work within the oil and gas sphere. To some degree, this makes the sector an oil and gas "spin-off" that has taken a part of the diverse knowledge platform and directed it into a new application. Note that the largest identifiable off-diagonal elements of knowledge flows in the inter-sectoral matrix are between oil and gas and environment.

Movement of People

A majority of both small and large firms mentioned internal human resources as key to innovations for purposes largely internal to firms (capacity-building innovations). These innovations serve to enhance the firm's overall ability to offer an advanced suite of services. This is illustrated by some firms' statements that taking on complicated projects was predicated on the pre-existing ability to access requisite local talent from the right knowledge culture. "You don't need a lot of people, you need the right people who believe similar things but come from different perspectives, who can take those random ideas and create something unique."

Talent acquisitions from other sectors appear to serve as explicit acts to expand the reach of a firm's shared memory of previously solved problems (Minsky 1988), from within and outside the industry, in order to enhance its ability to solve new ones. This may reflect a form of "hidden knowledge flow," both within and across sectors, where the recruiting of external talent carrying relevant knowledge generated the firm's internal contribution to the innovation process. Said a geological engineering firm executive: "The whole geology team at [large project] is [our] people, even some of them have moved on to some of the other operators. They all go to our clients, almost all ... There is some political return, when you're dealing with your own people and they know what you've done for them. If they need help, where do they go?"

Further, small firms tended to maintain a roster of contract talent for engagement in specific projects, while larger firms tended to proactively seek out and acquire unique talent in the anticipation that their expertise would non-specifically enhance overall firm-based capabilities.

Consortia and Associations

A number of innovations aimed at generating a collective advantage were identified above. Most of these are realized through formal, if transitory, consortia. Both financial and technical knowledge is shared in these consortia and they distribute knowledge and understanding among the participating firms. This is a standard mode of operation in the oil and gas industry. These consortia reach out to engineering firms and environmental and financial actors. The Petroleum Technology Alliance of Canada (PTAC) is an institutionalized structure to promote consortia for technology development that provides a forum for a proponent of a technology initiative to recruit other firms to participate in a project. PTAC projects normally relate to "conventional" reserves.[6] There is also an organization for heavy oil and oil sands activity. We also found evidence of consortia in the advertising/multimedia sector.

Associations are also significant venues for knowledge exchange. Many of these are national and even international, but ones centred on Alberta are also common. Given the different industrial structure of the two major urban centres in Alberta, Calgary and Edmonton, several of these can be thought of as primarily Calgary organizations. They serve as conduits for a significant fraction of the local tacit knowledge flows that were identified in our interviews. There are also important local organizations such as Innovate Calgary, Calgary, Inc., the Advanced Technology Association, and the Wired City project that bring together mostly smaller companies from the "other" group above. The majority of these are in manufacturing and organizations can be explicitly intersectoral, focusing more on common business problems of younger and growing firms.

Social Networks

The local knowledge circulation seen to be so prominent in innovations is dependent upon social networking. Informal networks were mentioned widely in the interviews. Interviewees from oil and gas, engineering, environmental, and multimedia/advertising sectors all mentioned the interdependence of their industry-based social, knowledge networks. Even though each sector's internal networks were distinguishable, there was also evidence of interaction. At one end of the spectrum, the Calgary "Plus-15" system of linking major downtown buildings above street level functions to promote "accidental" contact. As this was put in one interview: "You bump into someone on the Plus-15 and it reminds you [that you] should talk and you set up a meeting." Similarly, there are arranged gatherings in social settings, as revealed by this example linking oil and gas and environmental expertise: "There's currently a group of us who get together at the [local restaurant] once a month, swap ideas on water treatment for oil sands, drink lots of wine, that sort of thing. It's quite a social forum, but it's a technically useful social forum."

There are organizations that clearly link to one sector, but have broad influence. The Petroleum Club, for example, is an important venue for informal business and civic activity, reflecting the prominence of the oil and gas platform both economically and in civic affairs. National oil and gas organizations played an important, but less frequently mentioned role, as noted by a junior exploration/developer:

> There [are] a number. The Society of Petroleum Engineers, they have a local chapter. [I participate] a lot more in the Canadian Heavy Oil Association, and so I have lots of friends through that. And I have lots of friends through former work connections. Many of us get together, we do a lot of technology swap over lunch, there are a whole lot of projects to get done, so we've got to do some sharing … to all get our jobs done, so we have a fairly interesting tight-knit network.

Conclusion: Related-knowledge Diversity and a "Diversification" Problem

Our interview results provide qualitative evidence to support the statistical data on employment profiles. Calgary enjoys diversity of talent. The city's oil and gas extraction is a knowledge-based activity that depends on its rich mix of managerial, technical, and financial knowledge to manage resource extraction in Western Canada and globally. The firms are innovative and there is extensive evidence of rich local knowledge flow. The city has grown rapidly and the strengths identified here have been central to that growth. However, the weak point is the degree of reliance on a single commodity. The degree of Calgary's reliance on this dominant commodity as the source of its economic vitality limits any guarantees of future sustainability or resilience. In common with other centres in Western Canada,[7] a major concern of policy discussion is "diversification." The rich background-knowledge diversity of the oil and gas related-knowledge platform does not, by itself, promise future economic stability and long-term growth. Diversification into other markets is the centrepiece of the federal government's involvement in the Alberta economy. Its economic agency for the west, Western Economic Diversification Canada, has a mandate to "promote the development and diversification of the western Canadian economy" (WD 2009). The Alberta government is also making diversification a major goal of economic policy. For example, Alberta's premier mandated the minister of advanced education and technology to "enhance economic diversification and build a knowledge-driven future by identifying focused priority sectors" (AET 2010). Notably, promoting "value-added" production has been a pillar of provincial policy for both petroleum and agriculture. The diversification conversation has also been given a high degree of local prominence. At the city level, economic diversification is one of the goals of the city-funded Calgary Technologies, Inc. (now a part of Innovate Calgary) and a concern of Calgary Economic Development. The interviews revealed, however, that there is no agency or leading spokesperson, no "champion," articulating a strategy for promoting further diversification that would overcome the potential vulnerability of this related-knowledge platform with respect to either low commodity prices or major shifts away from carbon energy.

Part of this contribution to the present work was an attempt to assess the views of key spokespersons for the major elements of the regional governance system. This was achieved in collaboration with The Centre for Innovation Studies (THECIS) through a "summit" conference, "The Natural Resource Industries as Engines of Economic Diversification" (THECIS 2008). Invited participants included leaders from industry, government, and academia across Western Canada. This group considered the opportunities and barriers to diversification based on innovations arising from a natural resource knowledge platform.[8] Among the advantages seen to be available to the oil and gas related-knowledge diversity platform were the following:

- Most of the educational, technical, and management capabilities to innovate and diversify exist already in the region or can be accessed relatively easily by enterprises that stay in the region.
- The resource industries are linked into an already diverse range of neighbouring industries that are rooted in the region, offer wide

scope for innovative entrepreneurial activity, and have the potential to achieve global reach.

- The region supports a growing pool of experience in creating successful globally competitive high-value-added enterprises.
- Resource enterprises have extensive managerial experience and financial experience related to creating value from entrepreneurial risk.

In the context of the results obtained in this study of the nature of innovation and knowledge flows in Calgary's economy, as well as the analysis provided by the "Banff Consensus," it appears that the success or failure of the entrepreneurship needed for diversification-based growth and stabilization is largely a matter of attitudes in the private sector. Will these attitudes favour rent-seeking innovation or can they favour productive innovation (Baumol 1990) by opening up new directions? Important obstacles were recognized at Banff. Among these the one most relevant to oil and gas is "a 'rip-and-ship' mentality ... A rush to take commodity prices at the lowest levels of added value ... still dominates."

Hopefully, the recent shift of technological emphasis (Vanderklippe 2011) from technologies of exploration, which have already produced spin-offs to telecommunication and global positioning systems, to technologies of extraction driven by the focus on oil sands and heavy oil will stimulate new spin-off directions. As well, the problems of carbon management are stimulating attention to carbon capture and storage that is likely to have as much relevance to power generation, among others, as to oil and gas extraction. In this respect, there exists an untapped reservoir of knowledge arising from the current related-knowledge diversity platform in Calgary that can be directed to the exploitation of these new opportunities to provide a more sustainable and resilient future for the city-region.

NOTES

1 Thanks to the Calgary team: Julie Alati-it, Kelly Bergstrom, Christine Cheung, Patrick Feng, Richard W. Hawkins, Stefan Mendritzski, Ray Op'tLand, Terry Ross, Sheila Taugher, and Nathan Voisey.
2 LQ is defined as the ratio of a sector parameter (e.g., employment) to the value of that parameter (e.g. employment) for the CMA divided by the ratio of national sector parameter to the overall national parameter in national

statistics. It measures the intensity of localization of the sector parameter normalized to the national average. The most commonly used parameter is employment.

3 This term denotes here variety of specializations (e.g., IT, engineering, finance) that share a common element in the demand profile (Frenken et al. 2007). This applies in this case to the firms that serve the common goal of upstream oil and gas production and marketing.

4 In total, 121 interviews were conducted in Calgary. Full transcripts of interviews conducted with 34 senior firm executives characterized firm behaviour for the purposes of this study. Sectors represented in this subset include oil and gas, science and engineering services, environmental engineering, construction, independent consultants, advertising and multi-media, and financial services. Data on knowledge inputs to innovations were obtained from analysis of the transcripts of interviews with senior firm executives who had personal involvement with the innovations described. The coding method based on extraction of substantial text fragments is summarized in appendix 1.

5 The magnitude of the diagonal element in a knowledge-flow matrix may offer an interesting approach to characterizing a platform or cluster.

6 PTAC's mission statement is: "Facilitating Innovation, Collaborative Research and Technology Development, Demonstration and Deployment for a Responsible Canadian Hydrocarbon Energy Industry" (PTAC, n.d.).

7 Dependence on, e.g., potash in Saskatchewan or lumber in British Columbia stimulates parallel concerns over reliance on volatile commodity markets and the need for diversification.

8 A detailed report on the conference edited by R.W. Hawkins and C.H. Langford (THECIS 2008), subtitled *"The Banff Consensus"* to reflect the convergence of opinion that arose, is available at www.thecis.ca.

REFERENCES

AET. 2010. Alberta Advanced Education and Technology mandate, February. http://www.premier.alberta.ca/documents/Advanced_Education_Man date_Letter.pdf.

Alexander, I.E. 1988. "Personality, psychological assessment, and psychobiography." *Journal of Personality* 56 (1): 265–94. http://dx.doi.org/10.1111/j.1467-6494.1988.tb00469.x.

Arrow, K.J. 1962. "Economic welfare and the allocation of resources for innovation." In *The rate and direction of inventive activity*, ed. R.R. Nelson, 609–26. Princeton, NJ: Princeton University Press.

Asheim, Bjørn, and Høgni Kalsø Hansen. 2009. "Knowledge bases, talents, and contexts: On the usefulness of the creative class approach in Sweden." *Economic Geography* 85 (4): 425–42. http://dx.doi.org/10.1111/j.1944-8287.2009.01051.x.

Bathelt, Harald, Anders Malmberg, and Peter Maskell. 2004. "Clusters and knowledge: Local buzz, global pipelines and the process of knowledge creation." *Progress in Human Geography* 28 (1): 31–56. http://dx.doi.org/10.1191/0309132504ph469oa.

Baumol, W.J. 1990. "Entrepreneurship: Productive, unproductive, and destructive." *Journal of Political Economy* 98 (5): 893–921. http://dx.doi.org/10.1086/261712.

Calgary Economic Development. 2009. "Calgary: A global energy leader." http://www.calgaryeconomicdevelopment.com/files/Sector%20profiles/CED_Profile_Energy.pdf.

Cohen, Wesley M., Akira Goto, Akiya Nagata, Richard R. Nelson, and John P. Walsh. 2002. "R&D spillovers, patents and the incentives to innovate in Japan and the United States." *Research Policy* 31 (8–9): 1349–67. http://dx.doi.org/10.1016/S0048-7333(02)00068-9.

Conference Board of Canada. 2007. *Metropolitan outlook, Fall 2007*. Ottawa: Conference Board of Canada.

Cooke, P. 2001. *Industrial and corporate change*. Oxford: Oxford University Press.

Cooke, Philip, Mikel Gomez Uranga, and Goio Etxebarria. 1997. "Regional innovation systems: Institutional and organisational dimensions." *Research Policy* 26 (4–5): 475–91. http://dx.doi.org/10.1016/S0048-7333(97)00025-5.

Cowan, R., P. David, and D. Foray. 2000. "The explicit economics of knowledge: Codification and tacitness." *Industrial and Corporate Change* 9 (2): 211–53. http://dx.doi.org/10.1093/icc/9.2.211.

Feldman, M., and D. Audretsch. 1999. "Innovation in cities: Science-based diversity, specialization and localized competition." *European Economic Review* 43 (2): 409–29. http://dx.doi.org/10.1016/S0014-2921(98)00047-6.

Foran, Max, and Heather MacEwan-Foran. 1982. *Calgary: Canada's frontier metropolis. An illustrated history*. Windsor, ON: Windsor Publications.

Foray, D. 2004. *The economics of knowledge*. Cambridge, MA: MIT Press.

Frenken, K., F. Van Oort, and T. Verburg. 2007. "Related variety, unrelated variety and regional economic growth." *Regional Studies* 41 (5): 685–97. http://dx.doi.org/10.1080/00343400601120296.

Gibbons, M., C. Limoges, H. Nowotny, S. Schwartzman, P. Scott, and M. Trow. 1994. *The new production of knowledge: The dynamics of science and research in contemporary societies*. Thousand Oaks, CA: Sage.

Glaeser, E.L., H.D. Kallal, J.A. Scheinkman, and A. Shleifer. 1992. "Growth in cities." *Journal of Political Economy* 100 (6): 1126–52. http://dx.doi.org/10.1086/261856.

Granovetter, M. 1983. "The strength of weak ties: A network theory revisited." *Sociological Theory* 1: 201–33. http://dx.doi.org/10.2307/202051.

Jacobs, J. 1970. *The economy of cities*. New York: Vintage Books.

Langford, C.H., B. Li, and C. Ryan. 2009a. "Creative talent in relation to the city: The case of a natural resource-based centre (Calgary)." Presented at ISRN 2008 annual meeting, Montreal.

Langford, C.H., B. Li, and C. Ryan. 2009b. "Innovation from an oil and gas platform: Calgary." Presented at ISRN 2009 November meeting.

Langford, C.H., J.R. Wood, and T. Ross. 2003. "Origins and structure of the Calgary wireless cluster." In *Clusters old and new: The transition to a knowledge economy in Canada's Regions*, ed. David A. Wolfe, 161–86. Montreal: McGill-Queen's University Press.

Lefebvre, Mario, Alan Arcand, Greg Sutherland, Lucie Tremblay, and Maxim Armstrong. 2009. *Calgary: Metropolitan outlook 1, Autumn 2000*. Ottawa: Conference Board of Canada.

Lundvall, B.A. 2004. "The economics of knowledge and learning." In *Product innovation, interactive learning and economic performance*, ed. J.L. Christensen and B.A. Lundvall. Bingley, UK: Emerald Group. http://dx.doi.org/10.1016/S0737-1071(04)08002-3.

Marshall, A. 1890. *The principles of economics*. New York: Cosimo Classics.

Maskell, Peter, Harald Bathelt, and Anders Malmberg. 2005. "Building global knowledge pipelines: The role of temporary clusters." DRUID working paper no. 05-20. http://www2.druid.dk/conferences/working_papers/05-20.pdf.

Miller, B., and A. Smart. 2011. "'Heart of the new west'? Oil and gas, rapid growth and consequences for Calgary." In *Canadian urban regions: Trajectories of growth and development*, ed. L.S. Bourne, T. Hutton, R.G. Shearmur, and J. Simmons. Toronto: Oxford University Press.

Minsky, M. 1988. *The society of mind*. New York: Simon & Schuster.

Polanyi, Michael. 1966. *The tacit dimension*. London: Routledge & Kegan Paul Ltd.

Porter, M. 1998. *On competition*. Cambridge, MA: HBS.

Rogers, Y. 1997. "A brief introduction to distributed cognition." University of Sussex, Brighton. http://mcs.open.ac.uk/yr258/papers/dcog/dcog-brief-intro.pdf.

Rogers, Y., and J. Ellis. 1994. "Distributed cognition: An alternative framework for analysing and explaining collaborative working." *Journal of Information Technology* 9 (2): 119–28. http://dx.doi.org/10.1057/jit.1994.12.

Romer, P. 1986. "Increasing returns and long run growth." *Journal of Political Economy* 94 (5): 1002–37. http://dx.doi.org/10.1086/261420.

Spencer, G.M., T. Vinodrai, M.S. Gertler, and D.A. Wolfe. 2010. "Do clusters make a difference? Defining and assessing their economic performance." *Regional Studies* 44 (6): 697–715. http://dx.doi.org/10.1080/00343400903107736.

Statistics Canada. 2009. *The input-output structure of the Canadian economy 2005/2006*. Ottawa: Statistics Canada.

THECIS. 2008. "The natural resource industries as engines of economic diversi-
    fication, the Banff consensus." http://www.thecis.ca/userfiles/files/Banff_
    Summit_Consensus_Report.pdf.
Vanderklippe, N. 2011. "Patenting the oil patch." *Globe and Mail*, 8 January: B1.
WD. 2009."The department." *Western Economic Diversification Canada*. http://
    www.wd.gc.ca/eng/36.asp#B.

APPENDIX 1: METHODOLOGY

The ISRN project prescribes interviews with senior executives of firms,
community organizations, and government entities, as well as interviews with individuals identified as creative talent. These were carried
out from 2006 to 2008. In all, 121 interviews were conducted in Calgary.
A subset of the full transcripts of interviews conducted with 34 senior
firm executives was selected to characterize firm behaviour.

The methodology is parallel to that employed in the study of talent
attraction and retention in Calgary (Langford, Li, and Ryan 2009a).
In each, innovations identified in responses to questions about innovations over the past three years were tagged. The method tags
sufficiently large text fragments (conceptual units) to explain (on a
stand-alone basis) the category assignments. Next, text was tagged
to identify the suite of knowledge resources assembled in the process
of generating and supporting each innovation. In all cases, interview
subjects were key participants in the innovative processes they described. Each innovation identified was categorized as addressing firm
advantage or collective advantage; and one or more from the problem
types. For each innovation, related mentions of supporting knowledge factors were categorized into one or both of tacit or codified; and
one or more of internal (to the firm), local (to the CMA) or non-local.
The frequencies of the use of each type of knowledge factor alone,
and in combination, were recorded. Distinct multiple instances of a
given component type related to the same innovation were counted
individually.

From the sample of transcribed interviews,[1] 459 knowledge items
were identified, pertaining to 95 plausible accounts of an innovation with an average of 5 knowledge items per plausible innovation

---

1 Interviews were initially coded by all three authors, but the final majority of coding
    was accomplished by Ben Li.

(min = 1, max = 13). Of the 95 plausible innovations, 17 (18%) pertained to human resource practices required to attract and retain high-quality personnel, 19 (20%) pertained to the introduction or completion of suites of services, and 26 (27%) pertained to new collaborations, while the remainder pertained to new tools or new knowledge use. Among the 95 plausible innovations, 78% (74) were directed at securing firm advantages, while a minority (21) (22%) were directed at securing collective advantage.

PART III

The Specialized Characteristics
of Innovation in Medium-Sized Cities

7 Innovation in an Ordinary City: Knowledge Flows in London, Ontario

NEIL BRADFORD AND JEN NELLES

A key goal of economic development is the creation and perpetuation of vibrant and successful metropolitan economies. Policy orthodoxy on the best way to promote prosperity vacillates periodically between strategies that recognize the benefits of concentration and specialization and those rooted in industrial diversity. The range of policy interventions has expanded to include measures that enhance the cultural and environmental quality of place. Across the strategies and debates two central messages are clear. First, such activity is designed to stimulate and capture the benefits of the knowledge flows that fuel innovation. Second, the innovative capacity of a city or region has little to do with its size or even location. Rather, it depends on the breadth and quality of knowledge that is locally accessible, the degree to which it circulates widely and rapidly, and the extent to which actors can put it to productive use (Malmberg and Maskell 2002).

This chapter explores the implications of these two key messages about local innovation dynamics in the context of London, Ontario, a medium-sized Canadian city and middling economic performer that sits near the national average on almost every innovation and creativity indicator. As the innovation research literature seeks to understand better the *varying* forms of global–local knowledge interactions, we offer London as an intriguing case study. Recent scholarship has challenged the heavy, even exclusive, emphasis placed on localized knowledge creation through geographically proximate actors. Paying close attention to the role and impact of global knowledge flows in innovative places, scholars increasingly believe that successful places are those endowed with vibrant clusters of firms tapping both locally generated knowledge (buzz) and external sources of know-how (pipelines)

(Bathelt, Malmberg, and Maskell 2004; Gertler and Levitte 2005; Wolfe and Gertler 2006). This line of argument cautions against "over-embeddedness," where too strong relationships between actors in a locality can become an obstacle to adaptation and innovation (Trippl, Toedtling, and Lengauer 2009; Weterings and Boschma 2006). Yet, this corrective also opens consideration to another developmental trajectory. Local economies might exhibit signs of "over-connectivity." Here external sources of innovation are robust, while local relations among knowledge-intensive firms remain quite thin and episodic.

We argue that innovation dynamics in London, Ontario, exhibit just such a pattern of pronounced external connectivity. Our research reveals that many leading firms in London are innovative, while not embedded in any identifiable or thick local "regional network-based industrial system" (Saxenian 1996: 2). Rather, they are globally networked through what we term their "Global Niche Markets" or their "Custom Services and Solutions" strategies. Both of these business orientations rely on strategic interface with, and position in, global value chains to access the ideas and information that drive innovation. Such firms seldom engage in local technology consortia or learning communities, reflecting the marginal value they place on local knowledge flows. Indeed, our research found that global orientations and connections were often the reason why interaction with other co-located firms and the universities, community colleges, and laboratories remained a low priority.

Thus, the London experience suggests another innovation pathway beyond the familiar account of locally embedded innovation networks. Exploring this pathway, the chapter offers a four-part presentation. We begin by situating London in a broader discussion of economic development in so-called ordinary cities. These are "in-between" places without world-class clusters, hip cultural infrastructures, a high concentration of research institutions, or compelling built and physical environments. At the same time, such places innovate at an acceptable rate, lagging behind the new economy's hotspots, but staying ahead of marginal cities or regions "locked in" to structural decline. The second part of the chapter focuses on knowledge flows in London, detailing the behaviour and strategies of selected innovative firms. Next we discuss the implications of our findings for scholarly debates about firm-level knowledge flows and prospects for innovative urban and regional economies. We conclude by highlighting some recent business development projects implemented against the backdrop of the global economic downturn that may portend greater local focus as London repositions itself – in the memorable words of the *London Free Press* – "beyond the crisis."

Knowledge Flows in the Ordinary City: Think Globally, Act Locally

It is the nature of academic scholarship to concentrate on extremes. Stunning examples of success or failure provide compelling narratives and produce more clear-cut lessons than the middling cases. This pattern is no less true in the realm of metropolitan economic development. The remarkable innovative capacity of places like Silicon Valley and the industrial districts in Europe almost single-handedly reinvigorated the study of local and regional industrial dynamics and the "power of place." Marginalized regions have similarly captured the attention of scholars and policymakers in an effort to arrest their decline and emulate the examples of their innovative betters. The focus of this scholarship on hinterlands and hotbeds is understandable, but Amin and Graham (1997) warn against taking paradigmatic examples as the starting point of analysis. In this case, the focus on the extremes effectively shuts out the many cases that fall in the broad, if mundane, middle. These unexceptional places are mid-sized city-regions that lack clear cluster profiles, the cultural diversity of large cities, and the quaint identity or natural assets of many smaller cities (Bradford 2009; Sands and Reese 2008). Yet, the lack of scholarly attention to ordinary places in the innovation literature is unjustified. Not only do ordinary places vastly outnumber the global pace-setting and declining regions, but the economic development challenges that they face exhibit a rather different logic and therefore merit focused attention (Benneworth 2007; Tödtling and Trippl 2005).

Ordinary cities have just as much need to address the trends of globalization and international competition as their larger neighbours (Beaverstock, Smith, and Taylor 1999), but require strategies somewhere between the radical reinventions prescribed to "innovation backwaters" and the ongoing fine-tuning of established hotbeds (Bradford 2003). According to the Federation of Canadian Municipalities, there is little evidence to suggest that ordinary cities are unsuccessful or that their economies are not performing well enough to maintain a reasonable quality of place or living standards (FCM 2009). This performance happens despite the deficiencies in the local knowledge flows and dense networks that the regional innovations systems and creative cities literatures celebrate. From a theoretical perspective, the development patterns and industrial conventions in ordinary cities highlight gaps in dominant approaches, and other literatures have emerged to better position ordinary cities in the innovation debates.

Indeed, these literatures on the roles of external knowledge exchange address the realities of industrial organization and firm behaviour within ordinary places. The moderate success, or at least lack of failure, of innovative firms in ordinary cities may be partially explained by the concept of global pipelines. While the spatial proximity of firms to competitors and firms in related industries encourages the formation of trust relations and exchange of tacit knowledge that underpins the innovation process (Malmberg and Maskell 2002), innovation does not spring from local sources of knowledge alone (Wolfe and Gertler 2006). Connections to global "pipelines" (Bathelt, Malmberg, and Maskell 2004) of knowledge also matter for accessing expertise and know-how (Bunnel and Coe 2001). Similar networks based on trust and pre-existing social relationships are present within these externally oriented pipelines. Through such vertical relationships, firms are integrated into multiple selection environments, enabling them to tap the knowledge of successful firms and clusters elsewhere, thereby opening new pathways to innovation. A consensus has emerged around the importance of access to both global and local knowledge pools, but the optimal mix of each within innovative regions is actually indeterminate (Wolfe and Gertler 2006), and will likely vary by type of economic concentration.

This indeterminacy holds some interesting implications for ordinary cities. It suggests that it might be possible to sustain relatively innovative firms in regions that lack the strong industrial concentrations or thick inter-firm networks that have dominated the literature on innovative places. Instead, knowledge flows and the strength of the various dimensions of proximity depend in large part on the distribution of firms along global value chains and within global production networks. These concepts help specify the ordinary-city trajectory by conceptualizing location as only one dimension of increasingly globalized innovation and production. Global value chains and global production networks link spatial and organizational processes highlighting "the nexus of interconnected functions, operations and transactions through which a specific product or service is produced, distributed and consumed" (Coe, Dicken, and Hess 2008: 272). Such a relational view of the global economy and firm behaviour brings a different analytical lens to the study of economic regions. Firms perform several functions within global value chains and the distinction between labels like "producer" and "supplier" is increasingly blurred. Even within the same industries (such as advanced manufacturing or information technology), firms will play different (and often multiple) roles according to their positions in these global production networks.

This literature reveals the following insight for ordinary places: economic regions are the physical locations of firm structures, but frequently do not host the wide variety of other actors in production networks. Consequently, cities or regions with a large number of firms persistently integrated into global production networks may be quite successful, as their constituent firms nurture the extra-local linkages that matter most to them. Such firms may be innovative in spite of the apparent drawbacks of their locations as a result of their capacity to perform in these global production networks. With globally oriented firms in place, ordinary cities can sustain successful local economies despite their lack of industrial clustering and dense local networks.

The study of ordinary cities therefore strengthens our understanding of knowledge flows and innovation processes. While they are less headline grabbing than their declining or superstar counterparts, their particular trajectories expose gaps in existing theory, practice, and prescriptions. Ordinary city-regions challenge conventional wisdom about the relationship between innovation and proximity, inviting exploration of different social dynamics among firms, institutions, and public policy. We now turn to the study of London, Ontario, as an interesting "ordinary case," profiling the implications of its knowledge flows for local development.

London: Neither Hotbed nor Hinterland

In many respects, London, Ontario, is the quintessential ordinary city. Across a number of economic and demographic indicators, London's post-war performance and profile gravitated close to provincial and Canadian averages. For decades, corporations have relied on London as the test market for new consumer products from drugs to donuts. And while the post-war local economy's balance among traditional financial, manufacturing, and trade sectors provided a certain stability, it also left the city without an identifiable cluster identity to distinguish its economic brand (Spencer and Vinodrai 2009). At the heart of a Census Metropolitan Area of some 460,00 people, and located at the geographic midpoint between Toronto and Detroit on the NAFTA corridor, London exhibits an ordinary-city mix of innovation systems failures (Bradford 2009) and is regularly overlooked in Canadian urban-creativity tables. London consistently places in the lower half of the field of Canadian mid-size CMAs on measures of innovation such as cluster employment, educational attainment, employment, and employment in creative industries (Spencer and Vinodrai 2009). Its performance on indicators of technology, such as the Tech-Pole index, patents per capita, and patent growth, is also mediocre (Martin Prosperity Institute 2009b). These results are not surprising in an economy dominated by manufacturing (automotive and food processing) and health-care sectors (LEDC 2009). At the same time, it has not been an urban economy in long-term structural decline; as the Martin Prosperity Institute summarized in their mid-sized city technology, talent, and tolerance rankings: "London has some advantages but would benefit from improved performance on the 3Ts as it positions itself to compete in a creative economy" (Martin Prosperity Institute 2009b: 7).

The foundations of this unremarkable yet stable trajectory began to crumble in the 1990s (LEDC 1998). As continental and global restructuring took its toll on Ontario's economy, London lost several of its long-standing corporate head offices and many branch-plant manufacturers closed. London's economic performance, population growth, and median family income fell behind those of its key mid-sized municipal competitors in southwestern Ontario (Statistics Canada 2008). Moreover, the head-office flight drained away business leaders and philanthropic families heavily invested in the community, and signalled that London's labour market would not offer in the new knowledge economy the same opportunities for either senior advancement or movement across professions. The immediate impact of these structural shifts was buffered by the city's continued role as an important regional centre in southwestern Ontario for retail trade and services in health and education (Spencer and Vinodrai 2009). However, as we describe later, the so-called Great Recession that began in 2008 has framed a deeper and more enduring set of economic challenges for London.

This general pattern of economic evolution is familiar to many other ordinary places. While problems specific to local industrial profiles vary, the general themes remain the same: the gradual departure of many branch plants (usually in similar industries) leaves a gap filled gradually by firms founded by regionally based entrepreneurs or returnees (Grant and Kronstal 2010). In London, this dynamic was underpinned by the importance of the city as a secondary regional centre – a logistical, education, and retail node for southwestern Ontario – ensuring that its flagging fortunes would not founder. Consequently, the London case can stand for many ordinary places in North America. Its relative anonymity in the academic literature, its quite forgettable urban milieu and sense of place, make London a good case from which to draw lessons about innovation and knowledge flows relevant to the murky, yet wide, middle between urban hotbeds and regional hinterlands.

Knowledge Flows and Exchange in London

The most significant finding of this study of innovation patterns and knowledge flows in London was that ties between firms in the local economy are quite infrequent. This echoes the findings of a recent technology audit that reported that London's 250 or so high-technology firms exhibited few cluster synergies. They operated without the "glue" of anchor companies that branded the sector and seeded start-ups, and lacked the "local buzz" flowing through collaborative networks in research, finance, and entrepreneurship (LEDC 2006). In order to capture the knowledge flows across and between sectors of the London economy our analysis focused on a broad swathe of innovative firms. These firms were selected from a 2008 list of firms classified as "innovators" produced by the London Chamber of Commerce. This field included both large and small firms in a wider variety of industries and backgrounds, including foreign multinationals and local spin-out firms. Despite the wide base of subject firms, there was very little evidence of local knowledge flows – creative or logistical – either between firms in the same industry or between sectors.

The Structure of Innovators in the London Economy

The wide range of firms surveyed in this project can be broadly categorized into three different classes. Each category of firms has a different market focus, geographical market, research intensity, and, therefore, a distinctive innovation profile: Global Niche Market, Custom Systems and Solutions, and Local Manufacturers or Service Providers.

Global Niche Market (GNM) firms, as the label implies, serve niche markets with global scope and are one of only a handful of firms active in a certain market that service clients and customers all over the world. All the firms produce or provide a standard line of products or services within a narrow market and offer only minor tailoring to end-user needs. Occasionally these firms have custom capabilities – and respond to calls for tender – but their main focus is on developing a set of standard lines that can be adapted to most client requirements. The product lines of firms interviewed in this category include rowing skulls, light aircraft, waste-water treatment systems, pharmaceutical drug delivery devices, and digital narrowcasting, and wireless audio technologies. All the firms in this class are *development*-intensive. That is, they rely on product and process development based on market feedbacks more often than on research-driven innovations. This innovation profile has been described as "little r, big D" (Bramwell, Nelles, and Wolfe 2008) in industry parlance, a play on the R&D of research and development. The only exceptions to this observation are a subset of the GNM firms that are built on science-based (typically life science or pharmaceutical) product lines. These firms have much more active research departments that are more central to their innovation process. This subset of firms is referred to as GNM-S to denote their scientific focus.

Custom Systems and Solutions (CSS) firms share quite a few attributes with their GNM counterparts. These firms typically operate in fairly niche product and service ranges and their core business revolves around producing a truly unique solution for each client. While some large multinationals occupy this category, the majority of CSS firms in the London region are small and medium-sized enterprises originally founded in southwestern Ontario. The firms in this class engage in a wide range of industries from manufacturing-intensive custom fabrication and welding to very high-tech molecule discovery processes, custom silicon moulding, hardware-based component solutions, and interactive marketing systems. These businesses are based on expertise in a very small area that can be adapted to several different types of markets and industries, often as intermediate inputs. Again, their market reach tends to be global, though due to the customized nature of product lines their focus is often limited to clients in the Western Hemisphere.

Local Manufacturing and Service (LMS) firms are those oriented towards local and regional markets. In this case localized refers to southwestern Ontario in general, but some are much smaller in scope than that. These are typically service providers, but can include manufacturers of intermediate products for industries in the area. On the service side these firms include, for example, an educational firm and a server hotel (or co-location centre). Like the CSSs these firms were most responsive to their customer and client needs, but do not do much development beyond maintaining capacity to serve these markets. These firms are very innovative, but their innovations are more related to the identification and exploitation of a localized market niche (e.g., secure data storage) than to continuous product or process innovation.

This brief survey highlights the diversity of industries operating in London, Ontario. Although they can be grouped according to similarities in their production and innovation strategies, with few exceptions, their core businesses are quite different. As we will see, local knowledge flows and productive synergies do not necessarily flow from this type of strategic similarity.

Innovation Profiles and Knowledge Flows

On a broad level it is possible to identify some general characteristics of London's innovation profile. One commonality to all of the firms interviewed is that in-house development teams were almost unanimously cited as the primary sources of innovation. This was followed closely by customer input. Significantly, these customer relationships were almost exclusively non-local and were typically described in terms of customers expressing a need and development teams adapting to such requests, rather than collaborative development of a product or solution. As a result, these relationships may be more accurately characterized as *feedback* loops for market intelligence than as innovative knowledge exchange. Research or development partnerships – between firms or with research organizations – are quite rare across the board. Finally, firms rarely (consciously) sourced labour from other firms in the region, meaning that local "embodied" knowledge transfer flow between firms was quite low. In sum, knowledge flows, and inter-organizational linkages more generally, *within* the London region are quite weak.

GNM firms in the region most often base their competitive advantage in the market on factors such as product quality and design, responsiveness to customer needs, and efficiency of production and/or service delivery. In order to maintain these market advantages they rely first and foremost on in-house development teams to drive innovation. Customers are often cited as important sources of product ideas and market feedback, but are rarely considered as partners in the innovation process. Thus, the role of clients is more as providers of feedback to which firms can be more or less responsive. Regardless, it is typically the function of in-house development departments to adapt to these market inputs. In a similar vein, suppliers were also cited as sources of innovation, generally to the extent that their own product innovations open up new possibilities for the GNM firm. Foreign-owned firms also cite parent companies and affiliates as important sources of innovation. As with suppliers, foreign parents or affiliates were most instrumental in developing products or processes that the London firm then incorporated into its development cycle. Therefore, the transmission of knowledge and innovation through globally linked firm networks can be characterized in most cases as passive – a simple transfer – rather than a collaborative endeavour.

These observations reveal that *internal* sources of innovation – through in-house R&D divisions – have typically been the most important for GNM firms in the London region. Where customers, suppliers, and foreign parents and affiliates were cited as factors, these were overwhelmingly non-local. As a result, knowledge *flows* most frequently between the London firm and its individually developed, industrially specific, and primarily non-local customer/supplier network, and does not typically extend to firms in the region or outside the sector. This isolated pattern of knowledge flow may be symptomatic of the very niche market in which these firms tend to operate. Whether there are actually no potential partners or these firms *perceive* few opportunities for innovative synergy with other firms in the region, most firms have very little interest in reaching out to engage new local partners. To many firms interviewed, beyond the convenient and relatively inexpensive location and an adequate workforce, their location in London was nearly irrelevant.

Firms that provide customized systems and solutions (CSS) focus on quality of service to customers, an innovative approach to problem solving, and quality of product and design as the sources of their competitive advantage. Like GSM firms, those in the CSS category innovate primarily through their in-house development departments. But because of the nature of their business – to customize components, systems, and solutions – their non-local customers are integral to their innovation process. With each unique relationship the CSS firm acquires experience and knowledge that may contribute to creative problem solving in other contexts. In this sense customers are much closer to the innovation process and solutions are often collaboratively developed. None of the CSS firms interviewed cited physical distance from customers as barriers to innovation. The attitude prevails – correctly or not – that the benefits of engaging local firms in knowledge exchange are imperceptible or not worth the effort.

LMS firms have a very different innovation profile. These firms are engaged with local markets and fill specific niches in the product and services needs of the region. In most cases, ongoing innovations within the firms had more to do with evolving business practices and marketing than in changing the product or services offered. These firms have an established mission and are primarily concerned with expanding their local business. As a result, most innovations come from the specialized staff in the firm rather than in-house development departments. To the extent that expanding their business relies on responding to customer feedback, customers are also important to the evolution of these firms. In these cases responding to customer needs involves relatively low-level changes to business processes or infrastructure rather than adapting existing products or services to specific requirements. Knowledge flows *between* firms in this category are also scarce. However, of the three classes, LMS firms are more likely to use external resources and partners in their innovation process. This is partly because these firms tend to be small and medium-sized enterprises with few specialized internal resources, and partly because their innovation processes are more related to business and marketing than to the development of proprietary products and solutions. Because partnerships of this nature do not require sharing trade secrets, these firms have been more likely to engage external actors to grow their businesses.

The potential for partnership with firms in the region – within or outside the sector – was generally regarded with scepticism and suspicion. In addition to the perceived lack of opportunities for fruitful partnerships in the region, some GNM and CSS firms argued that the level of knowledge exchange required to engage in innovation with other firms would be too risky to justify collaboration. This lack of trust typically extended to firms outside the region as well. A small number of firms mentioned that informal collaboration with co-located firms occurred, but stressed that this was only on general areas of the business and not on proprietary issues. Interviews revealed that instances of informal cooperation on broad matters such as "best practices" rarely extended to processes of "technology development" involving exchange or sharing of intellectual property (confidential interview, 2008).

One of London's great strengths is the quality of its higher education institutions (such as Western University) and research organizations (such as the Robarts Research Institute in health sciences and the Centre for Automotive Materials and Manufacturing). Given the richness of these local knowledge assets, it is surprising that few firms in our study have formalized research linkages in London. This pattern may be more a function of corporate attitudes towards academia in general, and not an indictment of the quality of research infrastructure in London. Many GNM and CSS firms have adopted the attitude that academia and public research organizations have little to offer their businesses. For the most part these firms focus on development and do not see a need to partner formally with academic researchers. A common refrain among these firms is that public research and universities are not fast enough, are too difficult to deal with, and ask too much in return for formal partnerships to justify the effort. In regard to the former, one CEO stated: "They [external R&D labs/universities] move at the speed of institutions and we have to move at the speed of industry" (confidential interview, 2008).

For those London firms that have attempted research connections, those with the most success have been forged with universities outside southwestern Ontario. The only exceptions to this trend were two firms with ongoing relationships with Western University (WU), only one of which actively engages university researchers. The other firm is a spin-off in which the principal is still employed teaching at the university. While many firms announced their intention to engage with research within the university, most had no concrete plans to do so in the near future, and many were wary of dealing with technology transfer officials. These attitudes may evolve in a more positive direction in light of the university's recent restructuring and rebranding of its technology transfer office in partnership with two affiliated local research institutes. The new office, WORLDiscoveries™, aims to increase the performance of the partners in commercializing research. None of the firms consulted in this project mentioned the restructured office, nor did the prospect of new leadership of industrial liaison functions register as an opportunity to revisit the potential for knowledge exchange with the local university. However, it is still relatively early, and as we report below, the new office may yet accelerate and intensify university–firm relationships.

Finally, local flows of labour between the firms studied tended to be relatively rare. While these were reported as slightly more significant on the manufacturing side, the workers involved in the innovation process, and senior management, had not typically worked in other firms in the region. Hiring in these areas was also described as a global process, and these firms rarely formally advertised those positions in general job forums. Reputational factors were cited as an important magnate for talent. Talented people, typically on the research or technical side, interested in working in a given field know who the most innovative firms are and contact them for employment, even when positions are not available. Senior management was much more difficult to source, as interview subjects frequently commented that they needed managers who understood the intricacies of their quite unique type of business – knowledge that is much harder to find locally, as well as globally. This global approach to talent also limits the degree to which knowledge flows between firms in the London region. People who chose to leave these firms usually also left the region.

Overall, this analysis reveals that local knowledge flows in the London case are relatively weak. For most firms, networks of knowledge exchange tend to be global with firms and researchers in related sectors. In terms of local attachments, firms lack opportunities or are unwilling to collaborate with other firms in the London region. Similarly, ties to local research capacity have also been weak. This can be partly explained by the role that these enterprises play in global production networks. To the extent that London's local innovation profile is emblematic of ordinary cities, interesting questions arise about the prospects for future economic development in such places. The final two sections of this chapter open a broader discussion of the dynamics of its growth prospects.

Ordinary Cities: From Innovative Firms to Resilient Economies?

The patterns of knowledge flow and exchange described in the previous section indicate that leading firms in London are innovative, globally connected, and locally insulated. What are the consequences of this pattern for the city-region's economic development? Answers to that question likely vary by local stakeholder. Firms experience different consequences from city-regional economies than those anticipated by the economic development officials responsible for their overall vitality. Considering these two perspectives, we can situate a differentiated theory of local economic development appropriate to analysis of innovation in globally oriented, but ordinary, cities.

Proximity and dense localized knowledge flows impact the innovative potential of *firms* and local *economies* in different, but related ways.

On the one hand, firms in densely networked regions benefit primarily from the exchange of *ideas*. The experiences – successes, failures, experiments – of firms in related industries or at different points in the supply chain can help inform the innovation processes and the management strategies of others (Audretsch 2002; Maskell 2001). Casual conversations between workers at different firms, even in different industries, can spark creativity (Saxenian 1996). Knowledge also flows between firms as talent moves around, sharing knowledge and experience. On the other hand, managers of city-regional economies, defined as the group of policymakers and authorities with responsibility for governance, are interested in the character of knowledge flows to the extent that they want to attract and retain high-value-added firms that bring investment, income, and employment to their regions. Encouraging knowledge flows and spillovers can increase innovation and the value of firms as sources of employment and tax income. Successful firms can also attract others in similar industries, and their growth engenders new investments. Finally, these economies may benefit from "native" firm growth as spin-offs and start-ups find a potentially fertile place to grow.

It follows that both firms and economic authorities have an interest in leveraging the benefits of geographical proximity and local knowledge exchange. However, in places such as London, where the most important knowledge flows vertically and externally, there is little practical or immediate incentive for individual firms to act on this interest. Indeed, those interviewed acknowledged the *potential* for benefits from increased formal networking in the city-region, but typically argued that, in their experience, the payback had not been significant. This arises from a combination of their global sources of innovation (i.e., customers and suppliers) and their more guarded and "closed" innovation process that places a high premium on "in-house knowledge" as a form of intellectual property protection on the local scene.

As for officials concerned with overall prosperity and development, the consequences of weak local knowledge flows are probably much higher and more immediate. One of the major benefits of industrial clustering and the knowledge spillovers associated with local buzz is the creation of new firms as spin-offs and start-ups emerge out of innovative partnerships and other dimensions of labour-force mobility. The creation of spin-off firms and start-ups entails a certain degree of job creation (Dahl and Jensen 2010) and enhancement of economic resilience. Spin-offs created from innovative firms, as opposed to research and higher-education institutions, also typically exhibit higher performance with respect to net cash flow, longevity, peak market share, and revenue growth (Ensley and Hmielesky 2005; Eriksson and Kuhn 2006). They can also be important in the development and consolidation of industrial clusters that seed new innovation capabilities across sectors (Feldman, Francis, and Bercovitz 2005). While it is difficult to predict the potential foregone when regional economies evolve with weak local knowledge flows, research suggests that there will be less endogenous growth through spin-offs and start-ups. In ordinary cities relying on external connectivity, the loss to the regional economy will be of greater concern for economic development officials than for firms, who continue to access knowledge through their global networks. Yet, the moderately successful performance of such a local economy and the absence of clear signals of structural decline can lead to a degree of complacency about innovation (Bradford 2009). If leading firms are reluctant to engage in knowledge-driven collaborations, it is difficult for policymakers to generate support for either their formation or ongoing business participation once created.

Consequently, policymakers in the ordinary city confront particular collective action challenges in stimulating knowledge exchange between actors that may not see, or even discount, the potential benefit of increasing the density and quality of local networks. The challenges become more urgent when the global environment suddenly turns against the locality and the pressures mount for broad economic transformation. Under such conditions, the regional economy-wide importance of striking a balance between "global pipelines" and "local buzz" comes into sharp focus even in the ordinary city. The concluding section of this chapter pursues this theme of greater balance in London's innovation system, highlighting a series of civic initiatives designed to strengthen local networks and knowledge exchange. London may represent an ideal starting point from which to relaunch discussion about the economic dynamics of ordinary cities, specifically probing how their innovation agendas might better leverage both their global connectedness *and* local embeddedness.

The Ordinary City in Hard Times: More "Local Buzz" in London?

In 2008, the Great Recession rocked southwestern Ontario's key secondary manufacturing sectors, notably automotive assembly and parts sector. London's economy was plunged into its worst recession since the Great Depression. The statistics tell a grim story: an unemployment rate skyrocketing to over 11 per cent, the second highest in the country; social assistance claims increasing 20 per cent; and more than 8000 jobs disappearing, with only part-time jobs showing any resiliency (Hunter 2009). Complicating London's challenges were economic studies (Martin Prosperity Institute 2009a; Conference Board of Canada 2010) from national think tanks that took a dim view of the city's economic prospects, especially in comparison with key mid-sized competitors such as Waterloo and Ottawa. The *London Free Press* launched a year-long series "Beyond the Crisis" convening local leaders and outside experts to bring forward new development strategies.

In surveying London's recent difficulties, Annalee Saxenian's classic account of industrial and institutional change in different regional settings offers useful perspective (Saxenian 1996). She highlights the particular adjustment challenges facing firms in regions lacking sophisticated local knowledge infrastructures when markets, technologies, and tastes all suddenly change. Her work, and others', reveals that firms located in highly networked learning communities can be more resilient in times of shock and able to sustain innovation than those that rely extensively on vertical, in-house knowledge flows and where inter-firm knowledge exchange is less frequent (Cooke and Morgan 1998; Henton 2001; Wolfe 2010).

In London, economic development agencies have typically eschewed the lessons of networked regions and pursued a traditional development agenda, focused more on external recruitment of branch plants and efficient property servicing than any sustained effort to nurture a local infrastructure of innovation. These strategies aligned with London's prevailing business culture and learning practices, and they were appropriate for the relatively stable markets and technologies that characterized the city-region's post-war development. However, these orientations can become barriers when structural change is required. For an ordinary city their persistence can be especially problematic, as they may jeopardize the foundations of even middling economic performance.

London has confronted the major economic challenges of the early twenty-first century with an industrial culture and local policy system that makes broad-based economic transformation – as distinct from stable growth – difficult. As we have described, London's vertical and externally organized knowledge flows limited opportunities for new technology-driven firms through start-ups or spin-offs. The local infrastructure of support, ranging from venture capital to business incubators or commercialization services, never took root in a local economy where learning dynamics remained within corporate boundaries of global pipelines. Further, such local networks may be important in enabling more traditional manufacturing firms to upgrade their capabilities or renew product lines through incorporating new technologies and organizational forms. In London's case such industrial renewal applies across several traditional sectors including auto parts, transportation equipment, and food and beverages. In each case, new demands for more sustainable and energy-efficient production requires creative restructurings that tap a diversity of knowledge flows including those available only through close interaction with co-located research institutions.

Are there signs that London's economic infrastructure may evolve to support such transformation through a better balance of global and local knowledge flows? Saxenian makes the point that economic crises, such as those on the scale of the 2008 downturn, can be the catalyst for local policy innovation mobilizing a range of investors and partners. She specifies a two-step process. First, city-region economic leaders must recognize the need to act differently. Second, they need to deliver on the "shared understandings," forging "places and spaces" where diverse economic knowledge holders join together for change (Saxenian 1996: 167).

In these terms, London's recent history is noteworthy (Bradford 2009). A new civic awareness was announced in 2005 with two novel policy development processes, each funded and facilitated by the municipal government, representing what Gertler and Wolfe term "local social knowledge management" (Gertler and Wolfe 2004). The first was a task force on London's potential as a "creative city" (City of London 2005). With a mandate to change "the way London thinks" and determine a "strategy to help London become a leader in mid-sized cities in North America," the task force focused on issues of cultural diversity, workforce development, and urban design. Describing London as an isolated and complacent city, it challenged local leaders to welcome immigrants and make regional economic connections. The second report was a hard-hitting economic survey titled *London's Next Economy* (City of London et al. 2005). It highlighted the city's inaction in the face of a "stealth-like erosion" of high-end jobs and talent, and challenged London's "comparatively lethargic business community" and its "modesty and conservative style." The follow-up 2006 technology audit completed by the local economic development agency (LEDC 2006) confirmed these critiques. The findings of this study further underscore the lack of cluster synergies.

Recognizing these gaps and challenged by the Great Recession, London's key economic stakeholders have initiated several economic partnerships to engage and embed firms locally. Leadership has come from civic entrepreneurs directly involved in the task force reports with three priorities in focus: research commercialization, regional alliances, and "transformative projects." In 2008, two technology-transfer networks were created bridging distinctions between old and new economies. WORLDiscoveries™, mandated to commercialize research from laboratories at WU and the city's health and medical science institutes, has produced six start-up or spin-off companies in its first year and works with more than twenty enterprises. A national Centre for Automotive Materials and Manufacturing was established in the university research park to build smarter, greener vehicles through application of lightweight materials and alternative fuels, exploiting local research strengths in renewable energy, plastics and polymers, and biofuels. In 2010, the City of London created a $50 million Economic Development Fund for similar knowledge infrastructure investments in a green technology centre and eco-industrial parks for wastewater treatment and wind energy.

London's evolving strategy has also geographically expanded to encompass the larger region. In 2006, the City and the university joined forces to help create the Southwest Ontario Economic Assembly (SWEA), linking public- and private-sector partners across the region to promote economic cooperation, conduct research, and forge a regional development strategy. Another key priority achieved through SWEA's regional focus was the establishment in 2009 of the Federal Economic Development Agency for Southern Ontario (FedDev), with a $1 billion budget over five years to invest in local manufacturing and knowledge infrastructures (Bradford and Wolfe 2010). While initial hopes to locate the agency's headquarters in London were dashed when Kitchener-Waterloo was selected, London's municipal leaders have been positioning the city's business and research communities to access the $200 million Advanced Manufacturing Fund. New municipal partnerships and investments in an Advanced Manufacturing Park and a Medical Innovation and Commercialization Network have been proposed for FedDev support (Martin 2013).

The SWEA process informs London's own "transformative project" focused on transportation (LEDC 2009). In 2008, the City announced plans to make London a regional transportation logistics hub for moving freight internationally by rail, highway, and air. The project would leverage London's location along the NAFTA trade corridor and its transportation advantages over Toronto-area competitors – far lower airport landing fees, and much less road traffic congestion. Nearly $14 million was targeted for upgrades to the London airport, expressway interchanges, and an inter-modal terminal for trucks and trains. The City has mobilized a "Gateway City" coalition that includes the federal and provincial governments, two post-secondary institutions, and London firms linked into global supply chains. Local synergies are emerging. Fanshawe College launched a $31 million program in transportation trade and logistics, the WU business school has developed research expertise in sustainable transportation infrastructure and inter-modal linkages, and FedDev Ontario invested $8 million in the project.

In sum, over the past several years, civic leaders in London have come together to strengthen the city's economic knowledge base. These mobilizations signal that it is no longer "business as usual" in the ordinary city. In September 2009, the City and the local economic development agency co-hosted the first ever multi-sectoral London Economic Summit to identify strategic priorities. Public, private, and community sectors were challenged again to recognize that their goals "must be aligned, coordinated and mutually supportive in order to achieve success" (LEDC 2009). Observing that other Ontario mid-sized cities were further ahead than London in working locally for global success, the summit's keynote speaker, the new WU president, emphasized the "need to be innovative in our approach" and "to speak with a single voice and shape our own future" (Chakma 2009). While such "summiteering" can be dismissed as simply talk, considerable municipal and community resources are being dedicated to building a learning infrastructure for local firms, one that will support partnerships for innovation through spin-offs and start-ups. If these investments produce a durable foundation for knowledge-based development, then London's recent departures may offer a pathway for an increasing number of ordinary cities forced to rethink long-settled trajectories. These places, and the firms that inhabit them, may well require *extraordinary* collective action to ensure their future prosperity, capturing knowledge flows through both global pipelines and local buzz.

REFERENCES

Amin, A., and S. Graham. 1997. "The ordinary city." *Transactions of the Institute of British Geographers* 2 (4): 411–29.
Audretsch, D. 2002. "The innovative advantage of US cities." *European Planning Studies* 10 (2): 165–76. http://dx.doi.org/10.1080/09654310120114472.
Bathelt, H., A. Malmberg, and P. Maskell. 2004. "Clusters and knowledge: Local buzz and global pipelines and the process of knowledge creation." *Progress in Human Geography* 28 (1): 31–56. http://dx.doi.org/10.1191/03091 32504ph469oa.
Beaverstock, J., R. Smith, and P. Taylor. 1999. "A roster of world cities." *Cities* (London, England) 16 (6): 445–58. http://dx.doi.org/10.1016/S0264-2751 (99)00042-6.
Benneworth, J. 2007. "The role of leadership in promoting regional innovation polices in 'ordinary regions': A review of the literature." NESTA working paper, Newcastle University, UK.
Bradford, N. 2003. "Cities and communities that work: Innovative practices, enabling policies." CPRN discussion paper F/32.
Bradford, N. 2009. "Innovation and creativity in the ordinary city? Talent matters in London Ontario." Paper presented to Innovation Systems Research Network annual meeting, Halifax, Nova Scotia, April.
Bradford, N., and D. Wolfe. 2010. *Toward a transformative agenda for FedDev Ontario*. Mowat Centre for Policy Innovation. School of Public Policy & Governance. University of Toronto.
Bramwell, A., J. Nelles, and D. Wolfe. 2008. "Knowledge, innovation and institutions: Global and local dimensions of the ict cluster in Waterloo Canada." *Regional Studies* 42 (1): 101–16. http://dx.doi.org/10.1080/003434007015 43231.
Bunnell, T.G., and N.M. Coe. 2001. "Spaces and scales of innovation." *Progress in Human Geography* 25 (4): 569–89. http://dx.doi.org/10.1191/0309132016826 88940.
Chakma, A. 2009. "Western active partner in London economy." President's News Archive, Western University.
City of London. 2005. *Creative City Task Force report*.
City of London, LEDC, TechAlliance, Stiller Centre. 2005. *London's next economy: A game plan for accelerating new business development in the London region*.
Coe, N., P. Dicken, and M. Hess. 2008. "Global production networks: Realizing the potential." *Journal of Economic Geography* 8 (3): 271–95. http://dx.doi.org/ 10.1093/jeg/lbn002.
Conference Board of Canada. 2010. *City magnets II: Benchmarking the attractiveness of 50 Canadian cities*.
Cooke, P., and K. Morgan. 1998. *The associational economy: Firms, regions and innovation*. Oxford: Oxford University Press. http://dx.doi.org/10.1093/acp rof:oso/9780198290186.001.0001.

Dahl, M., and P. Jensen. 2010. "Spin-off growth and job creation: Evidence from Denmark." Paper presented at the DRUID summer conference 2010, London, UK, 16–18 June.

Ensley, M.D., and K.M. Hmielesky. 2005. "A comparative study of new venture top management team composition, dynamics and performance between university-based and independent start-ups." *Research Policy* 34 (7): 1091–1105. http://dx.doi.org/10.1016/j.respol.2005.05.008.

Eriksson, T., and J.M. Kuhn. 2006. "Firm spin-offs in Denmark, 1981–2000: Patterns of entry and exit." *International Journal of Industrial Organization* 24 (5): 1021–40. http://dx.doi.org/10.1016/j.ijindorg.2005.11.008.

FCM (Federation of Canadian Municipalities). 2009. "Quality of life reporting system." FCM report.

Feldman, M., J. Francis, and J. Bercovitz. 2005. "Creating a cluster while building a firm: Entrepreneurs and the formation of industrial clusters." *Regional Studies* 39 (1): 129–41. http://dx.doi.org/10.1080/0034340052000320888.

Gertler, M., and Y. Levitte. 2005. "Local nodes in global networks: The geography of knowledge flows in biotechnology innovation." *Industry and Innovation* 12 (4): 487–507. http://dx.doi.org/10.1080/13662710500361981.

Gertler, M., and D. Wolfe. 2004. "Local social knowledge management: Community actors, institutions and multilevel governance in regional foresight exercises." *Futures* 36 (1): 45–65. http://dx.doi.org/10.1016/S0016-3287 (03)00139-3.

Grant, J., and K. Kronstal. 2010. "The social dynamics of attracting talent in Halifax." *Canadian Geographer* 54 (3): 347–65. http://dx.doi.org/10.1111/j.1541-0064.2010.00310.x.

Henton, D. 2001. "Lessons from Silicon Valley: Governance in a global city region." In *Global city regions: Trends, theory and policies*, ed. A.J. Scott. Oxford: Oxford University Press.

Hunter, T. 2009. "Enhance London: A strategy for economic and societal prosperity." Prepared for London Economic Development Corporation.

LEDC (London Economic Development Corporation). 1998. *Investing in prosperity: A commitment to innovation, initiative, and competitiveness*. London, ON: London Economic Development Corporation.

LEDC. 2006. *London technology audit*. London Economic Development Corporation.

LEDC. 2009. *Advantage London. London economic summit: Creating the action plan for the next economy*. London Economic Development Corporation.

Malmberg, A., and P. Maskell. 2002. "The elusive concept of localization economies: Towards a knowledge-based theory of spatial clustering." *Environment & Planning A* 34 (3): 429–49. http://dx.doi.org/10.1068/a3457.

Martin, C. 2013. "Lots of cash, but Joe Who." *London Free Press*, 10 December, pp. A1, A7.

Martin Prosperity Institute. 2009a. *London 3Ts Reference Report*. Toronto: Martin Prosperity Institute.

Martin Prosperity Institute. 2009b. *Ontario's mid-sized regions' performance on the 3Ts of economic development.* Toronto: Martin Prosperity Institute.

Maskell, P. 2001. "Towards a knowledge-based theory of the geographical cluster." *Industrial and Corporate Change* 10 (4): 921–43. http://dx.doi.org/10.1093/icc/10.4.921.

Sands, G., and L. Reese. 2008. "Cultivating the creative class: And what about Nanaimo?" *Economic Development Quarterly* 22 (1): 8–23. http://dx.doi.org/10.1177/0891242407309822.

Saxenian, A. 1996. *Regional advantage: Culture and competition in Silicon Valley and Route 128.* Cambridge, MA: Harvard University Press.

Spencer, G., and T. Vinodrai. 2009. *Innovation Systems Research Network city-region profile, 2006: London.* Toronto: ISRN.

Statistics Canada. 2008. *Tracking trends in Canada: 2006 census.* Ottawa: Statistics Canada.

Tödtling, F., and M. Trippl. 2005. "One size fits all? Towards a differentiated regional innovation policy approach." *Research Policy* 34:1203–19.

Trippl, M., F. Toedtling, and L. Lengauer. 2009. "Knowledge sourcing beyond buzz and pipelines: Evidence from the Vienna software sector." *Economic Geography* 85 (4): 443–62. http://dx.doi.org/10.1111/j.1944-8287.2009.01047.x.

Weterings, A., and R. Boschma. 2006. "The impact of geography on the innovative productivity of software firms in the Netherlands." In *Regional development in the knowledge economy,* ed. P. Cooke and A. Piccaluga, 63–83. London: Routledge.

Wolfe, D.A. 2010. "The strategic management of core cities: Path dependence and economic adjustment in resilient regions." *Cambridge Journal of Regions, Economy and Society* 3 (1): 139–52.

Wolfe, D.A., and M.S. Gertler. 2006. "Spaces of knowledge flows: Clusters in a global context." In *Clusters and regional development: Critical reflections and explorations,* ed. B. Asheim, P. Cooke, and R. Martin, 218–235. London: Routledge.

8 Biotech and Lunch Buckets: The Curious Knowledge Networks of Steel Town

PETER WARRIAN

Hamilton, Ontario, historic home to the Canadian steel industry, represents a curious case study with respect to the role of knowledge networks and the social process of innovation. It has a strong, unitary industrial identity that qualifies it as a classic in specialization, displaying all the potential gains and dangers of specialization tied to the life cycle of one industry. However, with the emergence of the health sciences complex in the west end of the city, a diversity theme has also emerged. This chapter suggests that the special case of Steel Town strongly confirms the general themes of this volume.

Hamilton is a mid-sized city with a long history of industrial manufacturing. The broad studies of globalization have pointed to a migration of manufacturing over the last forty years from larger metropolitan cities to a relocation in mid-sized cities. Much manufacturing has moved offshore, but what remains within North America is sited in mid-sized cities. In this respect, the historical uniqueness of Hamilton has now become more characteristic of the spatial location of manufacturing more generally. The strength of its traditional industrial base has made Hamilton, like other mid-sized cities, more vulnerable, however, to the downside of path dependency. The city has clearly been locked into the trajectory and life cycle of its dominant industry, steel.

The trajectory of steel has, for most observers, been a downhill slide for the last thirty years. This has made Hamilton "just like Pittsburgh" in common parlance. On the other hand, the recovery of Pittsburgh, associated with health and health sciences, has just as quickly been applied to the Hamilton case, with reference to the role of the McMaster University Medical School and Hamilton Health Sciences. The following case study suggests that both ends of this perspective are wrong.

The steel story is much more fluid and complex than most observers assume (Warrian 2010). Hamilton is not Pittsburgh. Thirty years ago, Pittsburgh had seven integrated steel mills. Today it has none. Hamilton had two integrated mills thirty years ago and still does. The conventional image has been one of a precipitous drop in steel industry employment. However, if one thinks of steel as a cluster or sector, most of the drop in employment in the mills has been compensated for by a transfer of functions and talent from the steel producers to steel distributors and steel fabricators. The latter have grown through technology transfers and re-employment of production and technical workers in the broader sector (O'Grady and Warrian 2011). The moderate overall drop in employment in the steel sector since 1991 has followed the trend line of productivity increases. For all these reasons, the story of Hamilton steel is not the same as that of Pittsburgh steel.

At the same time, there has indeed been a growth in new-economy institutions like health services and health sciences. In the narrow sense, this has been similar to the case of Pittsburgh. Conventional observers have characterized this as the decline of the old-economy institutions and the rise of the new-economy institutions, as if these were completely dissociated developments. Hamilton has appeared to be a sort of bipolar economy – the east end being the old economy of steel and manufacturing and the west end being the new economy of the university and health sciences.

However, the emergence of the new economy institutions of the west end has been closely linked to the old-economy institutions of the east end. In previous studies, the investigation of movements between old-economy and new-economy institutions has focused on knowledge spillovers, particularly specialization/general knowledge transfers. The literature has focused on the net flow of human resources from the old to new economy, for instance, in skilled trades, specialized operators, engineers, and managerial talent. What is unique to the Hamilton case is the impact of unions and old-economy labour market institutions on the demand side of the equation. Schematically presented, Hamilton Health Sciences at McMaster University is the source of ideas that is fuelling the growth of the biomedical sector, but it is revenue streams provided by union agreements that are actually making it possible for innovation in the health sector to be implemented by new firms.

The Dynamics of Innovation in a Bipolar Local Economy

The Hamilton economy has been the site of two innovations with global impact in the last thirty years: the Stelco Coil Box[1] and evidence-based medicine. In both cases these developments were directly related to the scale and configuration of local resources. In the Coil Box case, the key resource was Stelco Engineering, which for a generation led technological development for the Canadian steel industry. To this day, the engineering labour market in Hamilton is three times that of its better-known neighbour Waterloo, 40 km to the north. In the case of evidence-based medicine, the key resource was the McMaster University Medical School. The McMaster teaching model was built from its beginning on a foundation of clinical practice, as compared to the bioscience model characteristic of mainstream medical schools.

For many, the Hamilton city-region is a near textbook old economy/new economy split story. The economic geography of the city is polarized between the east-end home of the traditional steel manufacturing industries and the west end, home to the university, health sciences, and biotech industries. Even local actors believe that Hamilton East and West are non-communicating solitudes. Analytically and on the ground the story is much more complex and nuanced, not least owing to the labour market, skills, and circulation of knowledge variables.

An important linkage between Hamilton East and Hamilton West is in the formal governance structures, wherein there was a direct transfer of knowledge from the steel industry to health sciences. Steel industry executives were instrumental in building the Hamilton Health Sciences complex. The chairs of the HHS board have consistently been drawn from steel senior management. As well, steel managers have served on the key committees of the board. Establishing and managing HHS was and is an incredibly complex affair. What the steel executives brought to the table was expertise in how to manage multi-site, high-capital-intensity, highly unionized environments. At the strategic and managerial level there was a heavy transfer of skills and knowledge from the traditional industries to the new health sciences and biotech domains. At the leading edge of the Hamilton new economy are the life sciences and clinical innovative firms. Interviews revealed a surprising linkage to the old economy institutions.

Complexity on the ground in Hamilton is fascinating. In encouraging the emergence of innovative new biotech and health services firms,

Hamilton has faced a critical shortage of venture capital funding. Compared to a place like Toronto, there was a large gap in venture capital markets for new firms. As noted above, in the face of this challenge, Hamilton start-up firms have developed service-based business models as an alternative source of financing growth and product development. This means that in Hamilton, as contrasted with Toronto and its MaRS Discovery District, health service innovation rather than biotech or pharmatech per se has led the innovation curve. The anchor of these revenue streams ironically has been the at times much disputed "Cadillac" benefit plans in industrial union collective agreements of the United Steelworkers and Canadian Autoworkers. In other words, it has been old-economy labour market institutions that have provided the base for new-economy innovation in the health sector. For instance, in customized orthotics, it has actually been the workers themselves and their families who have been the actual human physical sites for the product development experiments. In other words, the benefit plans negotiated by steelworker and autoworker collective agreements constitute a subtle feedback loop in the Hamilton old/new economy circuit.

This is a unique finding. Traditional studies have examined the flow on the skill supply side from specialized old-economy institutions into new-economy organizations. It is entirely new to observe the impact of old-economy labour market institutions influencing the emergence of innovative new-economy institutions on the demand side.

The Strength of Local Knowledge Flows within Industrial Sectors

Steel has dominated Hamilton's economic life for a century. However, globalization of the steel industry has qualitatively shifted the dynamics of labour markets and knowledge networks. Within the steel producers there has been vertical delayering of business units, with divisions of Stelco being sold off. Also, for the past decade, processing functions traditionally performed by integrated steel mills have been transferred to steel service centres to better match the inventory cycles of the auto and construction industries.

Labour shedding by existing industrial firms has led to many layoffs, but ironically has created a large pool of industrial skills for new start-ups, other industrial employers, and even the public sector's maintenance and trades needs. The traditional steel industry as everyone in Hamilton once knew it – that is, two giant integrated steel mills in the east end of Hamilton – has disappeared. However, steel has not disappeared. The mills are still there and the new international owners have reinvestment plans that, if implemented, will result in Canada having more aggregate steel production in 2015 than it did in 2007. Some of this will be in Hamilton, some elsewhere in Canada.

As stated, the steel industry has been the dominant knowledge network in Hamilton's past and largely continues to be so. However, that history has been ever changing. The classic mass-production steel industry (1910–85) was largely the story of Stelco, the dominant firm. The industry had a unique organizational structure, rejecting the multi-divisional corporation shift in mid-century led by DuPont and General Motors. Instead, the industry remained a unitary hierarchy (Chandler 1962), which meant that all knowledge flows were vertical and the steel companies only learned what their engineers learned. The development of industrial corporate R&D centres like the legendary Stelco engineering division reinforced this trend. It was a model of indigenous technological development. Hamilton was the undisputed technical leader of the industry in Canada and had global innovations like the Stelco Coil Box, invented locally and then spread across the world.

With the onset of the technological revolution led by the Japanese steel industry, followed by Korea and India, knowledge flows through technology transfers and licensing caused a shift in global steel-industry knowledge networks. By the 1990s, the global steel industry had developed a broad market for traded knowledge. Dofasco, which had always trailed Stelco in technical resources, used these newly available resources, along with a different organizational model, to displace Stelco as the industry leader. Its focus was applied innovation and product development.

Knowledge networks changed because, having abandoned indigenous technological innovation, the steel companies now came to rely on international technology transfer and licensing. Dofasco proved itself much more adept at this game than Stelco. As a former Dofasco executive noted:

Because our major customers are quite bureaucratic – traditional, conservative, cautious – they don't really want to put a new steel in their cars that no one else can make, or they've not seen before, and they're not sure how it works. Typically, what happens is that new steels come out of Japan that no one has seen before and maybe they're stronger or thinner or more corrosion resistant or whatever, and the North American companies say we need this too and ask us to make it, so we may come up with our own design or similar design – many of these are patented so you can't use the exact same, so our research is trying to come up with grades that are at the leading edge but are not unique in the world.[2]

Dofasco, by contrast with Stelco's 1000-plus engineers, tradition-ally had only 50 metallurgical engineers. It still does. But from the late 1990s, under its strategic plan, *Solutions in Steel*, Dofasco reorganized to deploy 1000 people in the field for interaction with customers. They focused on application development. The Dofasco strategy was to inno-vate in applications with customers. The deep metallurgical technology developments they relied on came from Nippon Steel and others.

> We were able to get the two Japanese steel companies to partner with us, but in different areas of the plant – they were both nervous that we would give information to the other – so we had Nippon Steel working in our steelmaking plant and JFE working in our hot mill and the French would be working in our cold mill, and they'd all meet in the cafeteria or in the elevator at the same time – that was always tricky.[3]

As a result, the patenting Dofasco did was related to process engineer-ing in support of local-customer application requirements.

> Ninety-eight per cent comes from customers wanting us to replicate Japa-nese steel. If we went to them with a steel no one else could make, they wouldn't buy it. Historically, our view was that we should protect our IP [intellectual property] through know-how – you just don't tell people what you're doing, but in the last eight years, we've done more patenting. There are two philosophies: one is that when you move quickly, you don't have time to patent, and if you apply for a patent, you're telling people what you do and they can find a little way to get around it; so you just don't tell people – the Japanese are good at this – they call it "know-how" [tacit knowledge]. So that was our philosophy until about ten years ago, when we started protecting property because we got stung a couple of times, when we produced something that we didn't document and they used it and came after us, so now we protect it in our company; we don't have a lot of patents, but we've patented the unusual things, not the exten-sions ... We have partnerships with McMaster and various universities, but they're not really innovating new things for us as much as they are helping solve problems that we already have.[4]

During this period, there was interaction between the steel companies and university researchers and laboratories, notably the Steel Research Centre at McMaster University. However, the connection was limited. Stelco played a mostly passive part, funding a research chair, which stood unfilled for some years and had virtually no participation in ongoing continuing-education activities. Dofasco had a much more active involvement at McMaster, but in the end, their expectation of the university was a flow of well-trained graduates rather than steel technological innovation per se.

> We interact mainly with people at Mac where we have access to the facilities and labs and those are very valuable to us. They have some programs that we fund, but they're more longer term, knowledge for the benefit of the university rather than knowledge we're depending on for tomorrow. We've done the real R&D, which is really product development at home, but we like to have good skill sets around us and we like to have the university working in areas that are somewhat related.[5]

Dofasco proved to be the best at working within these consortia, such as the Ultra Light Steel Auto Body (ULSAB) and the Steel/Auto Partnership (S-AP) alliance of steel and auto companies that sponsored research and development of the next generation of auto steels. The drivers of Dofasco's commercial success in the past decade were the new technologies, such as hydro-forming and tailor-welded blanking, that arose from its participation in these consortia. Both were generic technologies they acquired through the research networks, but implemented much faster than their competitor/partners. CEO John Mayberry had a vision – in *Solutions in Steel*, published in 1998 – of steel innovation analogized to software. The deep metallurgy would be developed by Nippon and Usinor, like the heavyweight original equipment manufacturers in software – Microsoft, SAP, and Oracle. The Canadian producers would be like value-added resellers (VARs), developing vertical applications off these platforms for use in the local market. "With respect to customers, they are local, and are between Oshawa and the other side of Detroit, but we also sell steel into California and Mexico – so maybe 20% goes to the US and 80% stays in SW Ontario and [the] Detroit area."[6] Application of these new technology platforms for the auto and related manufacturing firms of southern Ontario became the steel innovation strategy and the basis of Dofasco's success.

In the first decade of the twenty-first century, the steel story shifted again. The global steel consolidation movement led to the takeover of all the Canadian steel producers by foreign steel companies. However, for the most part, this did not lead to Hamilton becoming just a convenient site for local steel branch plants and a further diminution of steel knowledge bases. Hamilton is re-emerging as an important site for indigenous steel innovation, but in this case the upward and outward flows are within the new global steel corporations. The new knowledge mechanisms are the benchmarking of operations and international product development methodologies. The former Stelco and Dofasco occupy different positions within these networks.

Stelco now occupies a classic branch-plant position. The US Steel takeover has largely meant a dismantling of its engineering and technology capacities, as these activities have been consolidated into the US head office. US Steel (Hamilton) now presents itself in its industrial brochures and website as having innovative products and plant capacities. The recent layoff announcements were accompanied by significant changes to its management structure. Quite simply put, US Steel has identified its strategic production assets as Birmingham, Alabama, Pittsburgh, and Gary, Indiana. US Steel Canada is now consolidated into the second-tier Great Lakes Division. It is a B Team asset.[7]

Dofasco, by contrast, is seen as a strategic asset for Arcelor's penetration of the North American automotive steel market. ArcelorMittal Dofasco, still emphasizes its innovative R&D capacities in its information material. There have been no layoffs of permanent employees at Dofasco and local talent has been circulated among Arcelor's global operations.

> At Stelco under Mott, there were lots of layoffs in the R&D and development side. It was a continuation of a long process. Stelco now looks more like the classic branch plant.
>
>   Arcelor sees Dofasco as a still important part of global R&D and innovation. They talk about sending Hamilton people around the world to other assignments.[8]

Industry-to-university links have a long history in Hamilton, particularly with the engineering faculty of McMaster University. There is broad and deep agreement in business, academic, and policy circles that a critical opportunity for the future of Hamilton's economy lies in deepening and developing linkages and capacities in the materials and manufacturing sector. As a McMaster faculty member said:

> The key word here … we need it … is this notion of partnership between the university, government and private industry. Private industry, and remember manufacturing is a very applied form of engineering – it is not basic science, it uses basic science for the purpose of creating some things that are really applicable, so industry by definition has to play a really important role because they give us the relevance of the research, they give us the interesting technical problems that we can work on, so the purpose is not simply to use industry as a source of funding.[9]

Currently there is a reconstituting of the link between Hamilton steel and local research capacity by building on the infrastructure of local universities and public research. The McMaster University Innovation Park became home in 2010 to the federal CANMET (Canada Metals and Metallurgy) laboratories, relocated from Ottawa. Arcelor has been in discussions with the Ontario government, McMaster University, and the University of Waterloo for the co-location of an Arcelor global steel research centre, focusing on issues of corrosion, manufacturability, and steel-composite materials, with the university and the relocated federal CANMET lab in the McMaster Innovation Park.[10]

Hamilton Bio-Medical and Health Sciences

Health science is an emerging, high-tech industry that is the most likely of the three partners, with universities and local institutions, to conduct high amounts of R&D and to create unique products with customers around the world. Interviewees have identified the unique McMaster Medical School model as a significant factor. In contrast to other medical schools, such as the University of Toronto's, McMaster focuses on clinical practices rather than laboratory research as one of its objectives. It should be recalled that the new global health-policy norm – evidence-based medicine – was invented by two McMaster faculty members.

Field research identified a number of innovative firms, which were neither pure biotech nor pharmatech firms, but were health services firms. This may serve to moderate expectations that research universities with large biomedical faculties will necessarily produce a fulsome stream of biotech spin-off companies to fuel regional growth and prosperity. However, this development at least in the Hamilton case may not be surprising.

New companies will be service-oriented more than biotech. It's investor driven. There are not investors in Hamilton willing to take risks on early-stage developments. We need a better funding model. For example, if there was more early-stage investment into early validation of genetic marker technology or basic pharma therapy, then firms could emerge sooner. It creates a vicious circle. Biotech and pharmatech opportunities haven't emerged because of the gap in the venture capital market.[11]

The healthcare/health sciences cluster in Hamilton currently focuses on testing and treatment. Two of the most successful new companies are both directly involved in the area of testing. One manufactures antibodies for use in haemostasis research, while the second conducts contract research for several industries, including mining and oil and gas. Several other companies provide products or services related to patient treatment. We found no pharmacological cluster in Hamilton, though a significant new company provides regulatory services to the drug industry.

Hamilton is home to medical professionals with a variety of backgrounds and specialties, which makes it a good place to recruit workers and to find partners to work together. One example is a spin-off from McMaster University which specializes in measuring the quality of the life impacts of different treatment protocols. Hamilton is known as a centre for health sciences, which also makes it easier for certain organizations to promote their services to an international audience.

Hamilton also has certain special assets that make it a particularly good place to do certain types of research. For example, it has a nuclear reactor, which plays an important role in the testing performed by spin-off companies. Also, as mentioned, Hamilton citizens tend to have generous employee medical benefits from unions and government, which means that specialized medical services can be funded more easily. Employee benefit plans are almost always regarded as a cost issue, but in this case they are an asset.

The dynamics of the health sciences cluster in Hamilton is affected by more than just the McMaster teaching model. It is also directly impacted by other factors that go to the heart of the theme of innovation as a social process. At the formal governance level, the present chair of the board of HHS is Don Pether of Dofasco. His predecessor was Bob Jones of Stelco. Interviews reveal that, for the past decade, the chairs of the board and all the key committees have been executives from the steel companies.[12] At the other end of the scale, interviews confirmed that it is the well-endowed health benefit plans of the USWA and CAW in Hamilton that have largely created the market for the new innovative health services firms to develop new customized prosthetics, specialized testing technologies and procedures, and so on.

Key informants also affirmed that this collective agreement–based financing has made up for gaps in the local venture capital markets. Innovators, even where they may provide a medical device or technology of some kind, have developed service-based business models in order to fund their next-generation development. It is certainly an irony and counter-intuitive that these "smokestack" industries and particularly their much criticized legacy cost structure is funding health innovation.[13]

How Should We Understand Hamilton's Knowledge Networks

Recent work by Phil Cooke has challenged our conventional assumptions about learning organizations and learning regions. The fault line in past academic and policy approaches is the asymmetric nature of knowledge and knowledge transfer. Knowledge is neither uniform nor ubiquitous. Within "learning companies," expectations have foundered on inertia and interests within the firm. Within "learning regions," asymmetries of knowledge and institutional configurations between firms have led to shortfalls in expected knowledge spillovers (Cooke 2005, 2007a, 2007b).

Cooke (2005) argues that a qualitative shift has taken place in the dynamics and drivers of globalization and local/regional capabilities. "Globalization 2" is characterized by a shift whereby the advantages of scale which favoured global corporations have now been counter-balanced by dependence on regional knowledge capabilities. As Penrose (1959) argues, the dynamic capabilities of the firm reside in knowledge networks, leading to the massive increased value of transferable knowledge to the wider economy of the firm. Cooke extends the argument to formulate the proposition that Globalization 2 is driven by the quest on the part of multinationals for exploitable knowledge in "knowledge regions" that are often dependent on public research funding resources. Increasing returns to scope and scale give the character of "spatial knowledge monopolies" knowledgeable

clusters and their broader regional innovation systems. Many industries show large-scale corporations having become supplicants to small
"scale" knowledge-intensive firms and clusters due to their asymmetric knowledge weaknesses.

> The takeover of Dofasco may be to the better in the long run. Arcelor is
> making available its global researchers on high strength steels and micro-
> structures for use in Hamilton and with McMaster. However, the diffusion
> out to the companies may take a long time, more of it because of cultural
> limitations and attitudes that are big restraints on the appetite for techni-
> cal innovations.[14]

This perspective lends support to the notion that the recent takeover
of Hamilton steel companies by new global steel titans, at least in the
Dofasco-Arcelor case, are taking place within the dynamics suggested
in Cooke's Globalization 2 model. Conversely, the aspirations of the
McMaster-Waterloo auto-steel research consortium confirm the other
side of Cooke's asymmetric knowledge dynamic.

The asymmetry goes to the nature of knowledge itself. The regional
innovation systems approach is not particularly predicated on learning,
but rather on knowledge and innovation. Knowledge exploration and
knowledge exploitation are strategic choices for organizations. Successful regional innovation systems have organizations that conduct
research that generates new knowledge in appropriate institutional
settings. And other organizations, mostly firms, commercialize this
knot of consumable innovations in the marketplace. Cross-boundary
mechanisms or bridging of social capital between research (exploration knowledge) and commercialization (exploitation knowledge) are
critical to success. Further, the latter can be entrepreneurial (ERIS) or
institutional (IRIS).

When the Hamilton steel industry's knowledge networks became
significantly non-local, the technology licensing and transfer mechanisms and participation in the ULSAB project can both be seen as examples of ERIS bridging mechanisms. The movement of key talent and
IP from Stelco to Hatch for subsequent commercialization can be seen
as part of the same trend. All these knowledge transfers were taking
place within the contractual framework of private firms. The present
move to use public infrastructure and university-government research
capacities, with the proposed Arcelor research centre partnered with

Table 8.1 Changing knowledge capabilities in the steel industry

	Steel 1	Steel 2	Steel 3
Anchor firm	Stelco	Dofasco	Arcelor
Market orientation	Local	Continental	Global
Bridging mechanism	Proprietary	Private consortia	Open source
Exploratory knowledge	In-house	Licensed	Public infrastructure
Exploitative knowledge	Individual customers	Regional	Global supply chains

McMaster, Waterloo universities, and the CANMET labs is clearly an example of an IRIS bridging mechanism.

Finally, the current changes under way in Hamilton, as local steel capacities have been globalized and configured within global supply chains for really the first time, would confirm Cooke's hypothesis that the configuration of local knowledge capabilities underpins the way in which globalization proceeds (table 8.1). This partly explains the more complex setting and potential excitement of the new McMaster Innovation Park and relocated CANMET labs in Hamilton. Synthetic and tacit knowledge transfers were also reflected in manufacturing firms outside steel and health; for example, in a welding technology firm:

> Our main source of competitive advantage is technology, application, design and quality. We introduced our system in 1990 way ahead of anybody else. That's how we made our mark, but the fact that we understand how to weld these parts better than other people do – we have a reputation for running systems that work and we control the technology from software to hardware, everything is done by us.
>
> We develop mainly in consultation with customers. We have our own IP based on what we see in the market and we get it by talking to customers, going to trade shows, seeing what the competition has, talking to key people in the industry to see what they want to see.[15]

In the health and biotech sectors, similarly, the teaching model of McMaster has been a factor. The successful startup companies are in testing (labs) and regulatory compliance domains, not hard science/discovery.

The McMaster teaching model is a good thing. The opportunity there is not as recognized as [it is with] MaRS. A lot of it is social networking, e.g., MySpace. Opportunities that come out of service models and knowledge networks are not just technology plays. It is not just a high-profit venture model. Innovators may be not-for-profit, innovating for a new social context, social networking, and social oriented technologies. There are different types of technologies coming out of McMaster. For example, currently HHS is working on techno-logistics for patient management in hospitals of pre/post-operative pain.[16]

Key Research Findings

Concentration vs Dispersion of Talent in Hamilton's New Economy

The Jane Jacobs model of urban knowledge networks holds that individuals in a vibrant and diverse urban environment interact with each other on a regular basis and help different types of companies improve the quality of their product and business processes. In the interviews we found that this was not necessarily the case. There were a variety of knowledge networks that were used to solve problems and develop new ideas. Those geographically local were often less relevant than many others.

Contrary to expectations that there would be a Jacobean effect of accumulation of dense, interactive networks of creative resources in the city core, interview results suggest instead a "Tim Horton's" effect.[17] The overwhelming number of innovative firms prefer the weak links of suburban Flamborough and interaction through local gathering places and trade conference attendance. One interviewee mentioned that the local Tim Horton's had been a fruitful recruiting ground for him. While standing in line waiting for a coffee, he would converse with others whom he knew, mention a particular need, and learn that the person that he was talking to either was currently between jobs or knew someone else who was available.

This interaction led to subsequent queries as to whether there was some regular local establishment or event – a coffee shop, a bar, a church, a softball team, an art gallery – where they often ran into others and spoke about their business. In some cases, the individual that we spoke to was particularly involved in his or her community and answered in the affirmative.

For example, an arts manager cites the importance of conversations at the softball diamond. Another interviewee who was heavily involved in his church mentioned that he has some work-related connections there. In general, though, the responses to this question tended to be negative. For most people, there was no common, casual location where they would just sit around and talk shop.

There was, however, a clear dividing line between different interviewees. Some seemed more urban and others more suburban. The former typically worked and lived downtown and felt a strong passion for the city of Hamilton. They enjoyed having shops and services nearby and appreciated the burgeoning arts, culture, and sports scenes. This was particularly true of those in the arts community, as well as some in the health-science field located close to McMaster University.

By contrast, the majority of private-sector managers and of successful businesses were located in suburban areas such as Ancaster, Burlington, and Flamborough. From a residential standpoint, these individuals enjoyed quiet and friendly neighbourhoods in the suburbs or the country that were a short, quick drive from their job. They also found that suburban business parks were able to provide the reasonably priced land and services that their organizations required.

The urban/suburban divide seemed to manifest itself in other ways too. Suburbanites tended to have stronger international networks and less connection to the immediate area around their business. Urbanites, by contrast, seemed more embedded in their local community and were more likely to cross-pollinate with other locals. The success of the suburban areas may point to a trend in which economic growth in the city of Hamilton will go from the outside in, with success in the suburban regions generating greater investment in the urban core – as opposed to the traditional model of major industries and population in the core radiating outwards.

Some managers were directly critical of city politicians and planners trying to force feed Jacobean development downtown and neglecting housing needs closer to the new industrial park. "They want everybody who works here to live downtown, so there's no residential planning in the area. Rather than putting up 15 acres of the 139 into high-rise condos – people live and work in one area – but no, someone on city council says, 'We have to develop the downtown core to attract people down there, so that's where the residential should go.'"[18]

Boundary issues were also manifested at the circumference. Emerging industrial sectors in Hamilton are not easy to identify, and when interviewed, the respondents are not connected to each other through established community linkages such as the chamber of commerce.

The Strength of Linkages between Local and Non-Local Actors

Linkages with non-local actors in the new global steel industry are now more similar to the knowledge and talent flows of the new-economy health networks of the west end. Both are critical to the innovation circuits of the emergent Hamilton economy. The patterns of interaction among network agents in Hamilton's urban economy include the following:

INTERNAL, INTERNATIONAL NETWORK
(LARGE ORGANIZATIONS) AGENTS
In large multinationals such as engineering, manufacturing, and processing, individuals having a problem to deal with, or a new idea to develop, will typically turn to others in their own organization, often sourcing talent from the corporate head office and other locations around the world. In some cases, there are regular visits and meetings used to exchange ideas, see what others are doing, and offer advice.

EXTERNAL, INTERNATIONAL NETWORKS AGENTS
Individuals who are not part of a large organization may also consult an international network of contacts. These are often co-workers from past jobs, friends from school, or colleagues met at a trade show or conference, or in a professional organization. Often a quick phone call or email is enough to get a solution to an issue. The Internet is a vital tool in this regard as well.

LOCAL INFORMAL NETWORKS AGENTS
Interviewees were queried for information on local, informal connections, with the idea that an organization benefits from cross-pollination of ideas from many different sources. However, there was much less of this sort of interaction than one might expect. Very rarely did an individual point to a specific set of activities they engaged in or companies that they interacted with who provided them with product ideas.

The cluster for whom the local context was most important was the artistic community, which in many cases lived up to the ideas of cross-pollination and idea generation that corresponds to the Jane Jacobs model of urban innovation. Artists, arts organizations, and even craftspeople such as woodworkers regularly interact with each other, develop events together, and, to some degree, strategize with each other.

Artists, for example, noted that galleries, shows, and discussions with other artists all influenced and improved their work.

In other clusters where local knowledge was shared, it was typically regarding how to better run the business: recruiting key personnel, finding land, and, for small businesses, acquiring sales and marketing skills. For example, an executive from a diagnostics company noted that attending her son's sporting events gave her a chance to speak to other small businesspeople and get new ideas.

TRADE SHOWS / ACADEMIC CONFERENCES / ART SHOWS

One of the most interesting commonalities between organizations in terms of knowledge networks was the importance of conferences and trade shows as sources of new ideas and relationship building. This was true for organizations in all sectors. The chief difference between the organizations was the nature of the event. For those in health science the events are academic conferences, where peer-reviewed research was discussed and papers presented. For artists, galleries or theatre events give an opportunity for socializing and inspiring new ideas. For manufacturers, the forums were typically trade shows, organized by a particular trade association, where companies displayed their new products. While the shows are ostensibly meant to display products to customers, or to share research, they were often even more valuable as opportunities to reinforce connections with distant personal contacts, to find potential collaborators, to see what your competitors are doing, or simply to prove, as one gritty moulder suggested, "that you're not dead," because everyone in the industry goes to the show.

One particular trade show that was strongly recommended was the Hanover Industrial Fair in Germany. This fair is an opportunity for university groups and other IP dealers to show off their products and auction them to companies who may have a use for them. One manufacturer uses it to find ideas for improving his products – machines for putting soft products into shipping containers:

> Let me say this into the microphone – send all of your people to the Hanover Industrial Fair and go to the R&D centre. Absolutely critical. You'll understand philosophically why the Japanese and Germans are so good at it ... This is the world's biggest industrial fair. They had two buildings of R&D stuff. One was just for cool design – from an iPod to a bicycle ... The University of Dresden had a tent there that would span 100 metres.[19]

Other reasons suggested for the lack of a central Hamilton meeting spot was that engineers are by nature just not as social as others, or that the individuals being interviewed tended to be senior individuals, and therefore older; they no longer had the type of lifestyle where they would pop down to the local bar after work.

Summary of Key Findings

This summary of the findings in the Hamilton case correlates with three significant themes in the literature.

With respect to the nature of global/local knowledge flows, previous research on the steel-auto cluster documented the critical linkages between Canadian integrated-steel producers and the auto industry, particularly research and development initiatives within the steel companies and with global consortia, such as ULSAB. Stelco's traditional indigenous research approach through its R&D arm, StelTech, was contrasted with that of Dofasco, which was based on a technology transfer strategy. The difference placed Dofasco in a more advantageous position to pursue an innovative product development strategy with its Ontario regional customer base. However, it also meant that the university research base at McMaster lagged behind more advanced developments at Carnegie Mellon and the University of Pittsburgh.

Global/local knowledge flows are now undergoing another major shift. The steel producers have themselves now been directly integrated into the global steel production networks (Sturgeon 2008) via mergers and acquisitions. The local knowledge networks and public research capacities are now trying to catch up through the relocation of the CANMET labs, negotiations over an Arcelor-McMaster research centre, and so on.

In the area of knowledge flows, most of the literature on knowledge spillovers via labour market developments focuses on the supply side. Studies have examined the transfer of human resources and skills through the dislocation/relocation of human factors between old and new economic sectors and companies. The role of unions and whether "unions make a difference" has primarily examined the net flow of skills and human resources on the supply side. The unique feature of Hamilton that is emerging is the impact of unions and old-economy labour market institutions on the demand side. In Hamilton, it is exactly the financial asset base of union health benefit plans that has generated the market for innovative firms to emerge in the health services sector. Among other things, biotech firms are developing health service business models to make up for deficiencies in venture capital markets in funding new devices and procedures. The HHS complex is a source of ideas, but it is union collective agreements that are enabling the innovation to actually be implemented.

Finally, with regard to the role of public research infrastructure, the evidence from this case study suggest that both public and private research infrastructure will be critical to the future of the Hamilton economy. The old-economy model assumed a linear model of research and innovation. Research took place primarily in the R&D labs of companies such as Stelco, which were then transferred internally for commercialization. Dofasco extended the scope of innovation by reaching out through technology-transfer agreements and international industry consortia. In the new globalization phase of steel, but also in biotech, fibre optics, and so on, we have witnessed the emergence of asymmetric knowledge networks. That is, innovation in the future will be non-linear and come from multiple centres of knowledge generation, with each specializing in different aspects of development.

Conclusion

At the macro-economic level, the traditional Hamilton economy is being merged into the larger economy of the Toronto city-region. As summarized by Wolfe in the introduction to this volume, the world economy is being restructured and consolidated into a set of world cities (New York, Paris, London, Tokyo) that are being linked in turn to major hub cities (Milan, Toronto, Frankfurt). The global and hub cities are characterized by the agglomeration and creative class dynamics described by Jacobs and Florida.

A clear indication in the Hamilton case is the shift of its leading steel fabricator into "digital manufacturing." A traditional producer of fabricated beam and other structural products for commercial construction was faced with the challenges of finding a niche within the new digitized world of CADCAM and Building Information Systems. Not finding in Hamilton the skills it needed in digital visualization, it acquired a design house in Toronto and a new client constituency among the architectural cluster in Toronto. It even has its own film studio to problem solve issues in design and erection with audiences of trades, construction supervisors, structural engineers, and architects who can review 3D graphical representations of the entire construction and erection process from the blueprint to the finished building.

Two important conclusions flow from the Hamilton case of the interface between the old and new economies which may be of wider interest to economic geography and innovation studies.

First, global/local knowledge flows now dominate both the old and new economies of Hamilton. As described above, within the steel industry itself, knowledge creation has fundamentally shifted. For the first seventy-five years of the twentieth century it was Stelco's traditional indigenous research efforts that led the technical development of the whole Canadian steel industry. In the 1990s Dofasco took over leadership with its innovative technology-transfer strategy (Warrian and Mulhern 2003, 2005). With globalization, Hamilton's steel plants have become directly integrated into the production and R&D networks of the new global steel companies (Warrian 2010). The new-economy institutions, led by McMaster University and HHS, are grounded from the start with international networks of scholarly, peer-reviewed journals, patenting and university and scientific journals, academic conferences, and so on.

Second, in previous studies, investigations of movements between old-economy and new-economy institutions have focused on knowledge spillovers, particularly specialization/general knowledge transfers. The literature has focused on the net flow of human resources from the old to new economy, such as skilled trades, specialized operators, engineers, and managerial talent. What is unique to the Hamilton case is the impact of unions and old-economy labour market institutions on the demand side. Schematically presented, HHS is the source of ideas. It is union agreements that are enabling the innovation for those actually to be implemented by new firms.

NOTES

1 The Coil Box was a technology invented by Stelco Engineering in the 1970s that provided a new method for controlling the temperature of hot steel coils and therefore allowed companies to better control the quality of the steel for their demanding automotive customers.
2 Confidential interview, former Dofasco executive, 16 October 2006.
3 Ibid.
4 Ibid.
5 Confidential interview, welding company executive, 21 October 2006.
6 Confidential interview, former Dofasco executive, 16 October 2006.

7 US Steel, press release, 3 March 2009.
8 Confidential interview, former Stelco executive, 20 November 2008.
9 Confidential interview, university administrator, 11 October 2008.
10 Confidential interview, university research manager, 2 December 2008.
11 Confidential interview, biotech network administrator, 24 April 2008.
12 Confidential interview, HHSC board member, 27 November 2006.
13 Confidential interview, biotech manager, 24 April 2008.
14 Confidential interview, government science consultant, 25 April 2008.
15 Confidential interview, welding company executive, 21 October 2006.
16 Confidential interview, biotech network administrator, 24 April 2008.
17 Tim Horton's is the classic suburban donut shop chain at which local personnel often meet to discuss business, technical, and recruitment issues.
18 Confidential interview, technology incubator manager, 16 October 2008.
19 Confidential interview, manufacturing executive, September 2007.

REFERENCES

Chandler, A.D. 1962. *Strategy and structure: Chapters in the history of the industrial enterprise*. Cambridge, MA: MIT Press.
Cooke, P. 2005. "Regional asymmetric knowledge capabilities and open innovation." *Research Policy* 34 (8): 1128–49. http://dx.doi.org/10.1016/j.respol.2004.12.005.
Cooke, P. 2007a. "Regional innovation, entrepreneurship and talent systems." *International Journal of Entrepreneurship and Innovation Management* 7 (2/3/4/5): 117–39. http://dx.doi.org/10.1504/IJEIM.2007.012878.
Cooke, P. 2007b. "Regional innovation systems, asymmetric knowledge and the legacies of learning." In *The learning region: Foundations, state of the art, future*, ed. R. Rutten. Cheltenham: Edward Elgar.
Kilbourn, W. 1960. *The elements combined*. Toronto: Clarke, Irwin & Co.
O'Grady, J., and P. Warrian. 2011. *Human resources in the Canadian steel sector: Final report*. Toronto: Canadian Steel Trade and Employment Congress, May 2011.
Penrose, E. 1959. *The theory of the growth of the firm*. New York: John Wiley and Sons.
Sturgeon, T. 2008. "From commodity chains to value chains." In *Frontiers of commodity chain research*, ed. J. Bair. Palo Alto: Stanford University Press.
Warrian, P. 2010. *The importance of steel manufacturing in Canada*. Toronto: Munk School of Global Affairs, University of Toronto.
Warrian, P., and C. Mulhern. 2003. "Learning in steel: Agents and deficits." In *Clusters old and new: The transition to a knowledge economy in Canada's regions*,

218 Innovating in Urban Economies

ed. D.A. Wolfe, 37–62. Montreal, Kingston: McGill-Queen's University Press for Queen's School of Policy Studies.

Warrian, P., and C. Mulhern. 2005. "Knowledge and innovation in the interface between the steel and automotive industries: The case of Dofasco." *Regional Studies* 39 (2): 161–70. http://dx.doi.org/10.1080/003434005200059934.

9 Innovation Linkages in New- and Old-Economy Sectors in Cambridge-Guelph-Kitchener-Waterloo (Ontario)

ANDREW MUNRO AND HARALD BATHELT

It is now widely believed that knowledge and innovation are key resources for regional economic success (Lundvall and Johnson 1994), and that learning is the decisive process that stimulates knowledge creation and innovation (Lundvall 1988; Malecki 1991). Under these circumstances, regional well-being depends on the capacity to stimulate processes of interactive learning, networking, and innovation at the local level (Cooke and Morgan 1998; Gertler 2004). In this context, the region around the mid-sized Ontario cities of Cambridge, Guelph, Kitchener, and Waterloo is regarded as one of the model economies for Canada – a claim based primarily on the region's success at innovating and transitioning from a traditional manufacturing economy to one that increasingly includes high-technology firms in the information technology (IT) sector. A regional success story of reconfiguration driven by spillovers from university-related start-ups and spin-offs has developed to explain this modernization process. One hypothesis that could be derived from this development is that innovation around technology-based university start-up firms in the region has not only benefited the growth of the IT sector, but also spilled over to the traditional manufacturing sector via material linkages, knowledge flows, and innovation networks.

This chapter investigates a number of interrelated questions regarding innovation practices and the social foundations of innovation in the region around Cambridge, Guelph, Kitchener, and Waterloo, as laid out by Wolfe (2009). Our analysis first investigates the contribution of university-related IT spin-offs/start-ups to regional innovation networks. Second, we analyse the nature of innovation processes in the traditional manufacturing sector. Third, this research investigates whether there are cross-sectoral linkages that drive regional innovation processes, particularly between IT-based university start-ups and traditional manufacturing firms.[1] Fourth, a central part of this analysis studies the nature and relationship of local and trans-local knowledge flows in innovation. As pointed out in the literature, firms that are connected through regional agglomerations of producer-user relationships can stimulate regional economic growth and competitiveness (Etzkowitz et al. 2000; Vohora et al. 2004). This has, for instance, been shown in the seminal works of Roberts (1968) and Cooper (1971), who studied technology-related spin-off phenomena along Boston's Route 128 and in Silicon Valley, respectively. While Route 128 and Silicon Valley have established themselves as the prototypes of advanced high-technology/IT regions (Saxenian 1985; de Jong 1987), relatively few other examples exist of regions that have benefited from equally strong local university spin-off activities and related innovation processes. Understanding these producer-user linkages can inform the question of whether innovation activities benefit more from a specialized regional environment that is characterized by industries with a similar or related knowledge base, or from a milieu that is highly diversified (Beaudry and Schiffauerova 2009; Boschma and Frenken 2011). Accordingly, this chapter is set in the context of debates about interactive learning in regional economies (Gertler 2004), local versus global knowledge flows (Bathelt et al. 2004), and the role of related versus unrelated variety versus specialization (Duranton and Puga 2001; Frenken and Boschma 2007).

The Kitchener and Guelph metropolitan areas, located about 100 kilometres west of Toronto, around which the initiative called "Canada's Technology Triangle" was founded in the late 1980s, have received a lot of attention by policymakers because of their success in shifting the economic focus from traditional manufacturing to new IT-related businesses and, as a result, maintaining high regional growth. The regional economies have achieved above-average performance levels, according to indicators such as job growth, unemployment rate, and average household income. Between 2001 and 2006, the Kitchener census metropolitan area (CMA) and the Guelph census agglomeration (CA) experienced an increase in population and jobs that was significantly higher than the national and provincial growth rates and similar to those in the Toronto CMA (Bathelt et al. 2010, 2011). This seems to support the view that the region benefited from university spin-offs and related innovation processes. Numerous IT firms, such as Open Text and BlackBerry (formerly Research in Motion), have successfully spun off from researchers and students at the University of Waterloo, establishing a growing high-technology sector in the region (Bathelt and Hecht 1990; Bramwell and Wolfe 2008), with high growth potential (BMO Capital Markets 2008; Florida and Martin 2009).

In the media hype around the supposedly "post-industrial" future of the Waterloo region (Perry 2009), it is often forgotten that the region has a strong manufacturing tradition (English and McLaughlin 1983; Holmes et al. 2005; Rutherford and Holmes 2008; Vinodrai 2011). Academic and policy analyses on innovation in the region, however, tend to focus on high-technology growth and ignore the often incremental innovation that happens in traditional manufacturing (Rutherford and Holmes 2007). Upon closer analysis, we can see that 44,100 of 76,700 manufacturing employees (57%) and 1048 of 2163 establishments (48%) in the region fall within traditional manufacturing in the plastics and rubber, metal-fabricating and processing, machinery, electrical equipment, and automobile supplier industries in 2008 (table 9.1). In the Kitchener CMA, the manufacturing sector had a share of 20.3 per cent of the total labour force in 2007, which was nearly twice as high as the Canadian average (figure 9.1). As a result, this sector's contribution, despite a substantial decline in the past decade, cannot be neglected in the economic success story of the region.

To explain this development, it is important to understand the social foundations – in the sense of inter-firm linkages – of regional innovation in both new- and old-economy sectors in the region, as well as to investigate the potential relationships between both. This was done in our study by analysing, first, the vertical and horizontal networks underlying innovation and, second, the institutional support for such activities. Our research in the region around Cambridge, Guelph, Kitchener, and Waterloo was conducted in two phases. In the first phase, we examined the role that university-related spin-offs and start-ups in the IT sector played in guiding technological change at the regional level, and investigated the resulting innovation linkages. In second phase, we similarly analysed the innovation processes and linkages in traditional manufacturing in the region. Throughout both phases, we explored the social foundations of innovation and how networks of relationships, such as supplier-customer networks, university-industry collaboration, cooperation with industry associations, and cross-sectoral linkages, influenced further innovation and growth in the region.[2]

Table 9.1 Number of firms and employees in the Kitchener Census Metropolitan Area (CMA) and Guelph Census Agglomeration (CA) by industrial sector, 2008

	Kitchener CMA		Guelph CA	
Industrial sector by NAICS code	Number of firms	Number of employees	Number of firms	Number of employees
311 – Food manufacturing	96	7392	32	1308
312 – Beverage and tobacco product manufacturing	13	131	6	408
313 – Textile mills	17	278	1	3
314 – Textile product mills	31	1004	3	21
315 – Clothing manufacturing	34	668	8	376
316 – Leather and allied product manufacturing	11	64	3	17
321 – Wood product manufacturing	79	1158	19	191
322 – Paper manufacturing	13	1221	13	795
323 – Printing and related support activities	104	1069	30	267
324 – Petroleum and coal product manufacturing	7	171	2	2.5
325 – Chemical manufacturing	53	1482	24	731
327 – Non-metallic mineral product manufacturing	49	2091	11	560
331 – Primary metal manufacturing	23	1111	3	69
334 – Computer and electronic product manufacturing	87	4837	11	296
337 – Furniture and related product manufacturing	114	2836	21	136
339 – Miscellaneous manufacturing	165	1929	32	163
326 – Plastics and rubber products manufacturing	93	5347	24	709
332 – Fabricated metal product manufacturing	329	8654	96	3004
333 – Machinery manufacturing	272	7671	52	2882

Industrial sector by NAICS code	Kitchener CMA		Guelph CA	
	Number of firms	Number of employees	Number of firms	Number of employees
335 – Electrical equipment, appliance and component manufacturing	*56*	*2255*	*11*	*967*
336 – Transportation equipment manufacturing	*75*	*7446*	*40*	*5199*
Total	1721	58,816	442	18,106
Investigated traditional manufacturing industries total	*825*	*31,373*	*223*	*12,761*

Note: Selected traditional manufacturing industries in italics
Source: Statistics Canada 2008

Figure 9.1 Percentage of the Canadian and Kitchener CMA labour force in manufacturing sectors, 1987–2007

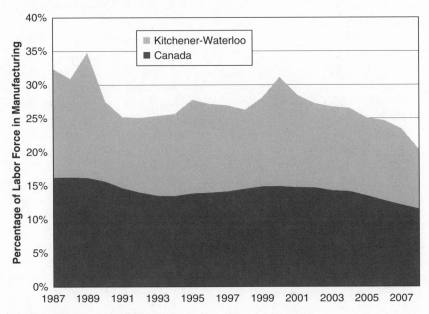

Source: Statistics Canada 2010

The argument developed in this chapter is structured as follows: In the next section, we present the regional context of our study. Then, the conceptual framework is discussed. The empirical part that follows analyses the nature of innovation linkages of both university-related IT spin-offs/start-ups and traditional manufacturing firms. At the end, a summary is provided and brief policy conclusions are drawn.

Regional Context

Regional modernization and success in innovation are sometimes connected to the presence of leading universities and research facilities. In the social science literature the advantages knowledge-based, technology-intensive firms can accrue from being in close proximity to a university have been widely recognized. Explanations often point out that high-technology ventures derive benefits from localized knowledge spillovers emanating from the two common tasks performed by universities: conducting basic research and creating human capital (Audretsch et al. 2005). While these knowledge inputs are appealing to firms due to their "public-good" character, access to these inputs appears to depend on the spatial proximity to specialized university institutes.

It was not until the late 1960s, when new industries such as those related to computer technology, microprocessors, and semiconductors emerged, that scholars began scrutinizing the technology-transfer mechanisms that led to the direct commercialization of university research, resulting in firm formation and subsequent regional economic growth (Landström 2005). Above all, it was the institutional and structural changes that began in the 1980s, such as the passage of the Bayh-Dole Act in the United States, and increased labour mobility – in particular of highly skilled individuals – along with a shift to more flexible modes of production and venture capital financing, that triggered research efforts concerning university entrepreneurship and technology-based growth (Rothaermel et al. 2007). Since a shift from a closed to a more open innovation system (Chesbrough 2003) has become visible, university spin-off firms are acknowledged in the literature as one of the key drivers of economic change and growth (Bercovitz and Feldman 2006). Today, most advanced national economies strive to generate economic wealth by exploiting and diffusing public research by means of university spin-offs (Clarysse et al. 2005). In many cases, however, such endeavours have limited success (Callan 2001). Although local universities – especially the University of Waterloo – played an important role in the modernization of the region around Cambridge, Guelph, Kitchener, and Waterloo (sometimes referred to as Canada's Technology Triangle), the strong economic performance in traditional manufacturing industries dates back a lot longer and is unrelated to university research.

The Kitchener and Guelph metropolitan areas were traditionally – and still are – characterized by a strong diversified manufacturing base (table 9.1). In the first half of the twentieth century, the region had well-developed economic strengths in the rubber, textile, leather, furniture, and food processing industries. Despite the differentiated industry structure, however, regional supplier linkages never seemed to be very strong. While the rubber industry was, for instance, originally established as a supplier sector to the shoe industry when Schlee and Kaufmann founded the Berlin Rubber Company in 1899, it later shifted towards other customer groups, especially towards the production of tires (English and McLaughlin 1983). In the post–Second World War period, manufacturing growth was driven by industries, such as fabricated metals, machinery, and electrical products. Furthermore, the region developed a strong basis in the automobile supplier and transportation equipment sector (Rutherford and Holmes 2008).

. Since the 1970s, numerous university spin-offs started up in the region. This was related to the foundation of the University of Waterloo in 1959 as a university with an engineering focus, with members of the university able to own patents from university research. Industrial leaders, such as Ira Needles from the rubber producer BF Goodrich, played an important role in the design of the university. They shaped the university's cooperative education program and its openness towards private-sector collaboration (Bathelt 1991a; Wahl 2007; Bramwell and Wolfe 2008). Compared to other Canadian universities in the post-war period, the University of Waterloo not only had a more pronounced focus on establishing university-industry linkages, but also developed a stronger focus on basic and applied research. At that time, the University of Waterloo became an important driver of a more research-oriented as opposed to a resource-led national production and innovation system (Niosi 2000). The university's initial advantage, however, decreased over time. Already in the 1990s, observers began speculating that the overall economic growth and success in the region was primarily due to the co-op program and a constant flow of highly qualified graduates, who found a job in the region's growing technology sectors, rather than being a consequence of university research and spin-off processes (Bathelt 1991a).

Aside from start-ups around the University of Waterloo, the region also attracted a number of multinational IT firms, such as Google, Hewlett-Packard, Microsoft, and NCR, which established branches or acquired existing technology firms. Although 87 per cent of the firms surveyed in information and communication technologies in a Communitech (2006) report had in-house R&D, most of this seemed focused on incremental development tasks rather than basic or applied research. In addition, local technology firms apparently had not developed extensive input-output linkages in the regional economy (Bathelt 1991a; Xu 2003; Bramwell et al. 2008).

In recent years, the regional economy continued to transform its manufacturing base while remaining strongly diversified. Traditional manufacturing sectors such as textile mills, clothing, and leather manufacturing lost between 50 and 60 per cent of their employees in the Kitchener CMA between 2001 and 2006; and chemical and electrical equipment manufacturing lost 20 per cent of their employees. This structural change was over-compensated by a 20 per cent increase of the employment in plastics/rubber products and computer/electronic products manufacturing. Furthermore, most knowledge-based producer-related services experienced substantial job growth. In the areas of professional, scientific, technical, and educational services, for instance, total employment increased from about 40,000 to 47,500 between 2001 and 2006. The most spectacular job growth in the Kitchener CMA in this period occurred in the scientific research and development services branch. Here, the number of employees increased by almost 300 per cent from 400 to about 1550 (Statistics Canada 2001, 2006a, 2006b).

Upon closer investigation, our understanding of the successes in regional economic development and innovation still appears limited. One of the key issues this chapter attempts to address is whether or not firms' innovation processes benefit from being located in city-regions due to robust regional benefits such as the presence of specialized suppliers, a local pool of specialized labour, and a shared specialized knowledge base; or whether a diversified economy brings more benefits by allowing firms in one industry to tap into knowledge regimes and technologies that are standard in other industries.

Existing studies have shown that the region around Cambridge, Guelph, Kitchener, and Waterloo does not form a true regional industry cluster of closely interrelated firms of a particular value chain (Bathelt 1991b; Xu 2003; Bramwell and Wolfe 2008; Bathelt et al. 2010). What we find instead is a highly heterogeneous and segmented regional economy characterized by limited commonalities. The region hosts a variety of larger and smaller establishments, old and young firms, and businesses with diverse manufacturing and service backgrounds. In our research, we rarely encountered cross-industry linkages between value chains in the region, and the influence of such linkages appeared to be of minimal importance for regional innovation processes, as will be discussed in more detail below. Overall, there is no simple explanation for the overall success of these different economic segments, as our study of the nature of innovation practices and the social foundations of innovation suggests.

Conceptual Framework

In this section, we outline our conceptual framework by focusing on the conditions for regional networking, clustering, and innovation. As the region under investigation cannot be described as a fully developed industry cluster, we focus on the nature of linkages and knowledge flows that could support further agglomeration and innovation in the future.

In the knowledge-based economy, creating and sharing knowledge has apparently become key to encouraging economic growth and innovation (Lundvall and Johnson 1994). Much of the literature on regional innovation and clustering has pointed out that firms that establish linkages to other regional firms in the same or a similar value chain have enhanced opportunities for interactive learning (Cooke and Morgan 1998; Morgan 2004), thus stimulating regional innovation processes (Lundvall 1988). Based on observations concerning the tendency of firms in complementary industries – defined by value-chain-related linkages – to form agglomerations (Porter 1990, 2000), cluster theory has developed as a network approach that examines material linkages and knowledge flows between co-localized firms (Pinch et al. 2003; Visser and Atzema 2008). In such settings, the co-location of firms provides numerous opportunities for inter-firm linkages (Storper and Walker 1989; Gordon and McCann 2000; Preissl and Solimene 2003). It supports frequent face-to-face communication, which enhances opportunities to resolve conflicts and exchange complex information more easily. Close geographic proximity does not, however, automatically ensure that firms engage in such exchanges (Amin and Cohendet 1999); yet, it can operate as a catalyst for channelling different corporate knowledge bases and organizational cultures, thereby increasing the potential to develop common understandings and interpretative schemes (Bathelt et al. 2004). Such a configuration may, in fact, be quite supportive to regional innovation.

Clusters and other economic agglomerations can be distinguished along several key dimensions. The vertical dimension typically consists of specialized suppliers and service providers that generate a division of labour, allowing each firm to focus on their core area of expertise (Malmberg and Maskell 2002). This reduces transaction costs, provides incentives for new firms to locate in the cluster, and stimulates the development of a specialized labour market (Scott 2006). The horizontal dimension consists of the firms that compete with each other in the same or related market segments, possibly establishing a precondition for vertical growth (Porter 1990; Maskell and Lorenzen 2004). These firms may not collaborate in joint research projects, but they have opportunities to closely monitor their rivals that develop similar products under similar conditions. This increases opportunities for mimicry and signals the need to differentiate products, all of which contributes to innovation.

Other organizations are also connected to an industry agglomeration and can provide significant input to firms, such as universities, government agencies, standards councils, and trade agencies (Parker 2001; Gertler 2004). Following Setterfield (1993), they establish the formal basis for institutions, in the sense of correlated behaviour of agents. The institutional dimension frames economic interaction, allowing specialized users and producers to discuss and solve particular problems, and develop reasonable expectations regarding each other's actions (Hodgson 1988; MacKinnon et al. 2009). If we consider that city-regions may consist of several industry clusters or cluster-like configurations, we can imagine that additional effects may emerge due to overlapping knowledge bases and labour markets. While we have already emphasized that our regional case study cannot be conceptualized as a highly specialized agglomeration of firms in a narrow set of industries, extended practices of vertical and horizontal cooperation and knowledge exchange are nonetheless important for innovation and economic growth within and across the regional industry segments.

In recent years, studies have argued that broad regional growth and innovation effects are more likely to occur if the regional economy draws from technological complementarities and related overlapping knowledge bases that enable firms to establish vertical networks and engage in knowledge exchange, even across industrial sectors – as the related-variety argument suggests (Frenken and Boschma 2007). A corresponding regional ensemble of firms may take the form of regional thickenings of a particular value chain or of fully fledged industry clusters with a well-developed supplier and institutional infrastructure (Porter 1990; Malmberg and Maskell 2002). In the case of a cluster, regional networks can develop and dynamic local knowledge flows, or "buzz," can unfold and drive innovation (Storper and Venables 2004; Bathelt et al. 2004). If local firms are not closely related to one another in terms of their used technology and knowledge base, possibilities for local networking and related growth triggers likely remain limited (Nooteboom 2000). In this case, growth has more likely few collective qualities, but is a result of individual firm successes that rely on bonds with partners outside the regional or even the national economy. Strong connections to global value chains and access to external markets and technology partners are then likely crucial to generating growth impulses and supporting processes of innovation in the regional context (Oinas and Malecki 1999; Owen-Smith and Powell 2004; Bathelt et al. 2004).

It is in this context that we designed a qualitative empirical study of the innovation networks and practices in the Kitchener CMA and Guelph MA to identify the nature and spatiality of knowledge flows and innovation linkages. Although we might have expected to see regional specialization and evidence of a related variety of firms in our case study, we found instead that the region is highly diversified (table 9.1) and that firms have developed strong supplier-producer-customer linkages nationally and internationally. Further, there is almost no evidence of relationships between the regional IT and traditional manufacturing firms.

University-related IT Spin-offs/Start-ups in the Region

The region around Cambridge, Guelph, Kitchener, and Waterloo is frequently portrayed as a dynamic technology region that draws from university start-up/spin-off processes, knowledge transfers, and corresponding regional networks. Our research, however, portrays a somewhat different picture of the development in this region and offers interesting insights into the underlying social dynamics of innovation. In contrast to what we expected, we did not find proof of strong value-chain-based or cross-sectoral networks and knowledge flows. In the first stage of this research, this aspect was systematically explored by means of semi-structured interviews with eighteen university-related IT spin-offs/start-ups (Bathelt et al. 2011).

We started off with the assumption that IT firms would be the most likely among the university start-ups/spin-offs to demonstrate evidence

Table 9.2 Local supplier linkages of firms in Cambridge-Guelph-Kitchener-Waterloo, 2007–9

	Firms with significant/high local supplies			
	Significant (≥ 10%)		High (≥ 50%)	
Firm type	Number	Share	Number	Share
IT start-up/spin-off firms	6 of 17	35%	1 of 17	6%
Traditional manufacturing firms	14 of 37	38%	11 of 37	30%

Source: Survey results

Table 9.3 Local and international sales of firms in Cambridge-Guelph-Kitchener-Waterloo, 2007–9

	Firms with significant/high local sales				Firms with significant/high international sales			
	Significant (≥ 10%)		High (≥ 50%)		Significant (≥ 10%)		High (≥ 50%)	
Firm type	Number	Share	Number	Share	Number	Share	Number	Share
IT start-up/spin-off firms	1 of 15	7%	1 of 15	7%	14 of 15	93%	13 of 15	87%
Traditional manufacturing firms	18 of 37	49%	8 of 37	22%	23 of 37	62%	14 of 37	38%

Source: Survey results

of regional horizontal and vertical relationships in innovation, within and across value chains, supporting broad "local buzz." However, the empirical results from our interviews were somewhat surprising. They showed, first, that there were fewer such start-ups/spin-offs than expected and, second, that most of these firms operated in specific cross-regional networks along market and technology linkages that adhere to their particular technological expertise. Local linkages with customers and suppliers and the existence of regional industry networks, such as those described in conventional cluster approaches, were quite limited in their extent or absent altogether. This is clearly indicated in tables 9.2 and 9.3. Only about one third of the IT firms had significant local supplier linkages (≥ 10% of overall supplies), and just one firm purchased its supplies primarily locally (table 9.2). In terms of sales linkages, only one firm showed a significant local orientation (≥ 10% of overall sales), while most had predominantly international customer linkages (table 9.3). About 90 per cent of the IT start-ups/spin-offs sold most of their outputs in the United States or overseas. This pattern clearly reflected the nature of knowledge flows in innovation, which primarily involved non-local producer-user linkages.

Regional industry associations/organizations were also of limited importance in stimulating innovation, primarily playing a role in deepening social networks and generic skill sets (Xu 2003; Colapinto 2007; Bramwell et al. 2008). Firms frequently turned to specialized Internet-based user groups as their initial problem-solving tool, and rarely found opportunities to collaborate with other firms in the region, typically stating that nobody else was working on the same type of products and problems they encountered. It appears unlikely that these firms would spur the development of specialized regional innovation networks.

Although the IT sector may be somewhat specific in terms of its ability to create international networks, it does not possess fundamentally different linkage patterns compared to other new technologies. In particular, we expected university start-up/spin-off firms to display a somewhat stronger regional orientation, especially in their early stages. This was, however, not the case. We found three reasons that help explain this: First, it seemed that firms in the area of specialized software solutions were able to establish a broader extra-regional customer base more quickly and easily than firms in other sectors. Second, the regional firms formed an extremely diversified pool, limiting the opportunities for local network creation in a mid-sized region. Third, acquisitions of firms by larger entities that took place in the region served to provide access to wider extra-regional corporate networks, and thus boosted market legitimacy for the respective units. Overall, we found that spin-off/start-up firms created a limited amount of specialized "local buzz" in innovation.

In our interviews, we encountered only a few IT start-ups/spin-offs that reported existing linkages in innovation with firms in other sectors, and none of these firms cited another industry outside their value chain as important for innovation. We were particularly interested in seeing if there were cross-industry linkages between the traditional manufacturing and IT sectors in the region, but only found two significant cases where IT firms indicated such linkages. One producer of electronic control boards identified some of the manufacturers interviewed, but did not name any of these firms as an important customer or partner in innovation. In a second case, a firm sold its computer numerical control (CNC) software to a local manufacturer, but distributed the software through an out-of-region machining hardware supplier. The regional connection was merely accidental, and the interviewee only knew about it through a social relationship. When asked about cross-sectoral relationships, firms often did not know in which way these could be important. The answers received were usually quite generic, as firms mentioned that a diversified economic base would provide a diversified labour market. However, nothing specific was mentioned about this – suggesting that such linkages were not very common.[3] The absence of noticeable cross-sectoral linkages clearly indicated that implicit claims about regional spillovers from regional IT spin-offs/start-ups to traditional manufacturing firms may be premature.

With respect to the conditions for regional innovation, we found that most firms were stand-alone units in the regional economy with strong international customer linkages, particularly to the United States. They had little ongoing research collaborations with R&D laboratories or regional universities, with the exception of those firms that had a hardware-related component to their product offering. Despite the lack of strong regional relationships, IT spin-off/start-up firms appeared to be clearly embedded in the community structure. The University of Waterloo provided important skill flows to the regional firms in the form of qualified graduates, but these were generic skill flows that did not directly strengthen innovative capabilities (Brady 2004). As Bill Gates of Microsoft emphasized during a visit to the region: "Most years, we hire more students out of Waterloo than any university in the world, typically 50 or even more" (CTV.ca 2005). In contrast to these labour market effects, our study indicated that limited specific knowledge was transferred to the region by entrepreneurial faculty members and graduates. Over time, existing university spin-offs/start-ups seemingly entered a stage of incremental innovation, with few strong R&D relationships to the university (or the region) persisting.

Traditional Manufacturing Firms in the Region

In the second phase of our research, we interviewed forty firms in traditional manufacturing industries in the areas of plastic and rubber products, fabricated metal products, machinery, electrical equipment, and transportation equipment (Bathelt et al. 2012). We were particularly interested in looking at the state of innovation activities in this sector and the potential for cross-sectoral exchanges of knowledge that could stimulate further innovation.

The firms interviewed differed widely in their role in the design process of the products that they fabricated or manufactured. Some firms primarily performed generic treatments, such as heat treating or painting, to the products. Similarly, contract fabrication shops, which did limited runs of machining or CNC manufacturing from a design provided by their clients, often had little involvement in the development of the products they produced. They provided feedback to their customers when parts became problematic to manufacture, but had little influence on the actual product design. Some of these firms, however, developed internal capacities to take the designs and provide input to their customers from a manufacturing standpoint. Over time, some firms became increasingly involved in early-stage design processes of the end products. Furthermore, we encountered firms in the upper tiers of the manufacturing value chain that both designed and manufactured products. Overall, we were surprised that two thirds of the firms were quite innovative: one third had developed new products and another third introduced new processes in the two years preceding our interviews.

Firms in traditional manufacturing differed substantially in terms of supplier-customer relationships and the kinds of knowledge-transfer-based innovative activities they engaged in, depending on where they were positioned in the value chain. On the one hand, firms with little knowledge of the end use of the products were rarely engaged in intensive interaction with their customers. On the other hand, firms with a stronger involvement in the actual design of the products, even in the case of short-contract fabricators, had the opportunity to gather more information about the operations of their customers and the kinds of capabilities needed. They used this input to develop specific innovation capabilities over time, often trying to extend their competence from manufacturing to research, development, and design.

Altogether, we identified two distinct cohorts of firms in our sample that were active in innovation. One cohort primarily engaged in custom manufacturing, converting raw materials into finished components using other firms' designs. They worked from a blueprint and gave feedback primarily on the manufacturability of the parts, rather than their end function. For these firms, most of their innovation came from cost-cutting measures and improving workflow and material management to reduce production delays. Another source of innovations were capital investments in new equipment to increase production capabilities and enhance automation to save on labour costs. The second cohort generally used their own designs to fabricate products, even if their outputs fed into other firms' product chains. Some of these firms were very innovative, with long-term research strategies in the development of new products. This was dependent on close interaction with users that were often not located nearby.

With an opportunity to observe the different cohorts of firms, we expected to see that the foundations of the underlying innovation networks would be found in regional value chains. However, it turned out that the firms interviewed usually did not identify either regional suppliers or customers as important to their innovation processes. Nonetheless, local/regional linkages through purchases of supplies and sales of products were higher than in the segment of IT start-ups/spin-offs. Almost 40 per cent of the firms purchased at least some raw materials (≥10% of overall supplies) from within the wider region (14 of 37; table 9.2), but their decision to do so was primarily based on logistical concerns and price factors, and not on innovation inputs. Compared to the IT firms, a higher share of traditional manufacturers had a primary regional-supply orientation (≥50% of overall supplies). In terms of sales linkages, half of the traditional manufacturers had at least some significant local sales and about one fifth were primarily locally oriented (table 9.3). Given that we interviewed numerous tier-2 and tier-3 manufacturers, as well as fabricators and metal treaters that would typically try to access the local market and provide customized jobs, this orientation towards local sales appeared low rather than high. In fact, a substantially higher share of firms was mostly reliant on customers in the United States or in Europe: almost 40 per cent of the firms had primarily international sales, which clearly had an impact on innovation linkages.

Interviewees overwhelmingly indicated that ideas for new products were developed from within their firms, sometimes through corporate ties connecting facilities in different countries. Most firms did not have important regional suppliers or customers that they collaborated with in innovation; thus, it was not surprising that these firms relied on internal problem solving. The firms interviewed also did not use consultants or have close relationships with research laboratories. When they ran into problems in their production processes, how they responded varied depending on the type of firm. More than half of the traditional manufacturing firms interviewed were metal fabrication shops. When they encountered problems, these were typically related to product designs and challenges in manufacturing. In these cases, firms typically solved problems in-house, but sometimes turned to their machine suppliers to help them improve manufacturing performance. A second group of firms provided coating and heat-treating services. The firms relied on particular processes and were limited in the number of sources they could consult to solve problems and innovate. Typically, they innovated by adding new capacity, in either volume or process rate. Original equipment manufacturers (OEMs) were the third group of firms focusing on in-house problem solving without engaging intensively in inter-firm innovation networks. Some were branch plants of larger multinational corporations and consulted with other corporate units around the globe to see how to solve problems in innovation. For the metal fabricators and treaters, the driving force in innovation was the customers' demand to lower costs or boost turnaround times. For the OEMs, the expectation of customers was related to improved product performance or design and feature enhancements. Regardless of the firm type, the customers were seen as the main push behind innovation, and often the key source for design input. Although sales in traditional manufacturing were not as internationally oriented as in the IT sector, such linkages were critical to the success of these firms.

As in the case of IT start-ups/spin-offs, we did not find strong evidence for cross-sectoral linkages in innovation. Most commonly, the firms indicated that they benefited from a joint labour pool in the region and the ability to draw skilled employees from firms in other industries using similar processes. But the flow of employees between firms was not seen as a significant input for innovation. A couple of firms indicated that they benefited from having a diverse pool of suppliers in the region, but suggested that the benefits came from the ease of having access to them, rather than from specific innovation inputs. Only one firm made explicit reference to business or management benefits, suggesting that, since so many firms in the region exported around the world, discussions with agents in unrelated sectors were useful in solving export challenges and selecting appropriate business strategies. Almost all answers appeared vague, however, indicating that cross-sectoral linkages were not common.

Generally, the results of the second phase of this research on traditional manufacturing paralleled the key observations about regional innovation processes from the first phase. Again, we did not see substantial patterns of local supplier-customer linkages that were important to innovation processes. Firms relied on international linkages, often with partners in the United States, and used regional suppliers primarily for generic business services, labour, and raw materials. Regional industry organizations, inter-firm labour flows, and the local universities and community colleges were only occasionally mentioned as being significant for innovation. Most firms indicated that they had little employee turnover. And many participated in the regional high school and community college apprenticeship programs, but had no distinct regional innovation networks otherwise.

Conclusions

The transformation that has taken place in the region around Cambridge, Guelph, Kitchener, and Waterloo from an economy based on traditional manufacturing to one with a substantial proportion of IT-related businesses is often attributed to knowledge transfers and growth triggers based on university spin-off processes and related networks. This would suggest that firms are regionally linked through cluster-like relations or other forms of inter-firm networks in innovation. In contrast to this expectation, our research shows that local firms are not closely related to one another in their technological and knowledge base. As a consequence, the possibilities for local networking and knowledge flows between firms are limited. More often, firms engage in international linkages to provide the necessary growth impulses, both within corporate networks and through inter-firm linkages. Therefore, restructuring successes in our case-study region are primarily due to individual-firm competencies, rather than being the consequence of collective action.

In addition, we found less than a handful of examples of some sort of relationship between traditional manufacturers in the region and IT firms, and where they existed the firms did not indicate that they were relevant for their innovation processes. Of course, this should not be taken to suggest that there are no cases of cross-industry linkages in the region, nor do we mean to suggest that such relationships have not been significant for some firms outside our sample; but our results show that cross-industry linkages between value chains are rare and that their influence on regional innovation processes appears small or minimal.

This is a different story of Canada's Technology Triangle and the University of Waterloo from that portrayed in the media and conveyed, in part, through academic publications. Successful regional restructuring and modernization appears primarily as a result of individual strengths, shared generic knowledge assets, and a strong sense of community in marketing the region's attributes, rather than being the effect of collective endeavours in innovation or networking. Although the fragmented regional economy may experience strong growth in some future periods based on diversification advantages, it may underperform due to a lack of economic cohesion and few collective synergies in innovation in other periods.

What we find is that knowledge about the economic success and the social foundations of regional innovation still appears incomplete. Considering that the University of Waterloo is often portrayed as one of the key examples of spin-off and start-up processes (Bathelt 1991a; Xu 2003; Bramwell and Wolfe 2008; Bathelt et al. 2010), we might have to lower our expectations regarding regional technology transfer through university–industry interaction. It might be more precise to conclude that the region's success is primarily based on weak ties and generic knowledge, rather than on strong, formalized ties that hold together the fabric of innovation. The role of local universities as a source of spin-off/start-up firms or as partners in leading-edge innovation appears overstated and does not, in itself, explain the successful modernization path.

At the same time, our study shows that traditional manufacturing firms are not necessarily less resilient in the face of economic crises than new creative or high-technology industries. Firms in the former industries can also be innovative and flexibly adjust to new market situations if they engage in ongoing innovation activities involving incremental improvements and adjustments, the acquisition of new resources, and related diversification and renewal processes. While the traditional manufacturing sector has proved to be more innovative than expected, this is not related to strong regional producer–user interaction, local innovation networks, university-industry collaboration, or cross-sectoral triggers.

There is little evidence suggesting that collective capabilities in innovation are key to the successes of these firms. Our observations, therefore, do not fit neatly into the current debates over the comparative benefits of regional industrial specialization versus relatedness or diversity for innovation and growth in city-regions (Duranton and Puga 2001). Instead, our research suggests that other more generic factors are important, such as a highly skilled labour market, successful corporate "role models" and community leaders, or a strategically important location within the southern Ontario transportation networks and markets; but these factors also do not fully explain strong economic growth and successful modernization in the region around Cambridge, Guelph, Kitchener, and Waterloo. Rather, it appears that regional suc cess primarily relies on individual endeavours of firms, internal corporate networks, and firm-specific competencies, as well as intensive linkages with the United States and other foreign markets – all of this within a diversified regional economy and labour market, beyond a certain minimum-threshold size. In conclusion, trans-regional connections appear to be the most significant linkages for innovation processes in the region. And neither Jacobean nor Marshall-Arrow-Romer models of regional diversity seem to describe adequately the role of global inputs in local innovation processes. Furthermore, despite being located in a medium-sized city-region, agents generate global linkages on their own (through their own networks), rather than relying on Toronto as a gateway to global financial and knowledge regimes, as could be expected from the global-city hypothesis (Sassen 2001).

What does this mean from a regional policy perspective? First, it will be important in the future to support firms in the region around Cambridge, Guelph, Kitchener, and Waterloo in maintaining strong transregional pipelines through corporate or inter-firm linkages that are key to innovation and economic success. Second, it is nonetheless important to provide platforms for regional technology transfer and knowledge exchange. This could potentially strengthen the adaptability to and robustness against future economic crises by helping local agents evaluate alternatives and variations to existing problem solutions. Third, we cannot expect that cross-sectoral networks will automatically form or play a role in innovation. Such linkages would have a greater likelihood to be useful if carefully orchestrated by regional policymakers or industry associations in areas where technological complementarities exist.

ACKNOWLEDGMENTS

This chapter is based on research presented in Bathelt et al. (2010, 2011, 2012). We wish to thank Meric Gertler, Dieter Kogler, Ben Spigel, David Wolfe, and the ISRN members for their support and critical inputs that helped shape this book chapter.

NOTES

1 The IT industry was selected partially because of its prominence in the region and the expectation that it could act as a platform technology (Cooke 2007) with the potential to induce innovations in other industries and create linkages between different sectors.
2 In the first phase in 2007/8, we conducted semi-structured interviews with 18 IT start-up/spin-off firms from the University of Waterloo. The total number of IT start-ups/spin-offs identified in the region was 42 (Bathelt et al. 2010). Of these, 32 firms were contacted, resulting in 14 rejections (44%) and 18 interviews. In the second phase in 2008/9, another 40 firms in traditional manufacturing industries were interviewed. Of a total of 642 traditional manufacturers identified in the selected industries (see also table 9.1), 310 were contacted and asked to participate in our study. Of these, 270 rejected, while 40 agreed to, a personal interview. The unusually high rejection rate of 87% is indicative of the difficulties we encountered in engaging firms in this research in the midst of a severe economic-financial crisis. Additionally, we conducted 8 interviews with university technology-transfer officers, economic developers of the cities, and leading representatives of business organizations. The latter interviews were conducted for the purpose of triangulation and getting an overview of the overall start-up policies and innovation dynamics in the region (Bathelt et al. 2011).
3 Interviewees were also not aware of specific cross-sectoral linkages between other regional firms.

REFERENCES

Amin, Ash, and Patrick Cohendet. 1999. "Learning and adaptation in decentralized business networks." *Environment and Planning D: Society & Space* 17 (1): 87–104. http://dx.doi.org/10.1068/d170087.
Audretsch, David V., Erik E. Lehmann, and Susanne Warning. 2005. "University spillovers and new firm location." *Research Policy* 34 (7): 1113–22. http://dx.doi.org/10.1016/j.respol.2005.05.009.
Bathelt, Harald. 1991a. "Employment changes and input-output linkages in key technology industries: A comparative analysis." *Regional Studies* 25 (1): 31–43. http://dx.doi.org/10.1080/00343409112331346267.

Bathelt, Harald. 1991b. *Schlüsseltechnologie-Industrien: Standortverhalten und Einfluß auf den regionalen Strukturwandel in den USA und in Kanada* (Key technology industries: Location behaviour and impact on regional change in the USA and Canada). Berlin, Heidelberg, New York: Springer.

Bathelt, Harald, and Alfred Hecht. 1990. "Key technology industries in the Waterloo region: Canada's Technology Triangle (CTT)." *Canadian Geographer* 34 (3): 225–34. http://dx.doi.org/10.1111/j.1541-0064.1990.tb01081.x.

Bathelt, Harald, Dieter F. Kogler, and Andrew K. Munro. 2010. "A knowledge-based typology of university spin-offs in the context of regional economic development." *Technovation* 30 (9–10): 519–32. http://dx.doi.org/10.1016/j.technovation.2010.04.003.

Bathelt, Harald, Dieter F. Kogler, and Andrew K. Munro. 2011. "Social foundations of regional innovation and the role of university spin-offs." *Industry and Innovation* 18 (5): 461–86. http://dx.doi.org/10.1080/13662716.2011.583462.

Bathelt, Harald, Anders Malmberg, and Peter Maskell. 2004. "Clusters and knowledge: Local buzz, global pipelines and the process of knowledge creation." *Progress in Human Geography* 28 (1): 31–56. http://dx.doi.org/10.1191/0309132504ph469oa.

Bathelt, Harald, Andrew K. Munro, and Ben Spigel. 2013. "Challenges of transformation: Innovation, re-bundling and traditional manufacturing in Canada's Technology Triangle." *Regional Studies* 47 (7): 1111–30.

Beaudry, Catherine, and Andrea Schiffauerova. 2009. "Who's right, Marshall or Jacobs? The localization versus urbanization debate." *Research Policy* 38 (2): 318–37. http://dx.doi.org/10.1016/j.respol.2008.11.010.

Bercovitz, Janet, and Maryann P. Feldman. 2006. "Entrepreneurial universities and technology transfer: A conceptual framework for understanding knowledge-based economic development." *Journal of Technology Transfer* 31 (1): 175–88. http://dx.doi.org/10.1007/s10961-005-5029-z.

BMO Capital Markets. 2008. *Canada's Technology Triangle:Waterloo Region and Guelph*. Toronto: Bank of Montreal.

Boschma, Ron, and Koen Frenken. 2011. "Technology relatedness and regional branching." In *Beyond territory: Dynamic geographies of knowledge creation, diffusion, and innovation*, ed. Harald Bathelt, Maryann P. Feldman, and Dieter F. Kogler, 64–81. London, New York: Routledge.

Brady, Diane. 2004. "RIM's Lazaridis." Interview with Mike Lazaridis. *Business Week Online*, http://www.businessweek.com/stories/2004-04-18/online-extra-rims-lazaridis-were-not-a-startup.

Bramwell, Allison, Jennifer Nelles, and David A. Wolfe. 2008. "Knowledge, innovation and institutions: Global and local dimensions of the ICT cluster in Waterloo, Canada." *Regional Studies* 42 (1): 101–16. http://dx.doi.org/10.1080/00343400701543231.

Bramwell, Allison, and David A. Wolfe. 2008. "Universities and regional economic development: The entrepreneurial University of Waterloo." *Research Policy* 37 (8): 1175–87. http://dx.doi.org/10.1016/j.respol.2008.04.016.

Callan, Benedicte. 2001. "Generating spin-offs: Evidence from the OECD." *Science Technology Industry Review* 26: 13–56.

Chesbrough, Henry W. 2003. *Open innovation: The new imperative for creating and profiting from technology.* Boston: Harvard Business School Press.

Clarysse, Bart, Mike Wright, Andy Lockett, Els van de Velde, and Ajay Vohora. 2005. "Spinning out new ventures: A typology of incubation strategies from European research institutions." *Journal of Business Venturing* 20 (2): 183–216. http://dx.doi.org/10.1016/j.jbusvent.2003.12.004.

Colapinto, Cinzia. 2007. "A way to foster innovation: A venture capital district from Silicon Valley and Route 128 to Waterloo region." *International Review of Economics* 54 (3): 319–43. http://dx.doi.org/10.1007/s12232-007-0018-1.

Communitech. 2006. *State of the industry: Technology in the Waterloo region.* Report. Waterloo.

Cooke, Phil. 2007. "To construct regional advantage from innovation systems first build policy platforms." *European Planning Studies* 15 (2): 179–94. http://dx.doi.org/10.1080/09654310601078671.

Cooke, Phil, and Kevin Morgan. 1998. *The associational economy.* Oxford: Oxford University Press. http://dx.doi.org/10.1093/acprof:oso/9780198290186.001.0001.

Cooper, Arnold C. 1971. "Spin-offs and technical entrepreneurship." *IEEE Transactions on Engineering Management* 18: 2–6.

CTV.ca. 2005. "Bill Gates draws a crowd at Waterloo University." http://www.ctv.ca/servlet/ArticleNews/story/CTVNews/20051013/billgates_waterloo_20051013/20051013?hub=Canada.

de Jong, Mark W. 1987. *New economic activities and regional dynamics.* Nederlandse Geografische Studies no. 38. Amsterdam.

Duranton, Gilles, and Diego Puga. 2001. "Nursery cities: Urban diversity, process innovation, and the life cycle of products." *American Economic Review* 91 (5): 1454–77. http://dx.doi.org/10.1257/aer.91.5.1454.

English, John, and Kenneth McLaughlin. 1983. *Kitchener: An illustrated history.* Waterloo: Wilfrid Laurier University.

Etzkowitz, Henry, Andrew C. Webster, Christiane Gebhardt, and Branca R.C. Terra. 2000. "The future of the university and the university of the future: Evolution of ivory tower to entrepreneurial paradigm." *Research Policy* 29 (2): 313–30. http://dx.doi.org/10.1016/S0048-7333(99)00069-4.

Florida, Richard, and Roger Martin. 2009. *Ontario in the creative age.* Consulting report. Toronto: Martin Prosperity Institute.

Frenken, Koen, and Ron A. Boschma. 2007. "A theoretical framework for evolutionary economic geography: Industrial dynamics and urban growth as a branching process." *Journal of Economic Geography* 7 (5): 635–49. http://dx.doi.org/10.1093/jeg/lbm018.

Gertler, Meric S. 2004. *Manufacturing culture: The institutional geography of industrial practice.* Oxford: Oxford University Press.

Gordon, Ian R., and Philip McCann. 2000. "Industrial clusters: Complexes, agglomeration and/or social networks." *Urban Studies* (Edinburgh, Scotland) 37 (3): 513–32. http://dx.doi.org/10.1080/0042098002096.

Hodgson, Geoffrey. 1988. *Economics and institutions: A manifesto for a modern institutional economics.* Cambridge: Polity Press.

Holmes, John, Todd Rutherford, and Susan Fitzgibbon. 2005. "Innovation in the automotive tool, die and mould industry: A case study of the Windsor-Essex region." In *Global networks and local linkages: The paradox of cluster development in an open economy,* ed. David A. Wolfe and Matthew Lucas, 119–54. Kingston, Ontario: McGill-Queen's University Press.

Landström, Hans. 2005. *Pioneers in entrepreneurship and small business research.* New York: Springer. http://dx.doi.org/10.1007/b102095.

Lundvall, Bengt-Åke. 1988. "Innovation as an interactive process: From producer–user interaction to the national system of innovation." In *Technical change and economic theory,* ed. Giovanni Dosi, Christopher Freeman, Richard R. Nelson, Gerald Silverberg, and Luc L.G. Soete, 349–69. London, New York: Pinter.

Lundvall, Bengt-Åke, and Björn Johnson. 1994. "The learning economy." *Journal of Industry Studies* 1 (2): 23–42. http://dx.doi.org/10.1080/13662719400000002.

MacKinnon, Danny, Andrew Cumbers, Andy Pike, Kean Birch, and Robert McMaster. 2009. "Evolution in economic geography: Institutions, political economy, and adaptation." *Economic Geography* 85 (2): 129–50. http://dx.doi.org/10.1111/j.1944-8287.2009.01017.x.

Malecki, Edward J. 1991. *Technology and economic development: The dynamics of local, regional, and national change.* Burnt Mill: Longman.

Malmberg, Anders, and Peter Maskell. 2002. "The elusive concept of localization economies: Towards a knowledge-based theory of spatial clustering." *Environment & Planning A* 34 (3): 429–49. http://dx.doi.org/10.1068/a3457.

Maskell, Peter, and Mark Lorenzen. 2004. "The cluster as market organisation." *Urban Studies* (Edinburgh, Scotland) 41 (5–6): 991–1009. http://dx.doi.org/10.1080/00420980410001675878.

Morgan, Kevin. 2004. "The exaggerated death of geography: Learning, proximity and territorial innovation systems." *Journal of Economic Geography* 4 (1): 3–21. http://dx.doi.org/10.1093/jeg/4.1.3.

Niosi, Jorge. 2000. *Canada's national system of innovation.* Montreal, Kingston: McGill-Queen's University Press.

Nooteboom, Bart. 2000. *Learning and innovation in organizations and economies.* Oxford: Oxford University Press.

Oinas, Päivi, and Edward J. Malecki. 1999. "Spatial innovation systems." In *Making connections: Technological learning and regional economic change,* ed. Edward J. Malecki and Päivi Oinas, 7–33. Aldershot: Ashgate.

Owen-Smith, Jason, and Walter W. Powell. 2004. "Knowledge networks as channels and conduits: The effects of spillovers in the Boston biotechnology community." *Organization Science* 15 (1): 5–21. http://dx.doi.org/10.1287/orsc.1030.0054.

Parker, Paul. 2001. "Local-global partnerships for high-tech development: Integrating top-down and bottom-up models." *Economic Development Quarterly* 15 (2): 149–67. http://dx.doi.org/10.1177/089124240101500204.

Perry, Ann. 2009. "Meet our post-industrial Waterloo." *Toronto Star*, 21 March: B2.

Pinch, Steven, Nick Henry, Mark Jenkins, and Stephen Tallmann. 2003. "From 'industrial districts' to 'knowledge clusters': A model of knowledge dissemination and competitive advantage in industrial agglomerations." *Journal of Economic Geography* 3 (4): 373–88. http://dx.doi.org/10.1093/jeg/lbg019.

Porter, Michael E. 1990. *The competitive advantage of nations*. New York: Free Press.

Porter, Michael E. 2000. "Locations, clusters, and company strategy." In *The Oxford handbook of economic geography*, ed. Gordon L. Clark, Maryann P. Feldman, and Meric S. Gertler, 253–74. Oxford: Oxford University Press.

Preissl, Brigitte, and Laura Solimene. 2003. *The dynamics of clusters and innovation*. Heidelberg, New York: Physica. http://dx.doi.org/10.1007/978-3-642-50011-4.

Roberts, Edward B. 1968. "Entrepreneurship and technology: A basic study of innovators; how to keep and capitalize on their talents." *Research Management* 11: 249–66.

Rothaermel, Frank T., Shanti D. Agung, and Lin Jiang. 2007. "University entrepreneurship: A taxonomy of the literature." *Industrial and Corporate Change* 16 (4): 691–791. http://dx.doi.org/10.1093/icc/dtm023.

Rutherford, Todd, and John Holmes. 2007. "'We simply have to do that stuff for our survival': Labour, firm innovation and cluster governance in the Canadian automotive parts industry." *Antipode* 39 (1): 194–221. http://dx.doi.org/10.1111/j.1467-8330.2007.00512.x.

Rutherford, Todd, and John Holmes. 2008. "Engineering networks: University-industry networks in southern Ontario automotive industry clusters." *Cambridge Journal of Regions, Economy and Society* 1 (2): 247–64. http://dx.doi.org/10.1093/cjres/rsn001.

Sassen, Saskia. 2001. *The global city: New York, London, Tokyo*. Princeton, NJ: Princeton University Press.

Saxenian, Annalee. 1985. "The genesis of Silicon Valley." In *Silicon landscapes*, ed. Peter Hall and Ann R. Markusen, 20–34. Boston: Allen and Unwin.

Scott, Allen J. 2006. "Spatial and organizational patterns of labor markets in industrial clusters: The case of Hollywood." In *Clusters and regional development: Critical reflections and explorations*, ed. Bjørn Asheim, Phil Cooke, and Ron Martin, 236–54. London, New York: Routledge.

Setterfield, Mark. 1993. "A model of institutional hysteresis." *Journal of Economic Issues* 27: 755–74.

Statistics Canada. 2001. *Industry – 1997 North American Industry Classification System: Class of worker and sex for labour force 15 years and over*. Catalogue no. 97F0012XCB01009. Ottawa: Statistics Canada.

Statistics Canada. 2006a. *Canadian Business Patterns, 1998–2005*. Catalogue no. 61F0040XCB. Ottawa: Statistics Canada.

Statistics Canada. 2006b. *Industry – 2002 North American Industry Classification System: Class of worker and sex for labour force 15 years and over*. Catalogue no. 97–559–XCB2006009. Ottawa: Statistics Canada.

Statistics Canada. 2008. *Canadian business patterns, June 2008: Establishment counts by CMA, industry sector (NAICS 2007, 2-digit) & employment category (Number of employees)*. Ottawa: Statistics Canada.

Statistics Canada. 2010. *Labour force historical review*. Catalogue no. 71F0004XVB. Ottawa: Statistics Canada.

Storper, Michael, and Anthony Venables. 2004. "Buzz: Face-to-face contact and the urban economy." *Journal of Economic Geography* 4: 351–70.

Storper, Michael, and Richard Walker. 1989. *The capitalist imperative: Territory, technology, and industrial growth*. New York, Oxford: Basil Blackwell.

Vinodrai, Tara. 2011. "The dynamics of economic change in Canadian cities: Innovation, culture, and the emergence of a knowledge-based economy." In *Canadian cities in transition: New directions in the twenty-first century*, 4th ed., ed. Trudi Bunting, Pierre Fillion, and Ryan Walker, 87–109. Toronto: Oxford University Press Canada.

Visser, Evert-Jan, and Oedzge Atzema. 2008. "With or without clusters: Facilitating innovation through a differentiated and combined network approach." *European Planning Studies* 16 (9): 1169–88. http://dx.doi.org/10.1080/09654310802401573.

Vohora, A., M. Wright, and A. Lockett. 2004. "Critical junctures in the development of university high-tech spinout companies." *Research Policy* 33 (1): 147–75.

Wahl, A. 2007. "Innovation station." *Canadian Business*, 8 October. http://www.canadianbusiness.com/article/16312--innovation-station.

Wolfe, David A. 2009. *21st century cities in Canada: The geography of innovation*. The 2009 CIBC Scholar-in-Residence Lecture. Ottawa: Conference Board of Canada.

Xu, S.X. 2003. "Knowledge transfer, inter-firm networking and collective learning in high technology cluster evolution: A network analysis of Canada's Technology Triangle." Master's thesis, University of Waterloo, Waterloo.

10 Knowledge Flows in the Consulting, Advertising/Design, and Music Sectors in Halifax

JILL L. GRANT

Halifax Regional Municipality – alternatively called Halifax or HRM – is the largest city-region in Atlantic Canada, with a population of around 360,000. While its population and income growth rates lag those of dynamic larger centres such as Toronto, Calgary, Vancouver, and Ottawa (Filion 2010), Halifax has outpaced other cities in Atlantic Canada. The Conference Board of Canada identified Halifax as one of nine hub cities driving the economy in the country: Halifax is the only hub that serves a region – Atlantic Canada – rather than a province (Lefebvre and Brender 2006). The city has benefited from its capacity to attract migrants from across Canada, but Filion (2010) suggests that its low rate of international immigration may render it vulnerable to shrinkage over time.

Like other Canadian municipalities, Halifax has been seeking to establish its credentials as a creative city (Florida 2002, 2005; Landry 2008) worthy of investment and talent migration (Gertler and Vinodrai 2004; HRM 2005). Halifax joined other municipalities in recent years in jumping enthusiastically onto the creative cities bandwagon in an effort to develop a climate that encourages local innovation and that woos talented workers from near and far. Branding itself early on as a "smart city," Halifax has experienced some important wins over the last decade in attracting high-tech firms like Research in Motion to open new facilities in the city-region (Grant, Holme, and Pettman 2008).

The Halifax economy has proved relatively diverse, with six industrial clusters identified in 2006 (see table 10.1). With major groupings of universities, hospitals, government services, federal research institutions, and technology firms, Halifax is the centre of innovation in Atlantic Canada and has an economy that has proved fairly resilient in the face of recent market downturns. Post-secondary institutions and the

Table 10.1 Industrial clusters in Halifax, 2001 and 2006

	Location quotient, 2001	Location quotient, 2006	Percentage growth, 2001–2006
Maritime	1.59	2.02	26.6
Biomedical	[1.06]*	1.14	25.9
ICT services	1.46	1.47	6.2
Business services	1.49	1.36	8.2
Creative & cultural	[1.00]*	1.14	15.3
Higher education	1.75	1.57	14.8
Logistics	1.20	[1.06]*	−5.7

Source: Spencer and Vinodrai 2006, 2009
* Note: Not qualifying as a cluster in the year indicated.

military draw a steady stream of young adults to the city (Grant and Kronstal 2010). The city-region's traditional strength in Maritime industries continued to grow in the early 2000s at the same time as new high-technology and creative clusters emerged. Halifax might well claim to be transitioning to what Scott (2007) calls cognitive-cultural capitalism: that is, an economy which generates clusters of high-technology, service, and cultural production activities bifurcated into high wage and low wage segments.

Contemporary thinking about economic performance suggests that innovation within city-regions depends on exchanges of information that allow creative people to collaborate as they develop and use knowledge (Wolfe 2009). Traditional models of innovation recognized the advantages of large and dense clusters of related industries as facilitating thick labour markets to attract talented workers and new businesses, and concentrated industrial and social networks to enable knowledge sharing within sectors (Wolfe, this volume). Jacobs (1969) offered an alternative view that argued that knowledge sharing between diverse sectors can generate innovation. These positions have generated debate within innovation studies on the relative benefits of specialization versus diversification (Beaudry and Schiffauerova 2009; Duranton and Puga 2000; Wolfe 2009). The contrasting perspectives help to frame the research questions about the sources of innovation and the nature of knowledge flows which guide the data reported here.

In this chapter we discuss selected components of the Halifax economy to illuminate some of the patterns of knowledge flows that may contribute to regional innovation and economic growth. Given the relatively small scale of Halifax, the strong and inter-connecting social and professional networks (Grant, Holme, and Pettman 2008), and the diversity of industrial clusters, we would require significant resources to investigate knowledge flows between and among all clusters. Consequently, to narrow the study we decided to focus on three groups which play important roles within the cognitive-cultural economy and which our previous research in Halifax had identified as innovative in the city-region.

As Scott (2000, 2006, 2007) notes, contemporary capitalism depends increasingly on creative industries to fuel consumption. Because the new cognitive-cultural economy rewards cultural and economic innovation, cities and businesses are pushed in the direction of promoting creativity (Landry 2008; Scott 2007). Halifax enjoys a reputation of being a city with a strong appreciation of arts and culture (Grant and Kronstal 2010). As cities increasingly recognize the potential contribution of cultural workers to economic development (Markusen and Schrock 2006; Wojan et al. 2007), understanding the ways in which knowledge flows might link the super-creatives into other components of the local innovation system becomes important. We therefore interviewed musicians and music intermediaries: an increasingly important creative industry in Halifax. While the total number of musicians in the city-region remains relatively small, Halifax has a dynamic music scene which pulls creative workers from across the country and beyond (Grant et al. 2009a).

We also decided to investigate two sectors which straddle creative industries and business services: advertising and design firms combine extraordinary levels of creativity along with business acumen to help link producers with consumers; built environment consultants (architects, landscape architects, engineers, and planners) offer creative design services to shape the built environment for cities anxious to attract talented workers. Both types of firms deliver business services within the region, although in the contemporary economic context they may range farther afield to find innovative projects and to serve clients no longer bound to hire locally. As Grabher (2002) noted in his study of London, UK, the project ecology within which many advertising firms now work involves collaboration between an array of colleagues, organizations, and clients across space: advertising engages cultural products such as music and works of art that contribute to understandings of time and place (Tota 2001). To what extent do such networks link talented workers in Halifax to those in other cities? Do the built environment consultants in Halifax contribute as knowledge brokers for continuous innovation (Hargadon 1998); do they involve highly mobile talent pools (Kloosterman 2008); or do they prove limited in their ability to contribute to innovation within the region (Alam 2003; Erbil and Akinciturk 2010)?

Although the sectors profiled here do not reflect the considerable diversity within the Halifax economy, they represent the types of activities which might be predicted to benefit significantly from knowledge flows and cultural products between sectors. In the end, however, we found that although the industries investigated rely on knowledge flows for their success, the Halifax interviews revealed relatively few concrete examples of knowledge flows across industries within the city-region: the sources of innovation in these sectors may depend more on talent flows, inter-regional project ecologies, and the social dynamics of the work environment.

Does the research in Halifax suggest that the city-region is heading towards a cognitive-cultural economy? The results are equivocal. While the cognitive and creative sectors are increasing in strength in Halifax and contributing to innovation and growth, their capacity to become dominant in the economy of the region remains unproven. Respondents pointed to the importance of the cultural sector in making Halifax a destination of choice for talented workers, but we found that knowledge flows between cultural sectors and business services remain relatively undeveloped. Advertising and design firms are drawing on other sectors, such as music, for cultural product inputs, and built environment consultants sometimes include public art in their designs, but the local economy appears to provide few mechanisms for mutual learning or exchange between sectors. As a relatively small city on the rim of a national network of cities, Halifax faces challenges to achieving the level of innovation which may be necessary to stimulate growth.

We begin by briefly summarizing the nature of the Halifax economy before discussing knowledge flows in the sectors profiled. In the final section of the chapter we consider the relationship between knowledge flows, innovation, and the cognitive-cultural economy.

Overview of the Halifax Economy

Halifax is a mid-sized city that serves as the hub of government services and economic growth in Atlantic Canada (Lefebvre and Brender 2006). The city-region differs from similar-sized cities in central Canada

Table 10.2 Some industry characteristics in Halifax, 2006

	No. in labour force	% of labour force	Average full-time income
23 Construction	11,590	5.5	$43,687
31–33 Manufacturing	11,015	5.2	$52,069
44–45 Retail trade	25,045	11.9	$34,778
61 Educational services	16,355	7.8	$51,708
62 Health care and social services	24,485	11.7	$46,462
72 Accommodation and food service	14,750	7.0	$25,189
91 Public administration	23,375	11.1	$59,645
All industries in Halifax	210,135	100	$48,092

Source: Spencer and Vinodrai 2009, 8

in that its manufacturing base is a modest component of the economy: this may have helped Halifax weather the consequences of trade agreements and globalization better than cities dependent on industrial production.

Originally settled in 1749, Halifax owes its history and character to its excellent port on the Atlantic Ocean. The port provides multi-modal service for eleven container lines (Port of Halifax 2009). While shipping volumes have decreased in the last decade, maritime services remain an important cornerstone of the local economy. The harbour plays a major role in the tourism sector in the city as well: the port welcomed its two millionth cruise-ship visitor in the summer of 2009.

Since its early days the city served as a centre of government and military activities. As the provincial capital, Halifax is simultaneously the site of several federal government departments for the Atlantic region. Canadian Forces Base Halifax employs over 14,000 personnel. As table 10.2 indicates, over 24,000 people work in health care services at the many hospitals in the city. With six universities and several colleges, the city has a large work force in educational services. Halifax hosts the largest shopping centres in Atlantic Canada and has become a major destination for tourism and hospitality activities.

Between 2001 and 2006 the city's population grew 3.8 per cent while the country as a whole expanded by 5.4 per cent (Spencer and Vinodrai 2009). As table 10.3 shows, Halifax has a well-educated population

Table 10.3 Halifax population characteristics

Characteristic	Halifax 2001	Canada 2001	Halifax 2006	Canada 2006
Population	359,185	30,007,085	372,855	31,612,890
% foreign born	6.8	18.2	7.4	19.8
% BA or higher	21.1	15.4	24.0	18.1
Employment rate (%)	63.0	61.5	64.5	62.4
% in creative occupations	37.4	29.2	38.0	33.2
% science tech occupations	7.1	6.4	7.0	6.6
% employment in clusters	21.7	22.1	19.0	22.1
Average employment income	$30,614	$31,757	$48,092	$51,221

Source: Spencer and Vinodrai 2006, 2009

more likely than the Canadian average to be engaged in a creative occupation. On diversity measures, though, Halifax remains below average. While almost one in five Canadians immigrated to the country, over 92 per cent of Haligonians were born in Canada: the city lacks the ethnic diversity of larger centres like Toronto and Vancouver. Incomes are lower than the Canadian average, but residents of Halifax are slightly more likely to be employed than other Canadians. Although many of Halifax's indicators have seemed relatively good for the last decade, Filion (2010) cautions that contemporary political, economic, and demographic processes may generate long-term challenges for mid-sized and smaller cities which struggle to attract international immigrants to drive growth.

As of the 2001 census, Halifax had five industrial clusters (see table 10.1): maritime, ICT services, business services, higher education, and logistics (Spencer and Vinodrai 2006). Clusters involve concentrations of specialized industries and workers co-located in a city-region at a scale above the national average. Associated with economic prosperity and growth, clusters are defined by a combination of elements that include specialization in employment, co-location of industries, critical mass, and breadth of industries (Spencer et al. 2010). The largest cluster in Halifax in 2001 was associated with higher education: Halifax has six universities and several community college campuses, making it a national centre for higher education. The smallest cluster in 2001 lost ground over the next five years: by 2006 logistics had declined in size (shedding 675 jobs to total 10,890) and despite its role in the economy no longer qualified as a cluster (Spencer and Vinodrai 2009). Two new clusters had appeared: biomedical and creative/cultural. While some Canadian cities and the nation as a whole lost industrial clusters over the five-year period, Halifax added a cluster. The maritime and bio-medical clusters grew more than 25 per cent in the five-year period between censuses, while creative/cultural and higher education components grew about 15 per cent. Employment in the biomedical industry increased by 445 persons to 2165, while the creative/cultural sector added 3615 jobs to reach 9140 people. These changes suggest that Halifax's economy was strengthening in some creative and scientific sectors over the period. The next section describes our interviews with people in some of the expanding creative industries and firms which we might see as central to the development of a cognitive-cultural economy as Scott (2007) describes it.

Methods of the Halifax Study

In order to understand the sources of innovation and the nature of knowledge flows in the Halifax economy we conducted interviews with participants in three kinds of industries in the summer of 2008. We were interested to discover whether knowledge flows were occurring across some key sectors and clusters in Halifax, and also in the importance of global versus local knowledge flows. Given their potential contribution to the cognitive-cultural economy, we focused on business services and cultural-creative sectors: built environment consulting (the architects, engineers, landscape architects, and planners who design the city), advertising and design firms, and the music sector.

Halifax is the regional centre for engineering, architecture, and planning consulting firms firms, and has some companies with a national profile. Since local sources suggested that some Halifax advertising and design firms were winning international awards for their creative products and innovative work, we added them. The final sector we examined was the music industry: earlier interviews with government and association representatives had identified the critical importance of the music sector to attracting and retaining talented workers to many industries in Halifax. We conducted a total of thirty interviews with thirty-one respondents to examine knowledge flows. Participants were fairly evenly distributed across the three industries profiled, with slightly more in the music category.

We summarize the responses from the interviews under several themes. First we report on knowledge flows, noting the patterns found in different sectors, and commenting on the variability. Then we consider the relationship between knowledge flows and innovation in the Halifax context.

Knowledge Flows in Built Environment Consulting

Halifax has built environment consulting firms that range from small (six or fewer employees) independent companies to large (150 or more employees) national consulting firms. Some are full-service firms with engineering, planning, landscape architecture, and architectural staff, while others concentrate on particular services. Acquisitions and mergers prove common in the industry. For instance, during late 2008 the large Western Canadian firm Stantec acquired Jacques Whitford Environmental, the largest Halifax-based consultancy, which had grown over two decades to become a national leader in the field, with offices in several cities and gross revenues of $230 million (Halifax Herald 2009). Expansion outside the region helps to establish a higher profile for firms, and opportunities for talented workers within the firms, but can be quite costly both in economic and in personal terms. For instance, when asked what other city would be more advantageous for his firm, one respondent said, "I think that if we wanted to really be bigger and have a further reach, then it would be Toronto, but there's such a price that you pay to move to Toronto" (Interview 1, Consulting).

Respondents indicated that they watch what their competition is doing in the region. They read each other's work, and sometimes borrow or build on good ideas. At times they may collaborate on projects or subcontract work to other firms that have special expertise or capacity. They participate in local professional networks that offer opportunities for exchanging and building knowledge.

Some firms have established particular niches: for instance, consulting for rural communities. Such firms may have less need to follow trends in other companies because they face limited competition for the specialized local work they do. Establishing local knowledge of place and people is important in consulting practice. If firms are not familiar with the local context in other places, then they look to partners with firms based there. As one consultant said, "If we're not local to an area, we would never walk into that area without partnering with somebody else" (Interview 4, Consulting).

Networking between firms is more important to some firms than to others. While some firms mostly work on their own, others regularly form alliances. One participant explained: "I think it's the nature of the industry – yes, we have relationships with other firms. They're pretty fluid and depend on work that we want to pursue and are able to do" (Interview 5, Consulting). Some firms are strategic in partnering with other firms, doing work alone when they can, but looking for temporary linkages where those prove advantageous. As one respondent noted, alliances are pragmatic: "We compete hard and we collaborate hard. Very often we compete and collaborate with the same firms ... We will collaborate with people who are major competitors if that's the way we both can serve a client or can best get the work" (Interview 14, Consulting).

In general, the consultants acknowledged that they are selling similar services to their competitors, so having good relationships with potential clients is essential to success. Most firms participate in local social and business networks and organizations. Networking activities provide a chance not only to learn about upcoming projects but also to build good social relationships that help to secure work.

A few firms have good connections with the universities to take advantage of the latest research, but most respondents did not. A small number have relationships with providers of special technologies, but for most firms opportunities to market those seem limited. In general, the respondents did not describe themselves as conducting research and facilitating innovation nor were they commissioning work from those who may be engaged in such activities. Respondents in consulting indicated that they ensured that staff attend conferences and read journals to try to stay current in the profession. Some worked within large companies that can draw on talent from other locations and that provide special training opportunities for their staff. Respondents described the Internet as an important source for research for firms of all sizes. We found limited evidence of cross-sectoral knowledge flows with other industries other than government and developers.

Most of the consultants focused their work within the Atlantic region and sought limited networking links beyond that. Some professionals with specialized expertise and national reach participated in professional networks beyond the region to maintain or upgrade their specialized knowledge: this provided opportunities to introduce professional innovations into the region. Some firms indicated that the city-region provided few opportunities to encourage design innovation or creativity and reported that they sought projects outside Atlantic Canada to find the creative challenges their employees craved. Although local governments and developers look to consultants in the built environment sector for the latest ideas about how to build communities, consultants in the sector saw their scope for innovation as limited by a relatively conservative economic and political climate in the city-region. While Vancouver and Toronto were sometimes mentioned as places of design innovation associated with the creative economy, respondents sometimes described Halifax as trailing the pack.

Knowledge Flows in the Advertising and Design Industry

Firms in the advertising and design industry vary from one person working at home to large national or international agencies with seventy or more employees in Halifax. Some of the local agencies had won important international awards for their work and served international as well as local clients. Respondents from advertising businesses presented themselves as extremely competitive. Those interviewed suggested that junior people shift employers with some frequency, whereas senior staff members seem more settled. Young staff may transfer knowledge from organization to organization as they learn the ropes, but more experienced staff members appeared less likely to share knowledge across the industry. Rather than saying that they seek to learn from their competitors, those in advertising firms contrasted their practices with other companies in the industry. Some explicitly criticized elements of the approaches that competitors used.

Because of the nature of the work, ideas in the design field become part of the public domain once they are used. Innovation is integral to the creative work done in the sector, but cannot easily be protected by patent or copyright. Accordingly, the creativity of the people in a firm is critical to its continued success. Companies use high salaries and good working conditions to try to retain talented workers and prevent them from being stolen by the competition.

Those interviewed from smaller graphic and communication design firms approached their work collaboratively, and talked about learning from each other. Most of them concentrated on the regional market, but one did website design for national and international markets. The smaller companies were committed to holding on to their employees, and used different hiring strategies to find staff than did larger firms: for instance, one respondent talked about seeking employees from outside the sector to avoid the usual strategy of stealing workers from competitors.

Although the agencies did not generally consider collaborating with each other, respondents did see some advantages in operating in a mid-sized city with competitors. As one respondent noted, clusters provide talented workers with options:

> Let's compare our agency to a competitor in Sackville, New Brunswick. The biggest challenge for the one in Sackville is if you're a talented person you want to be around a variety of firms that could take advantage of you, and not be in a town where there is only one. That is for career management, but you also want to be around like people. So there is a community effect in Halifax, which is making agencies here stronger and making those outside Halifax weaker. We are a bit of a magnetic pull for the talented people. This serves our firm very well. (Interview 2, Advertising)

Some advertising design companies indicated that at one time they needed to locate in the same community as their clients to win accounts. Twenty years ago, one said, it was advantageous to be in Fredericton if your client was McCain's, but that no longer is true. Still, a respondent from another firm thought proximity remained relevant: "I mean, we could potentially do all the work we're doing here from almost anywhere, so we're here because we believe that being closer to your clients is a good thing" (Interview 17, Advertising). Being in Halifax offered local advantages.

Respondents valued local network relationships highly, as one said: "A small business like this needs healthy good ongoing relationships [with clients]" (Interview 8, Advertising/design). The firms that worked locally seemed most interested in participating in local networking organizations like the chamber of commerce, while firms competing internationally took a dimmer view of the usefulness of such engagement.

Some firms had developed patented processes for addressing client needs, although such innovations were not offered by many respondents. More commonly, respondents highlighted the importance of face-to-face time with clients to gain a deep understanding of the needs and opportunities for an advertising or design campaign. Some saw the building of networks and client relationships as a long-term prospect, while others described it as fluidly related to particular jobs or projects. For those who looked for long-term relationships with clients, being located in the same place as their clients was helpful. For those committed to intense creativity in their products, the quality of the living and working environment proved central to their decision to live and work in Halifax.

Although parochialism may be diminishing (to the advantage of city-regions like Halifax), some respondents suggested that Toronto corporate head offices remain inclined to hire Toronto ad agencies rather than ones from Halifax.

> When you go to Toronto to pitch business they are inclined to think that the most talented agencies are right there in Toronto. That is a big disadvantage [for us] … The brand of Halifax as a centre of innovation, creativity, arts, culture, technology, the knowledge-based industries [… well,] the reputation is understated in Central Canada. They do not understand how good we are. (Interview 2, Advertising)

Several respondents in advertising talked about the role of research and of the utility of the Internet for gaining information and knowledge from other industries in other places that could be useful in their work. The industry works with communication technologies that often extend far beyond the local community. Larger firms have supply chains (subcontractors, etc.) located in other parts of Canada or beyond. Since many firms are working outside of Halifax and outside the region, travel is a mandatory part of the job. For some, gaining recognition through international design awards and trade shows has become an important sales tool and affirmation of their global reach and innovative capacity.

Those in advertising seemed the most likely of those interviewed to make connections with and to use knowledge from other sectors. For instance, they drew on the arts and on music for themes and content for some of their products. They valued research on marketing, consumer behaviour, and social psychology to inform their campaigns. Even in this industry, however, the knowledge flows appeared to occur primarily in one direction: being absorbed into ads rather than feeding back into other elements of the local economy.

The advertising and design industry in Halifax has firms that take divergent approaches to the social dynamics of their work and to the creative context within which they innovate. Older, more-established firms rely on developed social networks with clients and with the local business community; they engage creative talent from outside the region, but focus their work primarily within the Atlantic region. Newer firms engage less commonly with traditional business networks, but participate in global networks of suppliers and consumers; they establish different types of work environments to encourage and reward creativity and look to a global audience to affirm their capacity as innovators. The results suggest a bifurcation within the local industry between the innovative firms committed to playing in the global market, which necessitates and rewards knowledge sharing within international networks, and established firms socially embedded within regional networks wedded to traditional (conservative) approaches.

Knowledge Flows in the Music Sector

Halifax has a well-known indie musical scene that attracts musicians from across the country to play in its live-music venues (Grant et al. 2009a; Morton 2008). At the same time, though, the music *industry* may be undeveloped and under-served, with relatively few music managers, promoters, and agents (most of whom remain centred in larger cities like Toronto). Changes in the music industry that have undermined the viability of record sales have transformed the economic prospects for musicians. In the past, musicians had to move to major centres to become known and sign record deals. Prospects for becoming wealthy from a hit song have diminished in recent years, so that the industry is rapidly shifting gears and geography. Today, technology gives musicians different options of where and how to work. The Internet and social networking sites like Facebook and MySpace have become strategic venues for artists to reach and build audiences through downloads and touring. Halifax has benefited from the transformation while becoming known for its warm and nurturing musical community (Hracs et al. 2011).

As a medium-sized city, Halifax is a locale that lets musicians easily stay in touch with each other. People see each other's shows. They bump into each other on the street or in restaurants. They ask each other for help and favours. They play on each other's recordings or gigs. Knowledge flows fluidly between those with experience and newcomers to the scene. Respondent after respondent talked about the mutual support and encouragement that permeates the community. Individual musicians cross between music genres, often playing in several bands or helping each other out, as this respondent explained:

All you have to do is take twelve CDs and read the credits and you won't believe you are looking at the same thirty names over and over and over again. But they are playing different instruments, they are singing different things. It is different kinds of music. It's bizarre, but it's wonderful. It's what makes this industry unique and saleable from my perspective … It becomes fascinating as a sales feature: part of our brand. We work it from all angles, that this is a community. We're selling a community. (Interview 20, Music sector)

The primary source of innovation in the music industry is individual musicians working within a social context that encourages and rewards creative performance in somewhat unpredictable patterns. In the music sector technological and social innovation in the way that products are sold has changed the dynamics of the way in which products are produced (Leyshon 2009). Adjusting to the effects of music downloading has affected knowledge flows in dramatic ways. In Halifax the music industry adapted as musicians developed a collaborative atmosphere and an independent approach to producing and marketing music. The camaraderie in the Halifax music scene generates an intensely innovative atmosphere that draws potential musicians to the city to engage in face-to-face contact and creates a cultural "buzz" about the place (see Storper and Venables 2004).

Respondents talked about the important role that the local universities and colleges play in providing audiences and potential musical talent. Many musicians first come to Halifax to attend university or college. The Nova Scotia College of Art and Design (NSCAD University) has been an especially significant breeding ground for creative musicians to form bands and to try out their talents; several respondents also mentioned Dalhousie University as a place where bands germinated. The large numbers of students in the city create the core of the audiences that attend live venues where musicians have an opportunity to hone their craft (Grant and Kronstal 2010).

Public agencies play vital roles in facilitating the music industry and connecting it to other sectors. CBC Radio has provided a valued service in scouting and honing talent, giving many local artists a chance to produce work professionally and find an audience. Music Nova Scotia offers musicians vital opportunities for business training and funding: respondents suggested that its programs are unique in the country and have given local artists a significant edge in developing their talent. Many respondents spoke of the unique interest of the province under the leadership of former premier Rodney MacDonald (a fiddler) in supporting the music industry through new policy and funding initiatives.

While Halifax is a vital incubator of music, the city is too small a market to sustain many full-time musicians or music managers. To make a living in the business, musicians have to find other audiences: that usually requires touring. Those who hope to make it big in the industry may move to larger markets like Toronto or Montreal. Respondents often indicated that they find other ways to supplement their playing income by teaching, or by working in music promotion or management; some take supplemental jobs in other sectors. Some musicians have worked with other media sectors (film, television, and radio) and a few have seen songs used in national or international spots. Many musicians living in Halifax have committed to staying, and try to prove that it is possible to be successful while living in the city-region.

The Internet, email, and air travel are key technologies that facilitate the linkages that permit musicians to live in Halifax but tour elsewhere, and that allow music promoters and managers to work from the region. Through technology they can find mechanisms for distributing, promoting, and selling work, and they can arrange tours. Some artists rely on management services from Toronto because so few local agents and managers are available in Halifax. Some musicians have become successful in Europe through their Internet activities; they find it convenient to go on tour for several months a year while making Halifax their creative base. The changing technologies of music distribution have altered the dynamics of the industry in ways that favour smaller cities like Halifax, where the costs of living are reasonably low while the quality of life and the density of opportunities and connections in the music community are relatively high (Grant et al. 2009a). The small scale and collaborative character of the local scene fosters innovation and creative engagement in songwriting and performance that continues to attract musicians to the city-region.

Knowledge Flows, Innovation, and Economic Performance

While all sectors studied relied on knowledge flows for their economic success, we found that respondents reported relatively few traditional markers of innovation. For instance, while some engineering consultancy firms had established patents on processes or had innovated around a service, most respondents could offer few specific examples in response to inquiries. Some firms had diversified by establishing divisions to offer new services, but this often entailed bringing previous suppliers or subcontractors into the organization. The greatest degree of innovation appears to be occurring in the music industry as its workers struggle to create a new economic model after the collapse of the record/CD market.

Respondents revealed dense and overlapping social and professional networks that contributed to building trust, collaboration, and knowledge sharing in many sectors. They confirmed a mix of local and non-local ties and a range of economic actors participating in social and professional networks. However, they offered little evidence that linked economic performance to local knowledge circulation between industries or clusters. The basic hypothesis of the study, linking economic performance with knowledge flows between sectors, found no support within these particular industries in Halifax. We saw scant indication of knowledge flows across sectors except for advertisers' and designers' interest in music and the arts and their social connections with cultural-sector workers. We heard repeatedly that the vibrant arts and music scene in Halifax contributed to the city's appeal to talented and creative people, although respondents reported few concrete examples of knowledge exchanges. Rather than drawing on the economic strengths and innovative potential of creative industries, many local producers and firms appeared content to continue to treat cultural elements as attractive backdrops to everyday life.

Interviews indicated that participants in the sectors studied engaged in local and international knowledge flows within their industries, although in different ways (Grant, Haggett, and Morton 2009). Frequent and informal knowledge sharing and mutual learning proved especially important within the Halifax music scene, where musicians worked together to challenge and motivate each other (Morton 2008); musicians engaged with an international community that contributes to innovation in the sector. Respondents in the advertising industry indicated relatively little interest in sharing knowledge with local colleagues outside their own firms: they looked to clients for direction, to other cultural sectors for inspiration, and to firms in other cities for collaboration. In the built-environment consulting sector firms often competed with each other for projects; however, they said that when they believe it improves their odds of winning contracts, they collaborate with other firms and share information. Workers moving among companies transfer knowledge and skills. Larger or more complex projects generated partnerships with well-known firms in larger cities: such collaborations contribute global knowledge to the local scene.

The success of Halifax in attracting talented people challenges some of the hypotheses of the broader project. Anticipated flows across sectors which might reasonably have been expected to draw from each other were not found. Instead of innovation depending on knowledge flows among diverse organizations (as Jacobs 1969 predicted), innovation in the sectors studied depended on the creativity of individual actors and their collaborations within the industry: in some sectors (e.g., music) those collaborations were largely local, while for some firms in other sectors (e.g., advertising) they were regional or international. Despite the diversity of the economy as measured by numbers of industrial clusters, respondents described the local context as conservative and inhibiting innovation.

Given the results from the sectors profiled, Halifax's relative economic success seems almost counter-intuitive. Yet Halifax has become something of a niche destination for young people. The city-region benefits from its coastal setting, affordable housing costs, laid-back lifestyle, friendly people, and rich arts and culture environment: perhaps, as Glaeser and Gottlieb (2006) suggest, Halifax is among those cities which attract talent and growth because of consumption opportunities. Strong universities and colleges contribute to innovation and creativity in the region, and to the social ambience of the city. Talented and creative workers looking to avoid the hectic pace and competitive life of larger urban centres like Toronto and Vancouver see Halifax as an attractive option. Migrants recognize Halifax's economic weaknesses (compared to the strength of larger city-regions), but appreciate the comfortable scale and socially supportive workplaces they find in the city-region. As the regional hub, Halifax has many of the cultural and educational benefits of a larger city. Its vibrant creative sectors make it attractive to an educated population that need not depend on an especially robust economy to thrive. Across the sectors, respondents reported they could earn higher salaries if they lived elsewhere, but they made lifestyle and work-style choices in coming to and staying in Halifax. Thus, Halifax's appeal differs from that of other hub cities in Canada: while talented workers seeking diversity stimulate growth in the large cities of Toronto and Vancouver, and those following entrepreneurial ambitions head to Calgary, Halifax attracts Canadians making unique lifestyle and work-style choices. The city's recent economic success in some ways seems to defy the prognosis of contemporary theory which might predict decline for a region unable to lure immigrants (Filion 2010) and with a political climate many see as inherently conservative (Grant and Kronstal 2010).

Did we find evidence of a new cognitive-cultural economy emerging in Halifax that links cultural industries, design industries, and higher-order business services? Certainly the interviews suggest that Haligonians see the potential for the growth cognitive-cultural industries in the city. These sectors are gaining strength relative to traditional resource-based and manufacturing industries. To date, however, the links between business services and cultural industries remain relatively undeveloped.

We chose to profile music and advertising in part because Haligonians indicated that these were sectors where Halifax was experiencing international success. Does such success indicate that the city-region is developing a cognitive-cultural economy? The ability of some Halifax-based advertising companies to win international commissions and awards has garnered recognition for the city-region and helped the firms attract talented workers who might otherwise have located in larger city-regions. Rapid growth in some of these firms brought creative workers from larger centres that traditionally were the hubs of the industry and thick labour markets for workers in the field. Since technological innovations via the Internet make design and creative work possible anywhere, Halifax may have cost and natural beauty advantages that other places lack. Some companies that began in Halifax have stuck with their roots and have not relocated despite achieving international success. Respondents in advertising suggested, however, that the shortage of educational programs in copy writing and art direction limited the local talent pool and constrained growth of the industry in Halifax. While important to the city, the success of a few advertising and design firms is not sufficient to indicate the imminent arrival and explosion of a cognitive-cultural economy. Scale does matter for the growth of some industries because some creative workers may find the lure of larger scenes irresistible. The small scale of the city-region facilitates serendipitous social and business connections among actors in different industries that may lead to opportunities for linkages, but some respondents showed little inclination to take advantage of this potential. The extent to which exploiting such linkages could stimulate economic development and population growth in a smaller city on the periphery of the nation remains untested. Participants in the incipient cognitive-cultural economy in Halifax may have different priorities and values than their counterparts in larger city-regions. Lewis and Donald (2010) noted that in its emphasis on urban size and diversity the creative cities discourse marginalized smaller cities and set them up for failure. Similarly, Scott (2007) describes the cognitive-cultural economy as based in large city-regions. While cognitive-cultural capitalism will undoubtedly develop more quickly and fully in large centres, smaller cities may well incubate their own variants to respond to local sensibilities and priorities.

The small size of the regional market inhibits the growth of the music sector, which along with the arts is commonly cited as the pride of the city-region. Although musicians increasingly rely on the internet and touring to earn a living they need local venues and affordable housing to maintain a viable base in Halifax. Halifax has established a national reputation as a centre of popular music (Brooks et al. 2009; Grant et al. 2009a), but industry infrastructure remains rudimentary and funding for musicians is precarious. A cognitive-cultural economy cannot thrive without appropriate resource inputs and long term commitment from government. If a cognitive-cultural economy exacerbates the growing gap between elites and the disadvantaged in society it may undermine its own foundations: as Scott (2007: 1474) notes, the impacts of this shift in capitalism are far from benign.

While the research revealed some signs of an emerging cognitive-cultural economy in Halifax, local authorities have done relatively little to acknowledge such potential change, to invest in its growth, or to consider its implications. The city-region's cultural plan is a first step in that direction (HRM 2006), but meagre resources for investment in the cultural sector limits opportunities for growth. Current government policy focuses on issues such as infrastructure investment, construction, and regional development. While these contribute to long-term growth, local and provincial authorities who continue to sideline other important issues of social and cultural development may find they cannot protect Halifax's long-term economic performance and competitiveness.

The Halifax case indicates some problems with generalizing about innovation and growth in ways that may apply to smaller city-regions. Theories developed to explain processes in large urban regions may not account for conditions in smaller city-regions: as Lewis and Donald (2010) remind us, the residents of smaller cities may value different urban attributes than do the inhabitants of large centres. Social and inter-organizational processes that may facilitate innovation in particular industries (see, e.g., Bramwell et al. 2008; Wolfe and Bramwell 2008) may be missing in others. While the methods of our study may not have extended far enough to uncover knowledge flows between industries or may not have made appropriate connections between innovation and economic performance, we nevertheless conclude with reasonable confidence that Halifax has enjoyed its recent prosperity and appeal despite falling short on some of the key expectations of the creative cities discourse.

REFERENCES

Alam, Ian. 2003. "Commercial innovations from consulting engineering firms: An empirical exploration of a novel source of new product ideas." *Journal of Product Innovation Management* 20 (4): 300–13. http://dx.doi.org/10.1111/1540-5885.00027.

Beaudry, Catherine, and Andrea Schiffauerova. 2009. "Who's right, Marshall or Jacobs? The localization versus urbanization debate." *Research Policy* 38 (2): 318–37. http://dx.doi.org/10.1016/j.respol.2008.11.010.

Bramwell, Allison, Jen Nelles, and David A. Wolfe. 2008. "Knowledge, innovation and institutions: Global and local dimensions of the ICT cluster in Waterloo, Canada." *Regional Studies* 42 (1): 101–16. http://dx.doi.org/10.1080/00343400701543231.

Brooks, Caryn, Peter Cooper, Michael Corcoran, Sue Carter Flinn, Jamie O'Meara, and Chris Powell. 2009. "Cities of song: Get into the groove in these hot cities for music lovers." *CAA Magazine*, Summer: 26–33.

Duranton, Gilles, and Diego Puga. 2000. "Diversity and specialisation in cities: Why, where and when does it matter?" *Urban Studies* (Edinburgh, Scotland) 37 (3): 533–55. http://dx.doi.org/10.1080/0042098002104.

Erbil, Yasemin, and Nilufer Akinciturk. 2010. "An exploratory study of innovation diffusion in architecture firms." *Scientific Research and Essays* 5 (11): 1392–401.

Filion, Pierre. 2010. "Growth and decline in the Canadian urban system: The impact of emerging economic, policy and demographic trends." *GeoJournal* 75 (6): 517–38. http://dx.doi.org/10.1007/s10708-009-9275-8.

Florida, Richard. 2002. *The rise of the creative class, and how it's transforming work, leisure, community, and everyday life*. New York: Basic Books.

Florida, Richard. 2005. *Cities and the creative class*. New York: Routledge.

Gertler, Meric, and Tara Vinodrai. 2004. *Competing on creativity: Focus on Halifax*. A report prepared for the Greater Halifax Partnership.

Glaeser, Edward, and Joshua D. Gottlieb. 2006. "Urban resurgence and the consumer city." *Urban Studies* (Edinburgh, Scotland) 43 (8): 1275–99. http://dx.doi.org/10.1080/00420980600775683.

Grabher, Gernot. 2002. "The project ecology of advertising: Tasks, talents and teams." *Regional Studies* 36 (3): 245–62. http://dx.doi.org/10.1080/00343400220122052.

Grant, Jill L., Jeff Haggett, and Jesse Morton. 2009. *The Halifax sound: Live music and the economic development of Halifax*. http://www.utoronto.ca/onris/newsletter/2009_10/pdf/Grant09_HalifxSound.pdf.

Grant, Jill L., Robyn Holme, and Aaron Pettman. 2008. "Global theory and local practice in planning in Halifax: The Seaport redevelopment." *Planning Practice and Research* 23 (4): 517–32. http://dx.doi.org/10.1080/02697450802522848.

Grant, Jill, and Karin Kronstal. 2010. "The social dynamics of attracting talent in Halifax." *Canadian Geographer* 54 (3): 347–65. http://dx.doi.org/10.1111/j.1541-0064.2010.00310.x.

Halifax Herald. 2009. "Shareholders back Jacques Whitford sale." *Chronicle-Herald / Mail-Star*, 3 January: C1.

Hargadon, Andrew B. 1998. "Firms as knowledge brokers: Lessons in pursuing continuous innovation." *California Management Review* 40 (3): 209–27. http://dx.doi.org/10.2307/41165951.

Hracs, Brian, Jill L. Grant, Jeff Haggett, and Jesse Morton. 2011. "Tales of two scenes: Civic capital and retaining musical talent in Toronto and Halifax." *Canadian Geographer* 55 (3): 365–82. http://dx.doi.org/10.1111/j.1541-0064.2011.00364.x.

HRM (Halifax Regional Municipality). 2005. *Strategies for success: Halifax Regional Municipality's economic development strategy 2005–2010*. http://www.halifax.ca/EconomicStrategy/EconomicStrategy.html.

HRM (Halifax Regional Municipality). 2006. *HRM cultural plan*. http://www.halifax.ca/culturalplan/documents/CulturalPlan112007.pdf.

Jacobs, Jane. 1969. *The economy of cities*. New York: Random House.

Kloosterman, Robert C. 2008. "Walls and bridges: knowledge spillover between 'superdutch' architectural firms." *Journal of Economic Geography* 8 (4): 545–63. http://dx.doi.org/10.1093/jeg/lbn010.

Landry, Charles. 2008. *The creative city: A toolkit for urban innovators*. London: Comedia Earthscan.

Lefebvre, Mario, and Natalie Brender. 2006. *Canada's hub cities: A driving force of the national economy* (Ottawa, Conference Board of Canada). http://www.conferenceboard.ca/e-library/abstract.aspx?did=1730.

Lewis, Nathaniel M., and Betsy Donald. 2010. "A new rubric for 'creative city' potential in Canada's smaller cities." *Urban Studies* (Edinburgh, Scotland) 47 (1): 29–54. http://dx.doi.org/10.1177/0042098009346867.

Leyshon, Andrew. 2009. "The software slump? Digital music, the democratisation of technology, and the decline of the recording studio sector within the musical economy." *Environment & Planning A* 41 (6): 1309–31. http://dx.doi.org/10.1068/a40352.

Markusen, Ann, and Greg Schrock. 2006. "The artistic dividend: Urban artistic specialisation and economic development implications." *Urban Studies*

(Edinburgh, Scotland) 43 (10): 1661–86. http://dx.doi.org/10.1080/00420980600888478.

Morton, Jesse. 2008. "Why do I love this town? The music industry in Halifax." Master's project, Dalhousie University, School of Planning.

Port of Halifax. 2009. Choose Halifax. http://www.portofhalifax.ca/.

Scott, Allen J. 2000. *The cultural economy of cities*. London: Sage Publications.

Scott, Allen J. 2006. "Creative cities: Conceptual issues and policy questions." *Journal of Urban Affairs* 28 (1): 1–17. http://dx.doi.org/10.1111/j.0735-2166.2006.00256.x.

Scott, Allen J. 2007. "Capitalism and urbanization in a new key? The cognitive-cultural dimension." *Social Forces* 85 (4): 1465–82. http://dx.doi.org/10.1353/sof.2007.0078.

Spencer, Greg, and Tara Vinodrai. 2006. "Innovation Systems Research Network city-region profile, 2001: Halifax." Toronto: Munk Centre, University of Toronto.

Spencer, Greg, and Tara Vinodrai. 2009. "Innovation Systems Research Network city-region profile, 2006: Halifax." Toronto: Munk Centre, University of Toronto.

Spencer, Greg, Tara Vinodrai, Meric Gertler, and David A. Wolfe. 2010. "Do clusters make a difference? Defining and assessing their economic performance." *Regional Studies* 44 (6): 697–715. http://dx.doi.org/10.1080/00343400903107736.

Storper, Michael, and Anthony J. Venables. 2004. "Buzz: Face-to-face contact and the urban economy." *Journal of Economic Geography* 4: 351–70.

Tota, Anna Lisa. 2001. "'When Orff meets Guinness': Music in advertising as a form of cultural hybrid." *Poetics* 29 (2): 109–23. http://dx.doi.org/10.1016/S0304-422X(01)00030-4.

Wojan, Timothy, Dayton M. Lambert, and David A. McGranahan. 2007. "Emoting with their feet: Bohemian attraction to creative milieu." *Journal of Economic Geography* 7 (6): 711–36. http://dx.doi.org/10.1093/jeg/lbm029.

Wolfe, David. 2009. *21st century cities in Canada: The geography of innovation*. Ottawa: Conference Board of Canada.

Wolfe, David A., and Allison Bramwell. 2008. "Innovation, creativity and governance: Social dynamics of economic performance in city-regions." *Innovation: Management, Policy & Practice* 10 (2–3): 170–82.

PART IV

Innovation for Survival or Growth in Canada's Small Cities

11 Social Dynamics, Diversity, and Physical Infrastructure in Creative, Innovative Communities: The Saskatoon Case

PETER W.B. PHILLIPS AND GRAEME WEBB

Context

City-regions are increasingly seen to be the engines of economic growth and innovativeness for the nation state. There is a strong debate about whether specialization or diversity is the key to success (e.g., Jacobs 1985 and Krugman 1998). One of the key questions in this debate is to what degree the social dynamics, economic, social, and cultural diversity, and physical infrastructure of a city-region contribute to its success in the knowledge economy. Saskatoon presents an interesting case study, as both its history and many of its economic and social dynamics have been assessed using a mixture of tools (e.g., Dobni and Phillips 2000; Phillips and Khachatourians 2001; Phillips 2002). This chapter examines Saskatoon as a creative and innovative city, reviews the recent history, lays out the competing theoretical orientations to innovation and creativity, and tests the theory against the evidence provided by the Saskatoon case.

Background

This is an interesting time to be examining the structures and dynamics in the Saskatoon region. Before the global debt crisis of 2008 and the fiscal and economic meltdown of 2008–10, the demand for most commodities had been growing strongly for a number of years, at least partly owing to economic strength in Asia. Saskatoon, as a centre significantly specialized in the mining and agri-food sectors, benefited from this buoyancy in commodity markets. Other regions of Canada which are far more diversified have been more seriously affected by the recent global downturn. The continued strength of key commodity markets (particularly oil and gas, potash, and uranium) and the sustained research enterprise at the University of Saskatchewan and in the local research park have allowed Saskatchewan to maintain a robust level of growth. New and potentially more profitable resource deposits have been delineated and an array of technologies has emerged that make local product markets more profitable. In recent years more than $1 billion of new investment has been directed to the university and related research facilities to take advantage of these new areas, while oil, heavy oil, potash, uranium, gold, and diamonds have all been targeted for further commercial development.

The result has been sustained growth in Saskatoon in what has proved to be a rather uncertain and tentative global economic recovery elsewhere. There are a number of proxies for that vibrancy. The city has bettered the national rate of growth in population (growing by more than 14 per cent in the past decade, compared with 11 per cent for Canada and only 3.8 per cent for Saskatchewan, and recording the fastest growth of any Canadian city in the twelve months ending 1 July 2010); employment growth (+24 per cent in 2000–9, compared with only 14 per cent for Canada as a whole and just marginally behind Calgary and Edmonton); and property-market price rises (more than doubling in the 2006–8 period and holding relatively steady between 2008 and 2011).

Opinions are also buoyant. Mario Lefebvre, director of the Centre for Municipal Studies at the Conference Board of Canada, in February 2009, labelled Saskatoon one of Canada's three "new economic powerhouses." He wrote:

> I have been keeping a close eye on the economic situation of Regina, Saskatoon and Winnipeg. And what is about to unfold this year is no luck. These three cities have been posting solid economic performances for years now ... These three economies have diversified significantly over the past 20 years and while agriculture still matters, it plays much less of a role in the overall economic picture ... The success of these three cities certainly is not a new story. The Conference Board recognized years ago that something special was taking place ... And now, the country as a whole is noticing. In fact, the economic success of these cities has turned them into a magnet for people over the past two years.

There is debate about what underlies the economic dynamism in Saskatoon. Some argue that it is simply a cyclical capital-stocking effect driven by commodity markets. Others assert the infrastructure boom is a tangible, visible sign of a creative and innovative community. Saskatoon offers an excellent case for studying the balance between these perspectives. As a small to medium-sized creative city (it was Canada's Creative Capital in 2006) with a strong innovation record (with world firsts in uranium and potash mining, farm machinery fabrication, crop varieties, and biotechnology) flowing from a few world-class industrial clusters (especially mining and biotechnology), Saskatoon has been a major beneficiary of investments in soft infrastructure (research networks and projects) and hard infrastructure (e.g., buildings, labs and big science projects) funded by federal, provincial, and municipal governments and agencies. Meanwhile, as a small Great Plains city, Saskatoon has until very recently had a predominantly white, Midwestern look and feel about it, which at least superficially suggests any success may be in spite of rather than because of economic, social, and cultural diversity.

Saskatoon occupies a unique place in the list of cities in Canada. While the city's census metropolitan population in 2010 of 265,000 suggests it has a number of natural comparator cities in Canada – Windsor, Victoria, Oshawa, at over 300,000 population, and Regina, Sherbrooke, and St John's, at under 215,000 population – it is largely in a class of its own. Saskatoon occupies the top of the central-place hierarchy in a market area of more than 600,000 (as a primary wholesale-retail centre), has no significant higher-order government and is relatively remote (minimum 600 km) from any city of competing or larger size.

Theory

A range of theories attempt to explain innovation and creativity. Early theory simply looked at the influence of investments in hard infrastructure on growth. More recent theory asserts that investments in soft infrastructure and organic growth of networks and relationships in a community may be as or more important than built infrastructure. This soft infrastructure is often posited to depend on tolerant and diverse economies and societies.

Economists posit that hard infrastructure (what they call gross fixed capital formation) and soft infrastructure (sometime called human capital) are at the core of the innovative and growing economy. Traditional growth theory (the Arrow-Debreau-Solow-Swan model) asserts that with any given ratio of labour to capital, adding more capital (in the form of buildings, machinery and equipment) would raise the marginal productivity of labour, which would bump up real wages and draw in more labour. Ultimately, more capital would create more jobs and raise both incomes and the quality of life. But, in this conceptualization, diminishing returns eventually will set in as successive capital investments add less labour productivity, because technologies are assumed to be exogenously derived.

The so-called new growth theory endogenized technological change. Just as land, labour, and capital can be drawn into the market through price signals in the traditional growth model, the Romer-Lucas endogenous growth model assumes that people will similarly be willing to invest to change the production processes and recipes that define technological capacity. The theory provides a robust explanation of how decisions to invent and innovate are linked to both capital investment and human capital (which is posited but not well articulated), but the theory does not identify the necessary or sufficient conditions for individuals to engage in this endogenous activity of investment in invention and innovation.

A number of contextual theories of growth have stepped into the breach. Marshall (1890) provides a clear explanation for why one might want to invest in people and infrastructure around districts, milieux, poles, or regions where there is an agglomeration of workers and firms engaged in competitive and collaborative activities. A century later, Porter's diamond theory of competitiveness (1990) simply updates Marshall's concept to the twentieth-century vernacular. Thus, the rule of thumb might be to invest in physical and human capital in areas that specialize.

One logical conclusion is that the key goal of government should be to sustain and support specialization (what is often called building on strengths), but there are both conceptual and practical arguments that any viable community requires some degree of diversity. Kauffman (1995) asserts that invention and innovation are largely about realizing adjacent potential opportunities – hence, the more diverse a community, the more adjacent possibilities abound. Practical experience similarly adds weight to the value of diversity – Canada abounds with predominantly single industry/firm towns and cities (e.g., many provincial capitals and mining towns such as Sept-Îles and Uranium City) that have suffered body blows as either their core sector has suffered economic hardship or resources have been exhausted. Diversified centres offer some protection for government, citizens, and firms that their sunken investments may have sustained value.

Similarly, there is a strand of innovation theory suggesting that beyond the micro and macro effects of diversity, there is a range of meso effects of diversity on investment in innovation. Florida (2002), for example, posits that creative, talented individuals are attracted to centres that have the appropriate mix of hard infrastructure (technology) and economic, social, and cultural diversity (tolerance) that can support creative ventures.

What is not clear from the theory, and much of the analysis to date, is how one might balance specialization and diversity and which types of investment in hard and soft infrastructure might yield the highest economic and social benefits.

Soft infrastructure has been addressed by Bourdieu (2006, 110), among others, who defined social capital as "the aggregate of the actual or potential resources which are linked to possession of a durable network of more or less institutionalized relationships of mutual acquaintance and recognition – or in other words, to membership in a group – which provides each of its members with the backing of the collectively-owned capital, a 'credential' which entitles them to credit, in a various sense of the word." In other words, social capital accrues from relationships between two or more individuals, in the form of a "capital of obligations," which can then be spent by an individual or the group as a whole to achieve a desired end.

One view is that innovation depends upon interactive social learning between economic agents and that this is socially organized in spatially proximate regions (Morgan 1997; Asheim and Gertler 2005). The development of local "untraded interdependencies" (Storper 1997) is at least partly related to both soft and hard infrastructure in such things as university projects, programs and buildings, research laboratories, networking milieux, and cross-sectoral and interdisciplinary research networks (Scott 2004). Which of these many and varied structures are necessary and which are sufficient to generate untraded interdependences that induce innovation cannot be derived from the theory alone.

Taken together, these theoretical threads offer a range of testable hypotheses. Specifically, the economic and creative performance of city-regions could depend on three key characteristics: the strength of local knowledge flows within individual industries/clusters, the strength of local knowledge flows between individual industries/clusters, and the strength of knowledge-based linkages between local and non-local economic actors. The working hypothesis is that the economic performance of city-regions depends on the density of local networks, the relative mix of local and non-local ties as well as the heterogeneity, diversity, and tolerance of economic actors belonging to these networks. In this context it is not clear whether specialization or diversification will be the most effective strategy. This chapter examines how these factors affect the nature of the innovation process in Saskatoon, and specifically investigates how physical and social infrastructure may contribute to economic performance of the city region.

Findings

Data from studies in Saskatoon over the past decade is used to test the three hypotheses and to assess the role of infrastructure in influencing economic performance in the city. In brief, the analysis of the data for Saskatoon suggests that codified knowledge flows through global pipelines while contextual, intangible knowledge is embedded in relatively self-contained sector-specific local labour and knowledge markets. There is little or no evidence that there are any significant links across the creative platforms; those domains tend to remain relatively distinct. The recent and planned large investments in soft and hard infrastructure in Saskatoon are clearly part of the story. None of the investments in hard infrastructure, however, is unambiguously sufficient for spurring innovation – soft infrastructure and some indeterminate amount of organically developed social capital would appear to be needed to enable those investments. In short, the "if you build it they will come" mantra that has driven some of the hard investments in Saskatoon and elsewhere is a structurally incomplete model.

The following sections address the local evidence related to the role of soft and hard infrastructure and how Saskatoon has balanced specialization and diversity.

Strength of Local Knowledge Flows within Individual Clusters or Industries

The first hypothesis is that economic performance of a community depends on networks to support and sustain local knowledge flows and that physical and knowledge infrastructure can foster the knowledge circulation that underlies those networks. Ultimately, this could involve a mix of people, networks, institutions, and infrastructure, but it is not clear exactly how diverse a mix is optimal.

One place is to start with firms. As part of a national research project centring on the social dynamics of innovation in the city-region, we surveyed twenty-four firms from recognized and established innovation clusters in Saskatoon – predominantly from the mining and agri-food sectors. Generally these firms credited their competitive advantage to superior innovation. Of the twenty-four participating firms, half expressed a belief that they had superior research and development or had created innovative products. Another four indicated that their international connections or their network of alliances and partnerships were the basis for their success. There was a strong connection between the sector and their responses. Eight of ten firms involved in agricultural and biotechnology research believed that the quality of their research and development was a competitive advantage for their firm. To some extent, every firm reported that they *collaborated with other players in that process*. For some, particularly those offering services like contract research, this collaboration simply consisted of providing feedback from customers that influenced business practices. For others, collaboration included large-scale projects executed with other firms and organizations. Firms spontaneously mentioned collaborations with a number of major infrastructure facilities, such as the University of Saskatchewan, the Saskatchewan Research Council, National Research Council Plant Biotechnology Institute (NRC-PBI), POS Pilot Plant, Agriculture and Agri-Food Canada (AAFC), the local office of the Industrial Research Assistance Program (IRAP), the Innovation Place Bioresource Centre, and the Vaccine and Infectious Diseases Organization (VIDO). When firms were asked to describe the motivation underlying their collaboration, the most common answer was the need for efficiency and cost cutting; other answers revolved around gaining access to knowledge, innovation, and expertise in order to stay at the cutting edge of advancements in science and technology. This community-wide result correlates with the results of earlier work specifically focused on exploring the canola sector (Phillips and Khachatourians 2001) and, more broadly, with previous research that concentrated on the dynamics within the biotechnology cluster, which identified many of the same institutions and key individuals within those institutions.

Ultimately, however, knowledge flows are about people and ideas. This local labour market exhibits significant mobility and networking. Of 390 canola-industry employees surveyed in 1998, a full 39 per cent had linkages to the university, and 7 per cent had prior employment experience with AAFC and NRC-PBI. The 1998 survey also illustrated the critical role of the local education system in cultivating local talent (Phillips and Khachatourians 2001). It should be noted that the local labour market becomes significantly more mobile with employees having graduate degrees. Less than half the employees with master's degrees and only about one quarter of the employees with doctorates were trained in Saskatchewan. Thus, the local labour market depends critically both on local educational networks and systems and on attracting key personnel from away; hence, specialization through the education and labour market is perhaps the base for the sector, but a diverse group of scholars, scientists, and entrepreneurs have been attracted to the sector and are core to many of the world firsts of this community.

Moving beyond the labour market, there is some evidence that the Saskatoon biotechnology cluster generates both tangible and intangible benefits from co-locating with competitors and collaborators. Survey results conducted as part of an analysis of the social dynamics of the biotechnology sector suggest that most companies were highly dependent on face-to-face relationships to keep track of opportunities and threats, many of which appear to have some local component. In one way, this networking function at least partly addresses the specialization of the community – in-coming and out-going connections are designed to expand the adjacent potentials available for the community.

Furthermore, a 1998 survey of about thirty firms undertaking R&D related to canola globally investigated the factors influencing their decisions to locate their research efforts and provided interesting insights about the benefits and limitations of co-location (Phillips and Khachatourians 2001). Half of all the respondents, representing the majority of private companies responding, acknowledged the importance of proximity to either collaborators or competitors, especially the NRC and AAFC and key research universities. However, research conducted in 2003–4 on the dynamics of the biotechnology cluster in Saskatoon revealed that almost two-thirds of the companies interviewed indicated that once they were generating product, the value of co-location diminished. While relationships with customers are important, co-location is not critical – 82 per cent stated that relocation to be near key customers would not be considered. There was more diversity of opinion about whether it was important to be located near their suppliers. About 45 per cent reported relying solely on non-local suppliers and another 32 per cent reported they use both local and non-local suppliers; more than half (55%) reported that it is not important to be close to suppliers. Lastly, it is interesting to note whom the companies viewed as competitors. Approximately half of the actors recognize their competitors as global, while 41 per cent reported only local competition. In contrast to ambivalence about co-location with suppliers or buyers, more than three-quarters of firms reported that there was real value in being located near competitors. One could conclude that many firms at the R&D and early commercialization stage see co-location as improving their odds of competing.

Procyshyn (2004) used the results from the 2003–4 surveys in the Saskatoon biotechnology cluster to examine the role of eight central actors in the cluster. Each central actor was asked to identify their relationships with each of ninety-four other institutions in Saskatoon, noting whether they were engaged in exchanges related to R&D, fee-for-service work, financial capital, human capital, or contextual knowledge. A social network map was created and density measures for the entire network and each of the five "functional" networks were developed. On average, there were eighty-nine linkages per core institution, spread across the five functions analysed. The overall network density across all functions was 15 per cent (712 linkages identified relative to an absolute possibility of 8930). As one would expect, the density varied across function, from a low of 1.4 per cent for financial exchanges and 2.3 per cent for R&D links (which is consistent with a predominantly pre-commercial research based community) to a high of 7.6 per cent for networking, which suggests a quite highly linked community for industrial policy and promotional efforts, but relatively weak networks for financial intermediation.

More recently, research carried out in 2008 as part of an investigation into the social dynamics of innovation at the city-region level tested how firms were connected to innovation clusters in Saskatoon. More than 80 per cent of the firms reported that they had introduced a new product, service, or process in the past three years – most considered their innovations to be "new to the world." Overwhelmingly, the development of these new products, services, and processes was the result of internal R&D in reaction to customer and market demand; only one firm reported developing a new product, service, or process through a direct collaboration with another firm, though many firms cited collaborations which indirectly contributed to the development of the product (e.g., reliability testing of equipment under operating conditions). Participating firms were then asked about the advantages they saw their firm having over competitors. Participants commonly gave multiple possible sources of their competitive advantage. Half of the firms expressed a belief that they had superior research and development capacity, while others cited customer service and responsiveness as their source of competitive advantage.

Most firms assert that their firm's competitive advantage lies with having the right people on board. This advantage was gained either by building a good team through hiring, by the exceptional skills possessed by the firm's founders, or by intensive training. Willingness to respond to customer demand and market assessment were also commonly cited, usually in relation to efforts to develop a customer-service focus. Some firms attributed their competitive advantages to an effort towards product and process development and the building of new capacities. Perhaps most revealing, when firms were asked about their collaborations related to research, development, and commercialization, a plurality of firms, largely from the biotech sector, reported that they usually sought help from academics. For many, though, these connections were informal and were often described as simply picking up a phone and calling an acquaintance at the University of Saskatchewan or elsewhere who was thought to be able to lend assistance. Only two firms reported that they have dedicated science boards to advise their activities (cf. Zucker, Darby, and Brewer 1998). Mining and software firms, by contrast, tended to turn to consultants, parent or sister companies, or other firms in their sector in those situations. One innovative firm (in the biotechnology field) reported that they post their problems on a website and offer rewards for possible solutions.

Firms were also asked about how their relationships with the local government and various trade or industry associations affect their operations. Respondents often had difficulty thinking of the different trade associations and government structures that their firm was involved with. The key ones mentioned were all public-private partnerships: the Saskatoon and Region Economic Development Authority (SREDA), the Saskatchewan Trade and Export Partnership (STEP), the local chamber of commerce, the Saskatchewan Advanced Technology Association (SATA), the ag-bio industry organization located in Saskatoon (Ag West Bio), and the Prospectors and Developers Association of Canada (PDAC). Overall, it was rare for a respondent to indicate that traditional trade associations or any level of government had a very important influence on their business. Even if a firm indicated it worked with one of these groups, it often was unable to say what benefit, if any, they received from their engagement. Indeed, some organizations mentioned were derided for not doing enough. Positive benefits mentioned included information gathering, networking, client identification, lobbying, financing, market analysis, and training and recruitment. Only one firm (a very large employer) mentioned ever interacting directly with the municipal government.

In Saskatoon, it appears that much of the collaboration and cooperation that facilitates knowledge flows occurs through personal contacts. These collaborations, in particular, usually involve only brief, informal consultations. While formal connections are still at times important – especially in the context of labour-market mobility and relationships with key central actors – the impression from our research is that informal connections between individuals are a more important conduit of information in the average firm's day-to-day operations.

Firms were unlikely to refer to collaboration and cooperation with other firms as a major part of their business strategy. Only two firms cited collaborations with other firms as a strategic advantage, and in those cases they were referring to a parent or sister firm. While many firms reported that they collaborated for components of their innovation process, such as product testing, in many cases this is indistinguishable from merely purchasing the services of another firm. Tellingly, when asked to describe a product, process, or service developed in the last three years that was representative of their innovativeness – a question which most firms, understandably, used to exhibit their most impressive innovation – only one firm talked about a product that was developed in close collaboration with another firm or organization. Otherwise, each firm described a product created solely by internal R&D, though often with inspiration from outside sources.

In general, Saskatoon residents were open to exchanging information whenever this was not a direct threat to their company. That the default would be to share knowledge was seen as the natural order of things. Most respondents reported that they had a number of knowledgeable acquaintances outside their company that they could phone in order to seek help on specific issues. Likewise, most respondents stated they would also likely assist if the roles were reversed. Compensation for these brief consultations was never mentioned, except in those instances where the interactions began to run over an extended period. Generally, these consultations were based on pre-existing contacts, but this was not always required. And while these interactions may occur with individuals located anywhere, the majority of the examples given by respondents refer to connections with individuals in Saskatoon.

From this research, one might conclude that most knowledge sharing is done within a framework of social norms instead of market norms. The main difference is that interactions regulated by market norms require (usually immediate) compensation, while those regulated by social norms do not. These concepts represent two fundamentally different ways in which individuals think about social interaction; whether an interaction is perceived as being regulated by market norms or social norms radically changes the nature and outcomes of that interaction. Much research has, rather counter-intuitively, shown that the expectation of payment can radically decrease an individual's willingness to render a service to another individual in certain situations (e.g., Fukuyama 1996; Godelier 1999; Benkler 2006). Because of this, Saskatoon's environment of informal connections based on social norms may facilitate a greater level of knowledge transfer and willingness to assist other firms and individuals than another centre where paid consultations are the norm. In addition, in environments where social norms predominate, reputation becomes more important. It is possible that Saskatoon's small size and relatively specialized sectoral and labour markets contributes to its social-norms-dominated environment by allowing for easier tracking of reputational factors.

Strength of Local Knowledge Flows between Industries/Clusters

The theory suggests that communities can support more than one innovative cluster and that there can be spillovers between and beyond innovative agglomerations that contribute to local economic prosperity. Greater diversity can both create lower macroeconomic vulnerability and spur more innovation due to more adjacent potential opportunities.

One way to test this is simply by asking people whether the local economy supports mobility of knowledge between jobs and sectors. A 2008 survey of 115 creatives in Saskatoon tested whether the local economy enables mobility between sectors (Phillips and Webb 2014). Overall, respondents indicated that Saskatoon's economy did facilitate mobility in a meaningful way. When mobility between sectors occurs, respondents suggest that knowledge gained in other sectors was useful in their new job, although this knowledge crossover was by no means a major causal factor for most respondents. Perhaps most interesting about these numbers was that a comparison between the level of personal creativity of the respondents (i.e., a talent index) and the responses to these questions revealed that no significant correlation existed. In short, labour and skills may be mobile, but it is not clear that mobility differentially assists creative people to contribute in their own unique way.

The 2008 creativity survey also asked individuals to identify how the community supports creativity. In all, eighty people responded, saying there were specific aspects of the city that affected creativity. Of those who indicated specific institutions or features, twenty-six respondents reported that a specific industry or infrastructure facilitated creativity, while thirty-one reported that cultural aspects of the city supported creativity. Those citing institutions focused on the relatively large role that the scientific community plays in the city, including infrastructure at the university (including the Canadian Light Source and the federal labs), the biotechnology firms, and the nature of a competitive yet co-operative community. Those citing the community and cultural aspects of the city cited the cost and variety of amenities (e.g., sports venues), rural/agrarian/small-town virtues (e.g., friendliness, acceptance, and volunteerism) and access to affordable and engaging cultural events and facilities. The correlation coefficient between the constructed talent index for those individuals and their responses related to industry/institutions was .298 (statistically significant at the 99 per cent level), indicating that those who have higher talent measures see value generated by those institutional/industrial features that are unique to Saskatoon. There is no statistical correlation between the talent index and community/cultural features. This does not necessarily mean that cultural and community attributes are not important, simply that they are not differentially recognized by those who form the "talent" pool surveyed.

Research conducted on the social dynamics of innovation in city-regions in 2007–8 posed a number of related questions to the firms in Saskatoon mentioned earlier. Firms were asked whether and how they benefited and learned from other sectors. While some firms indicated there was some benefit from other sectors in Saskatoon – e.g., mining firms are well served by Saskatoon's metal-working and manufacturing infrastructure – the overwhelming response to these questions was bafflement at the idea of learning from a non-competing or collaborating sector. While some were able to find benefits, such as in human resources and marketing, more than one third indicated that learning interactions did not occur between sectors in the city and another third said it was minimal. A quarter of respondents said it happened to a noteworthy extent, while only one firm claimed that they learned from other sectors to a great extent. Larger firms, presumably because of their greater range of activities, were more likely to say that cross-sectoral learning was taking place. Where learning from other sectors was identified, it was usually in terms of a very closely related industry, such as a gold mining firm learning from a uranium mining firm.

Similarly, when asked to what extent employees worked across fields and the importance of recruiting workers from other sectors in the local economy, most firms reported that their workers were strictly confined to their particular sector and most did not work across fields in any significant way. In fact, 75 per cent of firms reported that recruiting workers from other sectors either never happened in their firm or happened only rarely. The firms which reported recruiting from other sectors often asserted that this enabled them to bring in new perspectives and skills. Some of the largest firms stated that they hired outside of their sectors simply because of Saskatoon's limited workforce.

Firms were further probed about where their primary sources of recruitment for creative and highly educated workers were, in terms of both location and background. While many firms stated their willingness to look outside Saskatoon, most recruiting happens directly from the local economy. As noted above, most locally sourced workers had links to Saskatoon-based educational institutions: eleven firms stated that the University of Saskatchewan was a primary source for recruitment, while seven mentioned the Kelsey Campus of the Saskatchewan Institute of Applied Science and Technology (SIAST) or other local educational institutions. Half of the firms reported that they had some sort of special relationship with a local educational institution. SIAST was the most commonly cited institution, with a quarter of the firms reporting that they had some sort of special relationship with the Universities of Saskatchewan or Regina. These relationships varied in extent from participation in job fairs to providing internships and guiding curricula. Less than a third of the firms said they commonly recruited directly from competitors, while most firms stated that they avoided the practice because they believed that it was either unethical or otherwise inappropriate in such a small and tightknit community.

Symptomatic of these findings, there was a low level of space-specific communication and information exchange within the city; the Saskatoon milieu has limited local buzz. Firms were asked a series of questions about their location in Saskatoon and how it affects their operations. Firms were located throughout the city, including the north and south industrial areas, along Airport Drive, on campus, downtown, and in Innovation Place. Of these locations, only those at Innovation Place, a provincially managed industrial research park on the northern fringe of the university campus, reported that their area helped to facilitate creativity and innovation. Innovation Place, which started in 1980 and in 2011 had 140 tenants and more than 3000 workers, was praised for its beautiful buildings and grounds, its range of amenities, and the vibrant nature of its research community – it was the only location in Saskatoon that was said to have "buzz" (especially for biotechnology firms). Most respondents reported that Saskatoon simply was not large enough to have areas inside of it with buzz distinguishable from the city as a whole.

This rich environment of information exchange appears to have limited effect on knowledge transfers between sectors. The proposition that firms might benefit from the existence of other sectors in Saskatoon was usually dismissed out of hand, especially when the question was framed in terms of learning and sharing knowledge. When pressed, some firms talked about access to common support services as a benefit to their firm, while some of the larger employers said they needed workers from other sectors, especially considering the recent economic boom and tight markets for skilled labour. However, the overall impression from the interviews is that the presence of other sectors in the local economy in terms of learning, overall benefit, or hiring was not an important consideration. In this context, specialization appears to trump diversity – diversity does not appear to generate identifiable adjacent potential opportunities.

Strength of Knowledge-Based Linkages between Local and Non-Local Actors

A third issue is whether localized communities of actors interact with the world. As noted earlier, both clusters and the community recruit labour, technical services, and codified knowledge from away. In one way, global links might be able to compensate for lack of local diversity and cross-sectoral flows.

While conducting research on the social dynamics of innovation at the city-region level, we asked employers where they sought for and found new employees and discovered that about one third of companies got new employees from non-local markets, with firms relying

Table 11.1 Sources of new employees in private firms

	Local	Non-local	% non-local
Management	11	6	35
Sci., tech., eng.	17	9	35
Design	3	1	25
Marketing/sales	11	9	45
Production	15	3	17
Freelance/contract	8	5	38

Source: Author's tabulation of ISRN II Survey, part D: Q3.

more heavily on external recruitment for marketing, management, and science, technology, and engineering research staff. Local markets were more important for production and design workers (table 11.1). The prevailing view of many is that while Saskatoon has a dynamic economy and attractive local labour market, there are gaps in the market that limit any individual's ability to ascend the corporate ladder locally – many feel they need to spend part of their career away to gain skills and experience not available in the local market. This is reflected in the nearly 40 per cent of University of Saskatchewan graduates who have migrated outside the province (Phillips 2000).

An earlier study by Phillips (2002) looked beyond the labour market to a broader range of local-global linkages and knowledge flows in the context of the global oilseeds research complex, a key part of the Saskatoon biotechnology cluster. The study illustrated a mix of links to global sources of knowledge and local systems that mobilized and integrated that knowledge (Bathelt, Malmberg, and Maskell 2004 call this a system of "global pipelines and local buzz"). While one might a priori conclude that Canada was the main canola innovator (based on its record as the lead innovator and early adopter of all the new global traits over the past fifty years), that analysis showed that a significant share of the applied research to develop the processes used in the creation of those varieties has been done in other countries, and much of the applications-based research (e.g., uses for new oils) is happening elsewhere. This suggests that Canada instead has operated in a specialized niche in this global knowledge-based industry – as an entrepôt undertaking and assembling the local, common-pooled, tacit knowledge with codified, proprietary technologies from the global community – but that the bulk of the activities up- and downstream of that stage in the production system are now and may continue to be done elsewhere.

While conducting research on the dynamics of the biotechnology cluster in Saskatoon we added a module to the survey examining the nature of local–global relationships as reflected through intellectual-property management strategies. About twenty firms and organizations responded: three-quarters had formal IP strategies; about 35 per cent relied exclusively on non-local IP management; and key IP valuation models were locally and non-locally controlled about equally. When this sample of firms was segmented, those that were more innovative firms (i.e., with an innovative index 6 or greater as calculated by Procyshyn 2004) had a number of common approaches: they had formalized IP strategies and a management team approach to decision making and their IP strategies were mostly managed locally, but the critical IP valuation process was generally handled non-locally by a multidisciplinary team.

One might conclude from this analysis, combined with the earlier points, that much of the local innovation in Saskatoon is inextricably interconnected with the global system. The linkages that were identified all resulted from highly specialized competencies and capacities in Saskatoon. Hence, in this context, specialization appears more important than diversity.

Infrastructure and Innovation

Public investment in Saskatoon and in many places globally has been targeted in recent years on developing a mix of highly specialized hard infrastructure and some soft infrastructure targeted on nurturing creativity and innovation.

Hard infrastructure is particularly important to Saskatoon's economic success. Investment in gross fixed capital has equalled about 21.4 per cent of Saskatchewan's GDP in the past decade (ranging from 19 to 23 per cent on an annual basis). Government investment represents about 2.3 per cent of GDP and private investment (more than 80 per cent in non-residential construction and machinery and equipment) has averaged about 19 per cent. Saskatoon's traditional share of provincial investment has always been higher than its proportion of population, and evidence suggests that share has risen sharply in recent decades. Federal, provincial, municipal, and industrial investors have responded to the economic opportunities and demographic pressures in the city in recent years. In the past five years alone, more than $1 billion of new investment has been directed to the University of Saskatchewan and related research facilities and major new capital has flowed into oil, heavy oil, potash, uranium, gold, and diamonds, all headquartered in and served from Saskatoon. Even though world economic growth has stalled and a global recession has begun, commodity prices remain relatively robust. Saskatoon remains buoyant and a prime investment target.

Government has been a major investor in the development of Saskatoon infrastructure. The University of Saskatchewan, the single largest employer in town, is the destination of much of the investment, either directly into university colleges or buildings or in Innovation Place, the research park that occupies university land just north of the main campus. In between the two districts, many of the major government and university infrastructure investments have been sited, including the Western College of Veterinary Science, AAFC, the POS Pilot Plant, VIDO, the International Vaccine Centre (InterVac), and the Canadian Light Source (CLS). Beyond the campus district, there is no single location with significant industrial or economic infrastructure. The mining industry is scattered about in the downtown, west side, and airport areas, but there is no identifiable linking infrastructure (except perhaps the airport, but the site benefits are likely minimal, as no point in the city is more than twenty-five minutes' driving time from the terminal and there are no significant commercial or meeting facilities at the airport).

Since its inception in 1997, the Canadian Foundation for Innovation (CFI) has been a major investor in Saskatoon. The two single largest contemporary investments in economic development in Saskatoon are the CLS and InterVac. Approved with six beamlines (designed for the purpose of imaging biological tissue and conducting radiation therapy research) at an initial capital cost of $173-million, the CLS synchrotron is the largest science project in Canadian history and was the product of an unprecedented collaboration of federal, provincial, and municipal governments and agencies, universities from across the country, and industry. As of 2011, eight additional beamlines were funded and developed, each at an additional cost of at least $10 million. In 2008, VIDO at the University of Saskatchewan was commissioned to build InterVac, a $140 million vaccine research and development facility, which will be the largest Containment Level 3 facility in Western Canada, developing vaccines for humans and animals against emerging or persistent diseases. The two projects together account for more than $400 million of investment in recent years. The tag line for the CLS – "field of beams" – evokes the hopes of Ray Kinsella in the book and hit movie *Field of Dreams* that "if you build it, he will come." The notion that infrastructure imbues some irresistible draw for science and investment is not unique to the CLS – the supply push notion of investment as a primer for development permeates the literature and myths of economic development. While much of the identifiable innovation in Saskatoon industry is linked to one of these hard-infrastructure specialized investments, not all the facilities can claim credit for significant innovative success yet.

Meanwhile, soft infrastructure has been a significant focus of government policy with respect to innovation in recent years. One long-term and significant source of targeted funding that nurtures research networks has been the commodity check-offs facilitated by federal and provincial legislation. Producers pay small levies on crops and livestock produced and sold in the province and use that to leverage university and industrial research and development. While the individual levies are small, the aggregate funds available in Saskatchewan for research exceeded more than $100 million over the past twenty years. Much of that has been coordinated and used by groups such as the Crop Development Centre, POS Pilot Plant, and various federal labs, all located at the University of Saskatchewan. Meanwhile, governments have also directly put significant money into soft infrastructure. The province has contributed by developing groups such as Ag West Bio, a non-profit Saskatchewan corporation funded by government but with an independent board of directors, which is mandated to "initiate, promote and support the growth of Saskatchewan's agricultural biotechnology industries" and "the commercialization of related food and non-food technologies." The city, meanwhile, created the Saskatoon and Region Economic Development Authority (SREDA), one of the first independent economic development agencies in Canada, to coordinate and lead economic and community development. Nationally, funding agencies put money into networked science and research, including through the Networks of Centres of Excellence, Genome Canada, and various new tri-council-funded networking initiatives, all of which Saskatoon has worked to use to generate new activity in the city.

Most of this investment in both hard and soft infrastructure is pushing the local economy into further specialization and away from diversity. The labs and facilities are usually single-purposed and the grants and capital pools are increasingly focused on clearly delimited outcomes.

While this undoubtedly has created some local buzz – respondents to various surveys report the important role of the university, NRC-PBI, AAFC, and Innovation Place – the newer investments are highly specific. There is some recognition that these new investments may need some special soft infrastructure to optimize their benefits – the university, for example, has proactively sought to attract research-intensive scholars to take up Canada Research Chairs on campus to use the CLS, while the university business office has aggressively pushed for more work with downstream users. The jury is out on how much soft infrastructure is needed to ensure hard-infrastructure investment will be able to contribute fully to the creative and innovative community.

Conclusions

The Saskatoon case offers some interesting perspectives on the role of social dynamics, diversity, and infrastructure in mid-sized creative and innovative cities. First, the data offers some (but not universal) support for the assertion that knowledge flows within sectors and clusters, linked to specialized structures and processes, matter for innovation. Both employees and firms indicated in a variety of studies reviewed that thick, specialized labour markets, formal knowledge transfer, and informal knowledge systems are important for knowledge-intensive economic development; yet it is less clear that these matter for less knowledge-intensive industries. Second, the data does not offer much support that there are any acknowledged externalities between parallel innovative clusters or between innovative clusters and the rest of the economy – if anything, they seem to be solitudes. Third, there is some support for the need for global pipelines, if for no other reason than to compensate for a lack of diversity within the local community. Firms (and highly skilled individuals) seem to acknowledge the need to remain globally aware and connected, thereby generating more adjacent potential opportunities.

Investment in infrastructure is undoubtedly an important and vital part of any industrial economy. The theory and evidence, however, does not provide any definitive direction to its role and impact on innovation. In the context of Saskatoon, we can see the value of specialized infrastructure as a necessary but not sufficient condition for innovation in the agricultural- biotechnology sector (though not as clearly in the mining sector), while there is some evidence that soft infrastructure, especially the specialized kind, may be critical to realizing the local buzz that makes innovation happen.

On the whole, the diversity hypothesis is not highly supported in the Saskatoon case. While diversity may have micro and macro benefits, at the meso level it is not clear that it has any particular effect, at least in the context of the Saskatoon economy, where the mining and agricultural-biotechnology firms use global knowledge and competencies to compete in global markets. In the context of this research on Saskatoon, we can see that innovation both within specific clusters and in the broader economy is supported by a general focus on innovation and by the capacity to leverage capital in support of the foundational infrastructure generated. Nevertheless, much of this capacity seems to be narrowly focused in specific clusters rather than widely distributed. This may simply be the result of a relatively small community. Spencer and Vinodrai (2006) concluded that the Saskatoon city-region has four clusters that employ about 15 per cent of the workforce, compared with Calgary, which has six clusters employing more than 40 per cent of workers. Given that Saskatoon's clusters are much smaller in both absolute and relative terms, it is likely that the local economy has not fully exploited economies of scale in each of these economic areas; hence, there are likely to be ample opportunities for innovation and growth within each of the core clusters, which reduces the incentive to seek more diverse collaborations.

Meanwhile, social capital, in institutions, in policy systems, and in the community of creative individuals who propel firms, government agencies, industry associations, and various collective efforts, is vital to the Saskatoon economy. It is the medium that generates untraded interdependencies that spur innovation and development. As noted above, most knowledge exchanges are socially mediated rather than market-based. Nevertheless, the strong theoretical focus on meso-level interactions was not witnessed in Saskatoon. Rather, social capital appears to be most actively focused at the micro and macro levels, targeted at specific initiatives that would spur development and innovation. There was limited evidence that social capital has facilitated inter-sectoral exchanges that spur innovation in adjacent or tangential sectors.

It is unclear what these results tell us about theory and policy. At one level, they tend to refute the diversity hypothesis, which would imply that governments should continue to focus their efforts to enhance specialization by "building on strengths." While this has an intuitive appeal in times of fiscal restraint, it may not be the only option. Rather, this focus on specialization may be a phenomenon of a middle-sized isolated resource community at a specific stage of industrial development. Given that there are no direct comparators for Saskatoon in Canada and that the economic and social landscape in the city has been rapidly changing over the past decade and is likely to continue for the foreseeable future, this may not be a stable, long-term experience. As the Saskatoon economy and society adapt to the recent transitions, we may see the need for and emergence of more meso-level diversity and interactions. As with most highly contextualized policy studies, there are few definitive conclusions and few strong policy recommendations one can make. A balanced portfolio of policies, programs, and capital investments in support of both diversity and specialization may remain the best available approach for most governments seeking to support innovation and development.

REFERENCES

Asheim, B., and M.S. Gertler. 2005. "The geography of innovation: Regional innovation systems." In *The Oxford handbook of innovation*, ed. J. Fagerberg, D. Mowery, and R. Nelson, 291–317. Oxford: Oxford University Press.

Bathelt, H., A. Malmberg, and P. Maskell. 2004. "Clusters and knowledge: Local buzz, global pipelines and the process of knowledge creation." *Progress in Human Geography* 28 (1): 31–56. http://dx.doi.org/10.1191/0309132504 ph469oa.

Benkler, Y. 2006. *The wealth of networks: How social production transforms markets and freedom*. New Haven: Yale University Press.

Bourdieu, P. 2006. "The forms of capital." In *Education, globalization, and social change*, ed. H. Lauder, P. Brown, Jo-Anne Dillabough, and A. Halsey. New York: Oxford University Press.

Dobni, B., and P. Phillips. 2000. *Saskatoon's ScienceMap^{TM}*. Saskatoon: University of Saskatchewan.

Florida, R. 2002. *The rise of the creative class and how it's transforming work, leisure and everyday life*. New York: Basic Books.

Fukuyama, F. 1996. *Trust: The social virtues and the creation of prosperity*. New York: Simon and Schuster Inc.

Godelier, M. 1999. *The enigma of the gift*. Trans. Nora Scott. Chicago: University of Chicago Press.

Jacobs, J. 1985. *Cities and the wealth of nations*. New York: Vintage.

Kauffman, S. 1995. *At home in the universe: The search for laws of self-organization and complexity*. New York: Oxford University Press.

Krugman, P. 1998. "What's new about the new economic geography?" *Oxford Review of Economic Policy* 14 (2): 7–17. http://dx.doi.org/10.1093/oxrep/14.2.7.

Lefebvre, M. 2009. *Canada's new economic powerhouse*. Conference Board of Canada, 17 February. http://sso.conferenceboard.ca/Economics/hot_eco_topics/default/09-02-17/Canada_s_New_Economic_Powerhouse.aspx.

Marshall, A. 1890. *Principles of economics*. London: Macmillan.

Morgan, K. 1997. "The learning region: Institutions, innovation and regional renewal." *Regional Studies* 31 (5): 491–503. http://dx.doi.org/10.1080/00343409750132289.

Phillips, P. 2000. "The geography of the Saskatchewan macro-economy." In *The Atlas of Saskatchewan*, ed. K. Fung, 211–16. Saskatoon: University of Saskatchewan.

Phillips, P. 2002. "Regional systems of innovation as a modern R&D entrepot: The case of the Saskatoon biotechnology cluster." In *Innovation, entrepreneurship, family business and economic development: A Western Canadian perspective*, ed. J. Chrisman, J. Holbrook, and J. Chua, 31–58. Calgary: University of Calgary Press.

Phillips, P., and G. Khachatourians, eds. 2001. *The biotechnology revolution in global agriculture: Invention, innovation and investment in the canola sector*. Wallingford, UK: CABI Publishing. http://dx.doi.org/10.1079/9780851995137.0000.

Phillips, P., and G. Webb. 2014. "Talent, tolerance, and community in Saskatoon." In *Seeking talent for creative cities*, ed. J.L. Grant. Toronto: University of Toronto Press.

Porter, M. 1990. *The comparative advantage of nations*. New York: Free Press.

Procyshyn, Tara. 2004. "Saskatoon's agricultural biotechnology cluster and the Canadian Light Source: An assessment of the potential for cluster extension." Unpublished MSc thesis, University of Saskatchewan.

Scott, A.J. 2004. "A perspective of economic geography." *Journal of Economic Geography* 4 (5): 479–99. http://dx.doi.org/10.1093/jnlecg/lbh038.

Spencer, G., and T. Vinodrai. 2006. Saskatoon city-region profile: Summary and highlights, Unpublished mimeograph, 19 April.

Storper, M. 1997. *The regional world*. New York: Guilford Press.

Zucker, L., M. Darby, and M. Brewer. 1998. "Intellectual human capital and the birth of U.S. biotechnology enterprises." *American Journal of Economics* 88 (1): 290–306.

12 How ICTs and Face-to-face Interactions Mediate Knowledge Flows in Moncton

YVES BOURGEOIS

Virtual Spaces, Physical Spaces, and Knowledge Flows

Space matters to innovation in different ways. On the one hand, information and communication technologies (ICTs) are reshaping our economic landscape. ICTs create new tasks, jobs, firms, and industries and, once mastered, decrease the costs and time of production, distribution, and communication. Increasing bandwidth, wireless capabilities, and Internet-enabled applications augment opportunities to produce, to exchange, and to consume products, experiences, and knowledge. Businesses use Skype for videoconferencing to enhance communication while trimming travel budgets. They use YouTube for online training, product demos, and tutorials. Many adopt Twitter to enable instant communication throughout the organization, to interact with customers, and to monitor trends and competitor activities. They implement cloud computing applications to synchronize collaborations between workers and with business partners. Although there is a cost to learn them and a risk of sapping worker productivity if usage is not well defined, many of these technologies are free for download in a scaled-down version and are thus accessible to all. Globalization as the relocation and outsourcing of production has followed ICTs in tow. As Internet and wireless technologies in particular penetrate the four corners of the world, writers like Friedman (2005) believe the ICT revolution holds promise to level the playing field between rich and poor regions, freeing economic activities from the chains of location.

Revolutions often come up short. A generation after the commercialization of the Internet in the 1980s and widespread deployment in the 1990s, demographic and economic growth have concentrated in urban centres even more. The world's population grew four times faster in cities than in rural areas between 2005 and 2010 (UNFPA 2010). In the more industrialized nations combined, cities grew by 0.7 per cent while rural populations declined 0.6 per cent. If workers could telecommute, why would they move to large cities where the cost of living was higher, and why would employers locate where salaries and real estate would cost more? This book and chapter argue that an important part of the answer resides in the importance of cities in facilitating innovation and economic performance to those who agglomerate there.

Whether we call them agglomerations, metropolitan areas, or city-regions, there is something about geographic proximity that confers advantages to them when it comes to innovation. Cities facilitate innovation via *specialized infrastructure*, such as airports, universities, and research labs. They provide *critical mass* that affords deep labour pools of specialized skills or diversity in complementary and connected forms of knowledge. Cities can crystallize *supporting institutions* that grease the innovative wheels, formal or informal associations and mechanisms that facilitate the production, flow and capture of knowledge and its spillovers. The speed and complexity of innovative and creative activities place a premium on dense urban environments as they multiply opportunities for face-to-face (F2F) interactions, which in turn facilitate the sharing of valuable new ideas and knowledge that are more difficult to codify (Storper and Venables 2004).

Statistics Canada surveys of innovation (2002, 2005) reveal that customers, suppliers, and competitors are the three main sources of innovation for businesses, cited as important by more than 80 per cent of firms. Moreover, half of innovative businesses consider customers a highly important source of innovations, and more than a third consider suppliers highly important. Given Moncton's and Atlantic Canada's smaller population and relative geographic isolation, does distance from major markets and suppliers put them at a disadvantage in accessing sources of innovation? Do ICTs provide means to overcome these challenges and facilitate the creation, dissemination, absorption, and transformation of knowledge in smaller, isolated city-regions like Moncton?

How do agglomeration effects play out in Moncton and how do they contribute to the innovative and economic performance of the city-region? How does knowledge flow within the same economic sectors in the city (*intra-sectoral flows*) and to what extent does it flow across different sectors (*cross-sectoral flows*)? When do local innovation capabilities rely more on knowledge that circulates well within the city (*local flows*) and when is knowledge from outside the region (*global flows*) more important? These are the main questions underlying this research.

Transit City

The tri-city area of Moncton, Dieppe, and Riverview, New Brunswick, has long served as a transportation hub for the region. Mi'kmaq Aboriginals used the bend in the Petitcodiac River to portage between it and the Shediac River during spring and autumnal transits. Moncton morphed from eighteenth-century Acadian and German agrarian settlements into a wooden-shipbuilding town in the mid-nineteenth century. Its strategic location on the Chignecto isthmus, connecting Nova Scotia and Prince Edward Island to the rest of Canada, attracted the emerging Intercolonial Railway at the end of the nineteenth century and rendered Moncton a railway town and distribution centre for much of the twentieth century. Despite the funnelling of road and rail through the region, there was little expansion of the city's manufacturing base to leverage increased access to central markets (Larracey 1991).

Transition City

Moncton weathered the perfect storm in the mid-1980s. Within a few short years it lost a tenth of its jobs, including its largest private-sector employer – the CN locomotive repair shops – as well as other mainstays such as the Eaton's mail-order catalogue warehouse, and an armed forces base. The CN shops alone represented 5000 jobs lost in a city with a workforce of 53,000 and in an economic region of 77,000 workers in 1987 (Statistics Canada 2011a). The shops closed for good in 1988. The mood was dour, the funeral organized, but no eulogy spoken as Moncton got up and carried on with steady population and economic growth. Since 1987, growth of employment in the Moncton economic region has consistently surpassed the national average, as illustrated in tables 12.1 and 12.2.

What some called the "Moncton miracle" (Farnsworth 1994) owed more to a massive effort deployed by community and provincial leaders to put the economy back on its feet. Then New Brunswick premier Frank McKenna and local politicians banged on every door on Toronto's Bay Street and, capitalizing on Greater Moncton's large number of bilingual workers, returned with call centres and back offices. These

Table 12.1 Total employment by sector, Canada and Moncton economic region, 1987–2009

	1987		1994		1999		2004		2009	
	Canada	Moncton	Canada	Moncton	Canada	Moncton	Canada	Moncton	Canada	Moncton
Total	12,333,000	69,900	13,058,700	76,400	14,406,700	85,900	15,947,000	98,200	16,848,900	105,400
Goods	3,633,900	14,900	3,397,500	14,400	3742500	16,800	3,989,800	19,100	3,736,400	20,800
Agriculture	464,500	1,000	437,200	1,500	406000	600	326,000	1,100	320,500	1,200
Forestry, fishing, mining, oil and gas	287,100	1,600	285,600	2,000	263,800	2,200	286,600	1,700	316,200	1,700
Utilities	114,800	x	127,000	x	114,300	x	133,300	x	147,800	x
Construction	726,600	3,800	724,600	3,700	766,900	4,200	951,700	5,100	1,161,400	7,700
Manufacturing	2,041,000	8,300	1823200	6,900	2,191,500	9,600	2,292,100	10,800	1,790,600	9,900
Services	8,699,200	55,000	9,661,200	62,000	10,664,300	69,100	11,957,200	79,100	13,112,500	84,500
Trade	1,982,000	13,200	2,061,100	14,800	2,218,200	15,500	2,507,100	15,600	2,639,800	14,900
Transportation and warehousing	634,000	6,000	644,900	7,400	737,000	7,700	799,400	7,900	820,300	6,500
Finance, insurance, real estate and leasing	765,800	3,400	832,700	4,300	859,900	4,900	960,600	4,600	1,099,000	7,100
Professional, scientific, and technical services	489,800	1,300	642,500	2,100	900,700	2,600	1,018,300	4,300	1,201,600	4,400
Business, building and other support services	272,600	1,100	365,400	2,100	504700	2,800	630,200	5,600	656,500	6,900
Educational services	776,600	6,300	927,200	5,300	970,700	5,300	1,035,700	7,700	1,192,700	8,000

(continued)

Table 12.1 Total employment by sector, Canada and Moncton economic region, 1987–2009

	1987		1994		1999		2004		2009	
	Canada	Moncton	Canada	Moncton	Canada	Moncton	Canada	Moncton	Canada	Moncton
Health care and social assistance	1,152,000	8,000	1,364,200	8,400	1,436,000	9,400	1,733,400	13,400	1,955,000	13,300
Information, culture and recreation	511,100	2,500	537,400	2,100	630,500	3,800	738,000	4,000	776,700	4,200
Accommodation and food services	716,700	4,300	799,100	4,800	913,600	6,600	1,012,400	6,900	1,055,900	7,400
Other services	633,100	3,600	651,900	4,800	716,500	4,900	696,600	3,900	788,300	3,700
Public administration	765,400	5,400	834,800	5,900	776,300	5,400	825,500	5,300	926,600	8,200

Source: Statistics Canada 2011c.
x = Data suppressed because numbers are too small.

Table 12.2 Average employment changes (%) by sector, Canada and Moncton economic region, 1987–2009

	1987–94		1994–9		1999–2004		2004–9		1987–2009	
	Canada	Moncton	Canada	Moncton	Canada	Moncton	Canada	Moncton	Canada	Moncton
Total	0.84	1.33	2.06	2.49	2.14	2.86	1.13	1.47	1.66	2.31
Goods	-0.93	-0.8	2.03	3.33	1.32	2.74	-1.27	1.78	0.13	1.80
Agriculture	-0.84	7.14	-1.43	-12.00	-3.94	16.67	-0.34	1.82	-1.41	0.91
Forestry, fishing, mining, oil and gas	-0.07	3.57	-1.53	2.00	1.73	-4.55	2.07	0.00	0.46	0.28
Utilities	1.52	x	-2.00	x	3.32	x	2.18	x	1.31	x
Construction	-0.04	-0.38	1.17	2.70	4.82	4.29	4.41	10.20	2.72	4.67
Manufacturing	-1.52	-2.41	4.04	7.83	0.92	2.50	-4.38	-1.67	-0.56	0.88
Services	1.58	1.82	2.08	2.29	2.42	2.89	1.93	1.37	2.31	2.44
Trade	0.57	1.73	1.52	0.95	2.60	0.13	106	-0.90	1.51	0.59
Transportation and warehousing	0.25	3.33	2.86	0.81	1.69	0.52	0.52	-3.54	1.34	0.38
Finance, insurance, real estate and leasing	1.25	3.78	0.65	2.79	2.34	-1.22	2.88	10.87	1.98	4.95
Professional, scientific, and technical services	4.45	8.79	8.04	4.76	2.61	13.08	3.60	0.47	6.61	10.84
Business, building, and other support services	4.86	12.99	7.62	6.67	4.97	20.00	0.83	4.64	6.40	23.97
Educational services	2.77	-2.27	0.94	0.00	1.34	9.06	3.03	0.78	2.44	1.23
Health care and social assistance	2.63	0.71	1.05	2.38	4.14	8.51	2.56	-0.15	3.17	3.01

(continued)

Table 12.2 Average employment changes (%) by sector, Canada and Moncton economic region, 1987–2009

	1987–94		1994–9		1999–2004		2004–9		1987–2009	
	Canada	Moncton	Canada	Moncton	Canada	Moncton	Canada	Moncton	Canada	Moncton
Information, culture, and recreation	0.74	-2.29	3.46	16.19	3.41	1.05	1.05	1.00	2.36	3.09
Accommodation and food services	1.64	1.66	2.87	7.50	2.16	0.91	0.86	1.45	2.15	3.28
Other services	0.42	4.76	1.98	0.42	-0.56	-4.08	2.63	-1.03	1.11	0.13
Public administration	1.30	1.32	-1.40	-1.69	1.27	-0.37	2.45	10.94	0.96	2.36

Source: Author's calculations from Statistics Canada 2011c.

were not the highest-paying jobs or easiest to sustain in the long term in a globalizing world, but they had an immediate palliative effect: they kept workers in Greater Moncton and attracted others at a time when employment rates were shrinking across the province.

Moncton's economy was slowly transitioning into a business service economy. Table 12.2 shows that in the critical years after the CN shops closure (1987–94), average annual employment grew fastest in business support services (12.99% growth per year compared to the Canadian average of 4.86%) and in professional, scientific, and technical services (8.79% average yearly growth compared to 4.45% nationally). In the recession years of the early 1990s, IT and culture industries as well as manufacturing stoked the local economy (16.19% and 7.83% average annual growth, respectively, compared to 3.46% and 4.04% nationally). Professional (13.08%) and business services (20.0%) grew strongly between 1999 and 2004, as did education (9.06%) and health care (8.51%). From 2004–9, the fastest growing sectors were public administration (10.94%), finance, insurance, and real estate (FIRE) (10.87%) and construction (10.20%), testimony to fast population growth and accelerated transition into a service-oriented economy.

Table 12.3 shows location quotients (LQs) for the Moncton economic region. LQs lower than 1.0 indicate a region's share of jobs in that industry relative to its overall workforce is of a smaller percentage than at the national level. LQs greater than 1.0 show higher concentrations of employment than the national average and are indicative of industry clustering. Table 12.3 reveals that between 1987 and 2009 employment has been clustering somewhat in the construction and manufacturing sectors. Historically, Greater Moncton has been a hub for commercial, retail, transportation, and warehousing, but employment in those traditional sectors has been growing more slowly than the national average, with LQs decreasing from 1.18 to 0.90 (trade) and from 1.67 to 1.27 for transportation and warehousing. While education and health care employment has been growing in Greater Moncton, growth in these sectors has been faster in the rest of Canada. The greatest clustering effects can be observed in the business support services (0.71 to 1.68), FIRE industries (0.78 to 1.03), and public administration (1.24 to 1.41). Moncton's share of IT and culture employment has grown more slowly than the national average over the last ten years.

Shift-share analysis attributes how much of a region's job creation or loss owes to different trends. The *national growth component* suggests how many local jobs could be expected if the sector grew as did the

Table 12.3 Location quotients by sector, Moncton economic region, 1987–2009

	1987	1994	1999	2004	2009
Goods	0.72	0.72	0,75	0.78	0.89
Agriculture	0.38	0.59	0,25	0.55	0.60
Forestry, fishing, mining, oil and gas	0.98	1.20	1,40	0.96	0.86
Utilities					
Construction	0.92	0.87	0.92	0.87	1.06
Manufacturing	0.72	0.65	0.73	0.77	0.88
Services	1.12	1.10	1.09	1.07	1.03
Trade	1.18	1.23	1.17	1.01	0.90
Transportation and warehousing	1.67	1.96	1.75	1.60	1.27
Finance, insurance, real estate, and leasing	0.78	0.88	0.96	0.78	1.03
Professional, scientific, and technical services	0.47	0.56	0.48	0.69	0.59
Business, building, and other support services	0.71	0.98	0.93	1.44	1.68
Educational services	1.43	0.98	0.92	1.21	1.07
Health care and social assistance	1.23	1.05	1.10	1.26	1.09
Information, culture, and recreation	0.86	0.67	1.01	0.88	0.86
Accommodation and food services	1.06	1.03	1.21	1.11	1.12
Other services	1.00	1.26	1.15	0.91	0.75
Public administration	1.24	1.21	1.17	1.04	1.41

Source: Author's calculations from Statistics Canada 2011c.

overall economy. The *industry mix component* suggests how many jobs owe to a particular sector's performance. The *competitive share component* is a residual that suggests local circumstances are aiding or undermining a region's competitiveness in that sector.

As table 12.4 reveals, most of the jobs created in Moncton's construction industry between 1987 and 2009 were the result of specific characteristics of the local economy (2333 jobs) rather than to the overall economy (1391 jobs) or industry's performance (883 jobs). Indeed, Moncton is in the midst of a strong demographic boom. The same analysis applies to FIRE industries, as well as accommodation and food services industries. Local advantages also drive employment growth in professional and technical services, as well as business support services, although the industry mix component is also high, as these are fast-growing sectors throughout Canada. However, growth in trade, transportation,

Table 12.4 Shift-share analysis for Moncton economic region, 1987–2009

	National growth component, %	National growth component, jobs	Industrial mix component, %	Industrial mix component, jobs	Competitive share component, %	Competitive share component, jobs
Goods	36.6	5,456	−33.8	−5,036	2.5	370
Agriculture	36.6	366	−67.6	−676	−31.9	−319
Forestry, fishing, mining, oil and gas	36.6	586	−26.5	−424	8.8	141
Utilities	36.6	x	−7.9	x	28.7	
Construction	36.6	1,391	23.2	883	61.4	2,333
Manufacturing	36.6	3,039	−48.9	−4,057	−13.2	−1,097
Services	36.6	20,139	14.1	7,764	50.8	27,950
Trade	36.6	4,833	−3.4	−452	32.0	4,230
Transportation and warehousing	36.6	2,197	−7.2	−434	28.1	1,686
Finance, insurance, real estate, and leasing	36.6	1,245	6.9	234	45.3	1,539
Professional, scientific, and technical services	36.6	476	108.7	1,413	151.0	1,963
Business, building, and other support services	36.6	403	104.2	1,146	155.2	1,707

(continued)

Table 12.4 Shift-share analysis for Moncton economic region, 1987–2009

	National growth component, %	National growth component, jobs	Industrial mix component, %	Industrial mix component, jobs	Competitive share component, %	Competitive share component, jobs
Educational services	36.6	2,307	17.0	1,069	52.9	3,330
Health care and social assistance	36.6	2,929	33.1	2,647	70.2	5,614
Information, culture, and recreation	36.6	915	15.3	384	52.5	1,312
Accommodation and food services	36.6	1,575	10.7	461	48.0	2,063
Other services	36.6	1,318	–12.1	–436	23.1	830
Public administration	36.6	1,977	–15.6	–840	21.1	1,139
		51,153		3,645		54,791

Source: Author's calculations from Statistics Canada 2011c.

warehousing, and public administration employment owes relatively more to growth in the overall economy than to the sector's performance or local advantages. Moncton manufacturing job losses owe more to the contraction in national manufacturing sectors (−4057) than to the region's lack of competitiveness in that sector, suggesting that the region may enjoy particular advantages, such as lower wages for a relatively skilled workforce, attracting Canadian investments, and competing with lower-wage regions.

Innovation City?

At first glance, lights are bright for Moncton. Its 6.53 per cent growth rate between 2001 and 2006 made it Canada's tenth-fastest growing city-region and New Brunswick's largest metropolitan area and commercial capital. The Greater Moncton census metropolitan area (CMA) scattered 126,424 residents among the cities of Moncton (64,128), the city of Dieppe (18,565), the town of Riverview (17,832), and smaller adjacent communities (Statistics Canada 2007). Statistics Canada (2011b) estimated Greater Moncton's 2006–11 growth rate would climb to 9.8 per cent based on projections for 2006–9.

A sober second look at demographic trends suggests this population growth rate cannot be sustained, unless inter-provincial and international migration increase manifold. In what could be called *delayed urbanization*, much of Greater Moncton's recent population growth has occurred from migration *within* New Brunswick, one of Canada's least urbanized provinces. Historically, urbanization has been a dirty word in New Brunswick and settlements built around natural resources remain prominent to this day. From 2001 to 2006 12.4 per cent of Greater Moncton's population had moved to the CMA from other parts of the province, while 6.0 per cent had moved from outside the province and 1.1 per cent from outside the country.

More workers mean more production, as well as greater scale and diversity of economic activities. However, the size of a city's economy does not translate into better economic performance or higher wages. While Moncton's population grew rapidly between 2001 and 2006, average earnings for full-time, full-year workers grew slowly, from 84.4 per cent of the national average in 2001 to 86.1 per cent in 2006 (Statistics Canada 2007). The next section details how earnings are tied to labour and multi-factor productivity gains accruing from innovation and technology investments.

If Moncton is to continue transitioning into a higher-productivity, innovation-centred economy, much remains to be done. Despite rapid growth Moncton remains the fourth smallest of Canada's thirty-three CMAs. There are three universities in or within a half-hour of Moncton, yet they offer predominantly undergraduate programs with fewer than 5000 students each, providing limited access to specialized workers or training and research facilities. Air travel is expensive and direct connections are largely limited to Montreal and Toronto, and to a lesser extent Ottawa and New York. Meeting North American and global suppliers and customers is costly and time-consuming. Innovation-enabling assets are in shorter supply in Greater Moncton. Are local players working to overcome these challenges in building a well-functioning innovation system? Do ICTs help bridge the distance with global suppliers and customers?

Innovation and Growth

Sustained economic growth depends exclusively on one of three factors: population growth, capital investments, and innovation. Much of Moncton's economic growth is due to migration from within the province. Whether northern New Brunswick economies develop or empty out, in-migration as a source of population growth for southern New Brunswick cities will grind to a halt, and growing international migration will likely prove insufficient if the experience of other small and medium-sized cities in attracting immigrants proves the rule.

In any case, population growth can fuel economic growth by increasing the supply of labour, but it cannot raise average earned wages. Rising standards of living can only occur through increased productivity resulting from capital investments (more tools) and innovation (better tools). When capital investments and innovations increase output per worker, productivity gains can translate into increased profits and higher average earned wages. Sustaining local economic growth will depend on Greater Moncton businesses investing in training, physical capital, and innovation.

A Tale of Three Industries

Substantial mobilization of effort was deployed in the 1980s and 1990s to reposition the city from its image of a railway and retail town into a savvy high-tech centre. Focus was given to marketing the city in business magazine rankings as being among the best places to do business in Canada. The qualified and bilingual workforce was cited. So too were cheap rents and wages. In their way these efforts bore fruit in convincing large companies to relocate back-office operations and call centres to Greater Moncton. Other companies providing communications and management consulting services grew endogenously to serve the growing local market. Overall economic diversification towards white-collar professional services did occur. However, competitive advantages built around low wages cannot be sustained through continued economic and income growth, and many of the call-centre jobs, for example, are moving to even lower-wage regions such as India. Can knowledge-intensive business services (KIBS) firms remain competitive in globalizing markets and increase the value added of the services they produce? Doloreux and Shearmur (2010) find significant innovation differences between Quebec KIBS sub-sectors and caution against using a generic KIBS label. Moncton's small size makes some sector groupings unavoidable; however, there are indeed key differences between financial services and strategic communications.

As with most other regions, Greater Moncton economic-development officials believe the region has special competitive advantages in information technologies that could usher in a new era of prosperity. Many economic-development dollars have been spent in trying to emulate Silicon Valley or Nokia's Espoo success. However, there was little evidence when interviewing economic-development officials that specific IT sub-sectors aligned with local strengths were being targeted when attracting investments. The risk of "branch plant effects" in high technology sectors (Cusmano, Mancusi, and Morrison 2010) was not a mitigating factor. The public-policy attention given to the IT sector in Greater Moncton makes it another sector considered.

The third economic sector analysed reflects the varied trajectories of development faced by firms in advanced manufacturing. Often less celebrated in the local economy because of Moncton's blue-collar label, many manufacturing firms have invested considerably in advanced technology and materials to compete globally, and their toils are documented here. These three sectors – information technology, advanced manufacturing, and KIBS – illustrate the dilemma confronting the local economy: whether to build upon long-established strengths that remain present and seek to modernize, or to raze the landscape by privileging

Table 12.5 Key characteristics on knowledge and innovation: Greater Moncton advanced manufacturers, KIBS, and ICT firms

	Advanced manufacturers	KIBS	ICT
Importance	Opportunities to add value to local resource	Creative industry; economic weathervane for region	Specialized skills; explicit economic development strategy
Challenges	Southeast Asia competition	Moncton has few headquarters, but many regional and back offices; professional services contracts often given to firms close to HQ	Establishing critical mass; stemming brain drain; momentum lost when successful firm is bought out.
Product innovation	Incremental	Social marketing tools now 30% of revenues: e-marketing	Enterprise resource planning (ERP), collaborative and wireless applications
Process innovations	Organizational: lean manufacturing	Organizational: developing team-oriented solutions	Organizational: developing team-oriented solutions

Source: Author's aggregation of interview responses.

new technological sectors. Table 12.5 summarizes key characteristics about the nature of knowledge production among Greater Moncton's advanced manufacturers, KIBS, and ICT firms.

Knowledge Flows into Greater Moncton: Owner Must Travel

What are the primary sources of knowledge that Greater Moncton businesses use to innovate and how does such knowledge flow into the region? Table 12.6 summarizes the various sources of knowledge for Greater Moncton's advanced manufacturers, KIBS, and ICT firms. While the answers can vary by type of industry and type of knowledge, there are some generalizations possible across sectors, and I shall begin with those.

First, if we consider the source of knowledge used to innovate, responses from Greater Moncton firms were consistent with the rankings from Statistics Canada's 2005 survey of innovation. Irrespective of industry, firms in the supply chain (clients, suppliers, competitors)

Table 12.6 Importance and origin of various sources of knowledge: Greater Moncton advanced manufacturers, KIBS, and ICT firms

	Advanced manufacturers	KIBS	ICT
Workers	Predominantly local; unskilled labour, some retrained from fish processing plants because of dexterity Dearth of specialized skill sometimes requires increased automation.	Mostly local, and some from other parts of Atlantic Canada. Halifax is more successful at attracting "creatives" and creative side of communications more easily done in Halifax. "Competition for talent fiercer than for clients."	Mostly local, experienced in call centres and community colleges Many repatriated from stints abroad. Hefty premium paid to workers to have them relocate.
Clients	Mixture of local, Central Canada, and New England states	Communications: 90% local, 10% Central Canada Financial services: Traditionally Atlantic Canada, increasingly national	Off-the-shelf (OTS) including downloadable software is global. Custom wireless applications and IT solutions is local.
Suppliers	Global, seen as linchpin for both product and process innovations	Low importance. Market intelligence is global; creative is global and local.	Low importance. Global
Competitors	Global	Traditionally Atlantic Canada, but increasingly national as company HQs centralize communications, legal and other professional services accounts.	Global for OTS None for custom ("lower-lying fruit")
Consultants	Low importance	Low importance	Low importance
Financing	Local, moderate difficulty	Local, difficult to leverage work-in-progress versus inventories in manufacturing	Mostly outside angels and VCs
Univ./gov. labs	Low importance	Low importance	"Surprisingly little"

(continued)

Table 12.6 Importance and origin of various sources of knowledge: Greater Moncton advanced manufacturers, KIBS, and ICT firms

	Advanced manufacturers	KIBS	ICT
Trade fairs	High importance. North America	"Field trips to replenish creative juices"	"Less than expected"
Local networks	Informal, few collaborations	Prospecting for clients "but not for ideas"	Social but few knowledge exchanges
City scale or diversity	Low importance	Small importance, but cultural amenities help retain workers	Important but inadequate. "My workers walk, bike, live downtown."

Source: Author's aggregation of interview responses

were by far the most important sources of knowledge that Moncton firms used for innovating. The notable exception was the use of consultants: although many firms cited having used them, few mentioned having learned anything "earth shattering," instead being told "what we already knew," as two manufacturing company CEOs remarked. Given the knowledge-brokerage role consultants often play, including the import of practices and techniques into the region, their relative unimportance was surprising. The cited frequent use of consultants from outside the region suggests a lack of specialized knowledge among area consultants. It may also be that area firms preoccupied with innovating are already actively drawing upon their existing networks to leverage knowledge.

Given Greater Moncton's position as a distribution and retail centre, proximity to customers was seen as an advantage, but mainly for those serving Atlantic Canadian markets. For exporters, proximity to customers was seen as a major disadvantage, and their ability to harness knowledge more limited. For most, proximity to suppliers, competitors, and consultants was seen as a disadvantage in their attempts to innovate.

While supply-chain relationships are the most important sources of knowledge to firms in general, some sectors rely more heavily on scientific research and development, such as biotechnology, IT, aerospace, chemicals, and advanced manufacturing. Internal R&D as well as collaborative R&D ventures with university and government labs are seen as key sources of knowledge, and they benefit from the brunt of public spending in innovation programs, such as the Atlantic Canada Opportunities Agency's Atlantic Innovation Fund and the New Brunswick Innovation Fund.

In Greater Moncton, these R&D collaborations with universities and governments, including their research infrastructure, were rather weak. While IT and advanced manufacturing sectors cited technical and scientific R&D as very important to their competitiveness, few were actively engaged in collaborative arrangements, at least not with Moncton partners. One IT firm did mention more active collaborations with l'Université de Moncton in years past, but not recently. An IT consulting firm technical director pondered that they "probably should [collaborate with Université de Moncton and National Research Council IT researchers], but never get around to it," suggesting the perceived value of these collaborations was poor or uncertain.

These weak collaborations are owed in part to the limited research capabilities of local universities. Newly created Crandall University is first and foremost a teaching-based institution to its 800 students. Nearby Mount Allison University is also a liberal arts, teaching-focused university. The 5000-student Université de Moncton is a teaching-focused university as well, although it has a scattering of graduate degree programs and research laboratories, including science and engineering and NRC IT research facilities. Where capabilities exist, their outreach remains limited, according to IT firms interviewed. Moreover, there is a perception by local businesses interviewed that Université de Moncton "programs, researchers and facilities will only work with the French community," as an advanced manufacturing company CEO observed, echoing a sentiment shared by several English-speaking business owners. Two advanced manufacturing CEOs and one IT firm director did mention R&D collaborations with universities, but specifically the University of New Brunswick in Fredericton because of its reputation, specialized resources, business-mindedness, and relative proximity. "It's important to be within two hours' drive with researchers, as we like to see what they're doing, and they need to see what we're doing. It's unfortunate it's not closer [in Moncton], but Fredericton is close enough," a manufacturer shared.

Thus, for Moncton IT, KIBS, and manufacturing firms, knowledge generally does not flow into the region through its fixed local research infrastructure, whether universities, federal labs, or private R&D facilities. While there was evidence of learning-by-interacting with customers and suppliers, much of this was either endogenous learning or knowledge circulation among local actors. Instead, knowledge flows into the region primarily through worker mobility, and to a lesser extent via electronic communications.

Generally, the mobility of workers that mattered most was that of owners and managers. The stewards of Moncton's innovative businesses must travel frequently outside the region. Trade shows, national conventions, and networking events in larger urban centres were often cited as providing venues for learning about new ideas, competitor moves, and market trends. However, those interviewed suggested that more important than fortuitous encounters during temporary knowledge clusters (Schuldt and Bathelt 2011) were scheduled out-of-town meetings and site visits with customers and suppliers alike. For those exporting goods or services outside the province, face-to-face relationships with supply-chain partners mattered most. To Moncton businesses serving mostly the local market, electronic communications were used for competitive intelligence.

Some KIBS firms organize employee "ideas contests," whereby winning employees get to present their ideas at trade events in larger cities. This offers a win-win situation. Employees get a paid trip. The employer shepherds new ideas from workers, motivates employees through contests, and has employees report back on new ideas they hear at the global trade events.

IT, KIBS, and advanced manufacturing owners and managers not only valued travel, but also treasured outside talent, as both a source of new ideas and a means to bridge skills gaps. In smaller metropolitan areas like Moncton, the labour market for specific skills can be quite thin. "Competition for talent is fiercer than for clients," lamented one communications company CEO. For advanced manufacturing firms, this shortage of qualified labour impels them to automate production, even more than they would like, because, as one advanced manufacturing owner shared, "it is easier to import machines than to find the right workers." The economic development implications are broadened by the fact that advanced manufacturing firms have to import most of this technology as there were no local producers. This situation not only curtails local manufacturing company growth, but also hinders spin-off opportunities and regional growth, substituting local workers with imported technology.

For communications firms in Moncton, the dearth of creative workers, relative to Halifax and especially Montreal and Toronto, limits the scale and scope of services they can provide. In light of these constraints, Greater Moncton communications firms actively market themselves as those who understand the region the best, thus seeking the "lower-lying fruit" of local markets. Local knowledge requires an understanding of the cultural nuances that makes a "big difference to consumers": subtleties in humour between linguistic groups, the "surf's up" Maritimer mentality, and so on.

"Lower-lying fruit": Local Markets and Global Technologies

Many Moncton firms interviewed could be split into two separate camps based on their deliberate choice between local or global markets. Export markets are often portrayed as a threshold of success, as displaying an ability of businesses to overcome local constraints. For several firms interviewed, their choice of serving local markets was strategic: their intimate knowledge of local markets was precisely their main asset, and investments in training and new technologies were pre-emptive strategies to stave off the entry of outside firms that may promise shinier products or tools.

There is often a presumed higher value of global connectedness and by extension global knowledge (Weterings and Ponds 2009). In many cases, the most valuable knowledge comes from the ability of knowing your market best. As one IT custom-solutions provider described, "We make no bones about picking the lowest-lying fruit. It's just as tasty." In other words, there may be a lack of scale economies when one targets smaller, local markets, but higher margins may compensate. More important, the need to understand the local market well and customize IT and other solutions accordingly may protect medium-sized markets from large-scale producers and leave such "lower-lying" fruit to smaller, local firms.

Innovations, in these cases, tend to be of the first-to-local-market kind. KIBS and IT services firms adopt globally produced and often Internet-accessible technologies, and customize solutions to local clients. Those clients, however, may be global exporters, and so technologies and services need to be state-of-the-art, which has been made easier by more accessible small-scale, high-quality professional video production and social media.

If KIBS firms were predominantly focused on local markets, IT firms were polarized in their choices between global and local markets. Of the IT firms with a primarily global focus, the largest had more than four hundred employees manufacturing and exporting gaming terminals and software around the globe, while two other firms had fewer than ten employees, yet were global leaders in their field of logistics and time management software. Equally innovative was an IT solutions provider that specialized in coupling wireless and mobile technologies, focusing primarily on the local market.

The type of market coveted was key in explaining the type of knowledge flowing into and circulating within the region. For off-the-shelf software producers, both market and technical knowledge were very global. They could succeed in serving global markets from Moncton by minimizing the need for F2F exchanges with clients. Customers remained crucial in refining the product and developing subsequent versions, but this feedback could be gleaned electronically, through automatically sent error codes and customer feedback mechanisms.

For custom software and IT solutions providers, F2F interactions were indispensable. Key to serving the "lower-lying fruit" markets was developing an intimate knowledge of customer needs, and there was no substitute to frequent F2F interactions and to "being there."

While the importance of local or global market knowledge varied, in all the IT firms interviewed technical knowledge was definitely global, their access greatly enhanced by ICTs. The knowledge they sourced was mainly from online technical references such as O'Reilley Media, and for some the very building blocks of their products were open-source codes, the epitome of codifiable knowledge. None of the IT firms profiled cited any active involvement with local universities and only a few had casual involvement with universities such as UNB or any outside the province. Some economic development officials spoke of an IT cluster, but there was little evidence of clustering, of local formal or informal knowledge exchanges to resolve bottlenecks, to share resources, or to plan strategically. In fact, there was displeasure that social activities organized by a newly created IT association were being used by some firms to poach workers from others. There appeared to be little trust between firms or anything that anchored the globally exporting IT firms in Moncton, beyond being owned by people who had grown up locally or started their company after having moved here. The two largest IT firms were bought by multinational firms – one flourishes while the other has downsized considerably. As Giblin (2011) found in Ireland, large anchoring firms are not sufficient for sustaining clustering effects.

Manufacturers are often more export-oriented because of the nature of the physical goods produced, and because the scale required to minimize production costs far exceeds local demand. Manufacturers were also more likely to have dedicated R&D staff to improve products and tackle process bottlenecks. However, manufacturers saw F2F exchanges with business partners – suppliers in particular – as their main source for innovative products and processes, highlighting the importance of meetings and international trade fairs as temporary knowledge clusters (Ramírez-Pasillas 2010; Schuldt and Bathelt 2011). As these partners were mostly outside the region, meetings and conventions were key knowledge exchange venues, and poor connections from the local airport was deplored as a major obstacle to innovation. Unlike KIBS firms, manufacturers could not build competitive advantages from an intimate knowledge of local markets, and there were few local competitors with whom they could interact and learn. Unlike off-the-shelf software producers, manufacturers could not interact with customers or suppliers electronically in meaningful ways. Neither F2F nor ICTs helped Moncton manufacturers overcome the barriers to innovation from their relative isolation.   .

In summary, those identifying their major strength as the *intimate knowledge of the local market* tended to customize solutions to local clients from globally accessible technologies. Those identifying the *technical knowledge they developed (products or processes)* as their competitive advantage tended to be global exporters. It may be accepted that innovators striving for world-first innovations target global markets more, but it is important to highlight that how knowledge flows into the region depends greatly on market strategies, and that serving local markets can involve deliberate choices and not be just the result of those firms that cannot export.

To What Extent Do ICTs Overcome Geographic Isolation?

There are agglomeration forces that impel the clustering of activities, namely, because some types of knowledge are best communicated face-to-face. There are also forces of dispersal that fling activities to all corners of the world, provided there exists a minimum amount of telecommunication infrastructure. Can ICTs help level the playing field in smaller, relatively isolated metropolitan areas?

For communications firms, leveraging Internet-mediated technologies and social media such as Facebook, LinkedIn, and Twitter are increasingly used to advertise, promote clients' products and reputation, keep clients informed of new or improved products, leverage their feedback on products and quality of service, demonstrate how to use products, recruit workers, and so on. They are being leveraged the world over, and ICTs can help level the playing field in isolated areas.

The new generation of ICTs, wireless, web 2.0 technologies is reshaping how knowledge flows into and across the Greater Moncton region. While businesses in smaller urban centres are keen adopters of new ICTs, most interviewed admitted this was a poor substitute for F2F interactions. ICTs helped compensate for distance to major partners and for the time and cost constraints of interacting with them face-to-face, but F2F remained the prized medium for knowledge exchanges.

Market intelligence, web production, technical referencing, and open sourcing were among the examples cited of how ICTs can level the playing field. Yet some techies admitted their reliance on online IT reference forums were contingent on physical proximity, on trust established during past F2F exchanges with former college mates and colleagues. While a-spatial in transmission, the importance and reliability of these knowledge flows were structured on past linkages that were informal (Trippl, Tödtling, and Lengauer 2009) and localized.

There was a perception by business owners that there were no or few suppliers or competitors with whom to interact locally beyond social networking events. In some cases there was anxiety that competitors would poach workers. There was considerable uncertainty as to the value of networking with local research facilities, although some businesses were engaged in collaborative work with universities elsewhere. Some university/private-sector partnerships heralded by university officials were called "painful" by the very researchers and business owners involved. There are few science and engineering graduate students and no English-language computer science programs, to name some shortages.

Fitjar and Rodríguez-Pose (2011) may be right in suggesting that in some peripheral regions, economic success can owe more to connectedness with larger, world cities than to proximity and exchanges among local firms. Many of the company owners and operators interviewed were very well travelled and connected globally, knowledge pipeline builders to the outside world (Bathelt, Malmberg, and Maskell 2004). Global knowledge flows were fairly lubricated for manufacturers and many IT firms. However, it was also clear that meaningful knowledge exchanges were very weak on a local level.

The greatest deficiency was indeed the weak innovation linkages between community, business, government, post-secondary education, and other stakeholders, which were certainly not at the level found in other centres. There are successful innovative entrepreneurs and products, but there is little evidence of a culture of innovation in Greater Moncton. The region has done well at addressing important economic challenges from the 1980s, but continued prosperity requires a new mindset. If Moncton is to sustain continued economic growth, and particularly higher average earned wages, creative ways to co-produce, capture, and transform productive ideas are needed.

ACKNOWLEDGMENTS

I am indebted to the Social Sciences and Humanities Research Council for its generous funding, as well as Robert MacKinnon and Tracy Chiasson of UNBSJ, and Rodrigue Landry and Hélène Gallant of CIRLM, for administering the grant. I extend my appreciation to Daniel Bourque, who provided much appreciated research support, and of course to the many interviewees who volunteered their precious time. I am grateful to David Wolfe and Meric Gertler for their admirable stewardship of this comparative project, and to David in particular for his invaluable comments on the chapter.

REFERENCES

Bathelt, Harald, Anders Malmberg, and Peter Maskell. 2004. "Clusters and knowledge: Local buzz, global pipelines and the process of knowledge creation." *Progress in Human Geography* 28 (1): 31–56. http://dx.doi.org/10.1 191/0309132504ph469oa.

Cusmano, Lucia, Maria Luisa Mancusi, and Andrea Morrison. 2010. "Globalization of production and innovation: How outsourcing is reshaping an advanced manufacturing area." *Regional Studies* 44 (3): 235–52. http://dx.doi.org/10.1080/00343400802360451.

Doloreux, David, and Richard Shearmur. 2010. "Exploring and comparing innovation patterns across different knowledge intensive business services." *Economics of Innovation and New Technology* 19 (7): 605–25. http://dx.doi.org/10.1080/10438590903128966.

Farnsworth, Clyde. 1994. The "Moncton miracle": Bilingual phone chat. *New York Times*, 17 July.

Fitjar, Rune Dahl, and Andrés Rodríguez-Pose. 2011. "Innovating in the periphery: Firms, values and innovation in southwest Norway." *European Planning Studies* 19 (4): 555–74. http://dx.doi.org/10.1080/09654313.2011.548467.

Friedman, Thomas. 2005. *The world is flat: A brief history of the 21st century.* New York: Farrar, Straus and Giroux.

Giblin, Majella. 2011. "Managing the global-local dimensions of clusters and the role of 'lead' organizations: The contrasting cases of the software and medical technology clusters in the west of Ireland." *European Planning Studies* 19 (1): 23–42. http://dx.doi.org/10.1080/09654313.2011.530529.

Larracey, Edward W. 1991. *Resurgo: The history of Moncton.* Moncton: City of Moncton.

Ramírez-Pasillas, Marcela. 2010. "International trade fairs as amplifiers of permanent and temporary proximities in clusters." *Entrepreneurship and regional development* 22 (2): 155–87. http://dx.doi.org/10.1080/08985620902815106.

Schuldt, Nina, and Harald Bathelt. 2011. "International trade fairs and global buzz. Part II: Practices of global buzz." *European Planning Studies* 19 (1): 1–22. http://dx.doi.org/10.1080/09654313.2011.530390.

Statistics Canada. 2007. *Moncton, New Brunswick (Code 305)* (table). *2006 community profiles.* 2006 census. Statistics Canada catalogue no. 92-591-XWE.

Ottawa. http://www12.statcan.ca/census-recensement/2006/dp-pd/prof/
92-591/index.cfm?Lang=E.

Statistics Canada. 2011a. *Table 282-0054 – Labour force survey estimates (LFS),
by provinces and economic regions, 3-month moving average, unadjusted for
seasonality, monthly (persons unless otherwise noted)*, CANSIM (database),
using E-STAT (distributor).

Statistics Canada. 2011b. *Table 051-0046 – Estimates of population by census
metropolitan area, sex and age group for July 1, based on the Standard geographical
classification (SGC) 2006, annual (persons)*, CANSIM (database).

Statistics Canada. 2011c. *Table 282-0061 – Labour force survey estimates (LFS), employment by economic region and North American industry classification system
(NAICS), annual (persons)*, CANSIM (database).

Storper, Michael, and Anthony Venables. 2004. "Buzz: Face-to-face contact and
the urban economy." *Journal of Economic Geography* 4 (4): 351–70.

Trippl, Michaela, Franz Tödtling, and Lukas Lengauer. 2009. "Knowledge
sourcing beyond buzz and pipelines: Evidence from the Vienna software
sector." *Economic Geography* 85 (4): 443–62. http://dx.doi.org/10.1111/
j.1944-8287.2009.01047.x.

UNFPA (United Nations Population Fund). 2010. State of world population
2010. http://www.unfpa.org/webdav/site/global/shared/swp/2010/
swop_2010_eng.pdf.

Weterings, Anet, and Roderik Ponds. 2009. "Do regional and non-regional
knowledge flows differ? An empirical study on clustered firms in the Dutch
life sciences and computing services industry." *Industry and Innovation*
16 (1): 11–31. http://dx.doi.org/10.1080/13662710902728035.

13 Networking Patterns and Performance of the Trois-Rivières City-Region's Firms in the Light of Sectoral and Place Characteristics

MICHEL TRÉPANIER, PIERRE-MARC GOSSELIN, AND ROSEMARIE DALLAIRE

Innovation, as much for large companies as for small and medium-sized enterprises (SMEs), is of great importance in the economy of knowledge and information of the twenty-first century (OECD 2001). To innovate, firms must exploit not only their internal resources and capacities, but also, in a complementary way, their external environment (Pittaway et al. 2004; Calantone, Cavusgil, and Zhao 2002). Thus, R&D, the adoption of new ways of doing things, new processes, or product innovations are often the result of the "interactive" exploitation of the resources and capacities both internal and external to the firm (Becheikh, Landry, and Amara 2006; Dosi 1988).

Existing research also shows that companies of a given industry innovate by exploiting the resources and capacities of the organizations in proximity to them (Asheim and Gertler 2007; Sternberg 2007). Innovation is fundamentally social by nature and is the result of collaborations and interactions between firms, but also, and especially, between the companies (competitors, suppliers, consulting firms, etc.) and organizations (educational establishments, government and university research centres, technology transfer centres, economic development services, etc.) which work on the level of a city, region, or country (Lundvall 1992; Doloreux 2004; Cooke 2001; Cooke, Heidenreich, and Braczyk 2004; Wolfe 2009).

Thus, innovation is created and maintained primarily and essentially by local/regional processes (Asheim and Cooke 1999; Uzzi 1997). And because it unfolds better in relations of geographical and social proximity (Boschma 2005), particularly in face-to-face interactions, which are more favourable and more effective for the construction and transfer of the tacit knowledge and know-how that are so important in innovation projects

Table 13.1 Factors influencing the innovative behaviours of firms (adapted from Becheikh et al. 2006)

• Firm's general characteristics	• Firm's industry (sector)
• Firm's global strategies	• Firm's region
• Firm's structure	• Networking
• Control activities	• Knowledge/technology acquisition
• Firm's culture	• Government and public policies
• Management team	• Surrounding culture
• Functional assets and strategies	

(Sternberg 2007; Uzzi 1997), the city-region becomes a critical spatial scale for both investigation/observation and intervention (Wolfe 2009, 15).

In the literature on innovation and city-regions, the characteristics of the city-region are seen as key factors for understanding how firms innovate and for explaining the success they have. Variables like firms' size (OECD 2006, cited in Wolfe 2009), the degree of sectoral specialization or diversity (Frenken, Van Oort, and Verburg 2007; Beaudry and Schiffauerova, 2009), the presence of a concentration of economic actors, technology users and producers, and supporting institutions (government services, colleges, universities, etc.) (Drejer and Vinding 2007), the model of governance (institutions and political structures) (Safford 2004), the capacity to attract and retain talented people (Florida, Mellander, and Stolarick 2008, cited in Wolfe, 2009) are all key determinants of the innovative behaviours of firms and consequently of their capacity to succeed in that matter. From this perspective, because innovation is closely linked to the capacity to succeed in the worldwide economy, the characteristics of the city-region become key determining factors of economic development.

On the other hand, it is a widely accepted fact, in "business studies," that the sector of activity of firms influence the way they innovate (Becheikh et al. 2006; Malerba 2002). This literature has shown the influence of a wide variety of variables, ranging from internal factors such as the firm's global strategy and culture to external factors, such as its region and government and public policies (see table 13.1). As such, we can see that there is an interesting overlap between "city-region" and "business" studies when it comes to the variables considered to be relevant to explain the innovative behaviours and success of firms. In fact, the firm's industry or sector is the only "external factor" listed in table 13.1 that is neglected to some extent in the literature on city-regions.

Here, when we say neglected we do not mean that sectors have not been studied in this literature, but rather that, contrary to what is happening in business studies, the sector does not usually play a significant role as an explanatory/differentiating factor of the firms' innovation practices and innovative performance.

In business studies, a more detailed look at a firm's sector reveals that it plays an important role in the determination of the specific characteristics of a firm's innovation process and also in its success/failure (Becheikh et al. 2006).

These conclusions are not inconsistent with some research results in the geography of innovation. In a recent paper examining "to what extent and how the characteristics of places affect the innovative performances and innovation patterns (e.g. use of networks) of firms," Isaksen and Onsager (2010, 228) reach the conclusion that "the firms' innovation patterns are quite similar irrespective of location." More specifically, after studying innovation patterns and performance of knowledge-intensive industries in three of Norway's quite different regions (different size, centrality, types of agglomeration economy, innovative performance, types of regional innovation system, geographical knowledge flows, and firm's innovation models), they observe that

> knowledge-intensive firms have more or less the same innovation partners, knowledge sources and knowledge channels irrespective of their location. The differences and the similarities in innovation patterns in the three quite different types of region indicate that both the characteristics of the industrial sector and the regional environment matter. (Isaksen and Onsager 2010, 241)

In a study focusing on the correlation between innovation performance and strategies of Canadian firms and city size, Therrien (2005, 865 and 872) reached similar conclusions:

> The variation between establishments by city-size is much lower than when establishments are compared between industrial sectors … the last result showing that establishments' innovation performance does not vary much between city sizes … Using an inclusive definition of innovation (introducing a product or process new to the firm), city-size does not matter.

But Therrien also points out that "city-size matters when using a measure qualifying the innovation (world-first versus Canada-first versus firm-first innovation)" (Therrien 2005, 872).

In sum, both the sector and the characteristics of place seem to play a significant role in the innovative capacity and performance of firms, and therefore in the economic performance of the city-region in which they are located. Instead of thinking about these from an "either/or perspective," it seems more productive to use them in a complementary manner. It is from this perspective that the present research explores the case of Trois-Rivières – a small city-region specialized, on the one hand, in a few traditional and low-and-medium-technology sectors and which, on the other hand, is trying to diversify and renew its economy by developing new high-tech sectors. In term of performance, the region is going through difficult times mainly because its mature and more important industrial sectors are losing markets and, consequently, jobs (Rochette 2011; Rochette 2010a; Radio-Canada Nouvelles 2000). Moreover, the development of new high-tech sectors is undoubtedly slower and more difficult than expected.

In light of the literature presented above, what can we learn from this particular case and how can we explain the situation that prevails in Trois-Rivières? Is it a question of size: the Trois-Rivières city-region is too small and too "closed in on itself" to offer the knowledge and ideas firms need to innovate? Is it a question of diversity or, more specifically, the lack of it: the supply and diffusion of new knowledge/ideas and opportunities is "weak" because cross-fertilization between sectors is impossible due to the low level of industrial variety and the absence of cross-sectoral interactions? Is it a question of sector dynamics and characteristics: the sectors in which the city-region is specialized are simply declining or the technology firms utilize is becoming outdated?

The Trois-Rivières city-region case allows us to explore these questions, which are important from both a scientific and policy point of view. As we will see, the deficiencies observed in the renewal of the city-region economy are linked not only to its size or degree of specialization (or lack of variety), but also to the specific characteristics/dynamics of the sectors in which it is specialized or is trying to be. In fact, both the characteristics of place and the characteristics of the sectors seems to play a significant role in understanding current economic performance and also as inhibitors of the renewal and diversification that can potentially result from the new knowledge produced and diffused within each sector.

With that in mind, our analysis will focus on how, in the innovation process of firms belonging to a specific sector, the interactions between those and the socio-economic actors of their region can or cannot constitute crucial sources for access to the various assets which are necessary for innovation. On this level, our analysis seeks answers to the following questions: where do the firms find the external resources and capacities they need to innovate? How do they use proximity resources and capacities to conclude their innovation projects? Do all the companies have similar patterns of interactions independently of the sector? The answers to these specific questions will then be used to explore the link between these patterns of interactions, the characteristics of both the sectors and the city-region, and their economic performance.

In answering these questions, we will take into account that time is a key variable. In fact, to better circumscribe and understand the relational playing field of companies within a given territory, it is necessary to take into account that the patterns of interaction of the companies and their propensity to collaborate are dependent on their history, their sector, and the often concomitant evolution of the organizations and institutions which compose the city-region (Andersen and Drejer 2008; MacKinnon, Cumbers, and Chapman 2002). From this point of view, the city-region is engaged in a continuous process of adjustment (Asheim and Gertler 2007; Lazonick 2006) to consolidate the integration and the regional embeddedness not of all the companies and organizations but rather of certain companies and organizations and the "disembeddedness" of others.

The study carried out on twenty firms of the Trois-Rivières region shows that the regional, national, or international location of the organizations with which they interact varies largely according to the branch of industry of which they are a part. In fact, the sector induces variations in the networking practices of the firms. For the companies working in mature sectors, the innovation process is marked by interactions of proximity and the prevalence of regional actors. By contrast, the companies belonging to emergent industries interact with actors external to the region in which they were born and continue to operate. In all sectors we found little evidence of inter-sectoral interactions. As we will see, the characteristics of place and of the sectors play a differentiated role in the determination of firms' innovative behaviours and performance: in mature sectors, sectoral factors are crucial for understanding the situation, while in emergent industries the characteristics of place are critical.

Methodology

This case study rests on the analysis and comparison of data gathered by means of semi-structured interviews in twenty firms of the Trois-Rivières region.[1]

The construction of the sample was based on the regional strategy of economic development of the Trois-Rivières region. Like all the regions of the province of Quebec, the Trois-Rivières region takes part in the ACCORD (Action concertée de coopération régionale de développement) program of the Quebec government, an initiative of regional development aimed at building a productive and competitive regional system in each region of Quebec by identifying and developing "niches of excellence," which could become their region's hallmarks. The ACCORD program calls upon the capacity of the firms of a given region to innovate and adapt themselves to economic and technological changes, to mobilize and prove themselves in specific fields/sectors related to their competences and their comparative advantages. It supports the regrouping of people: businesses and contractors of the same region which have a common perception of their field of activities, its potential, its forces, and its weaknesses, and which define a long-term strategy for the latter (MFEQ 2013).

Because of the place it grants to innovation and regional collaboration, this economic development program had, from a scientific point of view, great relevance for our study. The sample of twenty companies was thus built to be representative of the various figure cases of each of the five niches of excellence of the ACCORD program, which are, in addition, representative of the "economic fabric" of the region: value-added papers and associated technologies, hydrogen and electro technologies, technologies of metal transformation (magnesium and titanium), furniture making, and industrial bioprocesses. Meetings were held with experts in the National Research Council (NRC) of Canada's regional office, the Technopole Vallée du Saint-Maurice, and the Société de Développement Économique de Trois-Rivières to validate representativeness within each niche. Needless to say, the twenty firms interviewed are not statistically representative of the Trois-Rivières's city-region economy. However, given that the region's economy is relatively small, it remains reasonable to think that these twenty firms are representative of the typical situations and characteristics one can find regarding innovation practices.

Portrait of the Trois-Rivières City-Region

From the theoretical perspective described above, the history of the Trois-Rivières city-region and more specifically its economic history are extremely useful for the comprehension of the bonds and interactions that the economic actors and intermediate organizations maintain between themselves, inside and outside the region (Carrier and Gingras 2004; LeMay 2002) Founded in 1634, Trois-Rivières is located at the confluence of the Saint-Maurice and St Lawrence rivers. From the sixteenth century onwards, this geographical location positioned the city as a major nerve centre as much for the development of its own economic activity as for that of the Mauricie region. From an economic point of view, Trois-Rivières became industrial with the entry into production of the Saint-Maurice Forging Mills around 1733. The foundry manufactured cannonballs, cast-iron stoves, various utensils and kitchen accessories, and represented the area's main economic activity for a long period of time (Bloomfield and Bloomfield 1994). In the next century, with developments in the forestry sector and energy production in the Saint-Maurice valley, Trois-Rivières welcomed first sawmills and then paper factories. Thus, the foundations of the economic development of the Trois-Rivières city-region are to some extent a trio with its origins in the eighteenth and nineteenth centuries: wood and paper, metallurgy, and energy. The last element is important: electricity coming from Shawinigan, the second city in importance within the region, fed the growth of the metallurgy companies and the paper industry. Even today, these three sectors are essential components of the socio-economic fabric of the city-region.

Since the middle of the 1990s and, more recently, since 2006, the Trois-Rivières city-region has experienced an intense period of changes (Desjardins Études économiques 2007). The basic economic activities of the region which were historically the engines of its economic growth are in a difficult position at the moment, particularly because of the crisis in the forestry and paper sectors. As in many Canadian small city-regions to which these traditional manufacturing sectors are crucial, the rise of production costs, obsolescence of the equipment used by firms, reduction of the volume of wood made available, and increasing foreign competition all challenge the performance of the industrial pillars of the local economy. However, one observes strategic changes in the manufacturing sector, where firms, old or new, try with varying degrees of success to reposition themselves in specialized niches where the materials, processes, knowledge, and know-how are to a certain point those used in Mauricie's traditional sectors: the manufacture of windmills, components for transportation equipment, plastic and rubber products, electronic materials, and aeronautical components.

In parallel, emergent industries in the city-region attenuate the impact of the decline of the large industrial actors (MDEIE 2008; Desjardins

Études économiques 2007). Thus, the investment in the services sector and the housing industry, for example, doubled between 2001 and 2006 (MDEIE 2008). It is in the services sector that the greatest increase in employment is observed: between 1996 and 2006, nearly 10,000 new people found work, particularly in the professional, scientific, and technical services, services within companies, health care, lodging, and restoration (MDEIE 2008; Desjardins Études économiques 2007).

The innovation support infrastructure of the Trois-Rivières city-region is also strongly marked by the region's industrial and institutional history. Particularly important were the opening of institutions of higher education starting in 1960 and of research and economic development organizations from the 1980s onwards. As we will see, the overlap of these two stories explains a good part of the companies' networking practices and use of the resources of the city-region.

All in all, the last ten years have witnessed growth in the resources dedicated to innovation. For example, companies' R&D expenses grew from $334,000 to $695,000 between 1999 and 2004 on an establishment basis, an increase of 15.8 per cent (MDEIE 2008). This increase originates mainly from the manufacturing sector (MDEIE 2007). Between 1998 and 2003, the relative size of the region within the total provincial R&D spending rose from 0.6 to 2.1 per cent. The workforce deployed in industrial R&D also grew from 2.5 person-years per 1000 active people in 1999 to 3.4 person-years in 2004. However, for all these indicators, the Mauricie region still has lower figures than the provincial average (MDEIE 2007).

In research (fundamental, strategic, or applied) and in technology transfer, the most significant interventions have without a doubt been the creation of the Université du Québec à Trois-Rivières (UQTR), the CEGEP de Trois-Rivières, and the Collège Laflèche, as well as the creation of the College Technology Transfer Centres and government research centres. More recently, organizations supporting innovation such as the Technopole Vallée du Saint-Maurice, a regional office instigated by the NRC (IRAP program), an FPInnovations[2] regional office, and other organizations supporting innovation joined the institutions mentioned above. This innovation support infrastructure is thus composed of organizations whose research programs and services are closely related to the traditional industries of the Trois-Rivières city-region.

The Université du Québec à Trois-Rivières (UQTR) accommodates each year nearly 11,500 students registered in some 160 undergraduate and graduate programs in fields like pulp and paper, neurosciences and health sciences, engineering, ecology and biology, membrane bio-technology, photo biophysics, industrial electronics, dielectrics, business administration, and education. The Centre de recherche en pâtes et papiers (CRPP) was founded in early 1980s and combines fundamental and applied research on subjects of interest to the regional pulp and paper industry. For example, CRPP has selected research projects that directly concern mill operations. Founded in 1994, the Institut de recherche sur l'hydrogène (IRH) is interested in the development of scientific and technological knowledge concerning the storage, transportation, and safe utilization of hydrogen. Collaboration with industry and the training of graduate students and qualified workers are prime concerns. The Institut de recherche sur les PME (petites et moyennes entreprises) was founded in 1997 and is interested in the development of fundamental and applied knowledge on SMEs in order to contribute to their development. Outside those three major centres, UQTR also hosts more than forty research groups, including twelve Canada Research Chairs, and eleven industrial research chairs. Those research groups work in many fields, including membrane biotechnology, photo biophysics, industrial electronics, dielectrics, aquatic ecosystems, molecular oncology, molecular neuropharmacology, neurosciences, chiropractic, plant biology, educational achievement, mental health, history, management, visual arts.

The Centre spécialisé en pâtes et papiers du CEGEP de Trois-Rivières (CSPP) was created in 1989 and houses a pilot factory in which it is possible to reproduce pulp and paper treatment and manufacturing processes by using virgin or recycled fibres to manufacture papers and paperboards. It is also possible to start from a pilot factory or industrial pulp to develop new products with high technological content or new processes or equipment. In 2007, CRPP and the CSPP merged into a single centre, the Centre intégré en pâtes et papiers (CIPP), which now counts on high-tech equipment in the field of value-added papers and is researching new production processes. Created in 1985, the Centre de métallurgie du Québec (CMQ), is a collegiate centre of technological transfer affiliated with CEGEP de Trois-Rivières. Its focus is to support the technological development of Quebec-based manufacturing companies in the metallurgical sector. CMQ provides a knowledge base on metallic and ceramic materials as well as on metallurgical processes.

In 1993, the Collège de Shawinigan created the Centre national en électrochimie et en technologies environnementales (CNETE). This collegiate centre of technology transfer is conducting applied research and providing technical assistance to firms in the fields of bioprocesses, technologies of separation per membrane and electrochemistry. Also in Shawinigan, the Laboratoire des technologies de l'énergie (LTE) of Hydro-Québec was founded in 1987 and devotes itself to the discovery of new applications for the use of electricity or of combined forms of energy. It develops new technologies or adapts existing ones for the new market requirements, but also to support its customers' performance.

In the last fifteen years relations between innovation support organizations and firms have increased steadily. In fact, the research and transfer organizations presented above are used extensively by the companies within the region. It is to be noted that in 2002–3, nearly two-thirds of the $627,214 spent on R&D by regional companies in Quebec universities went to UQTR (OSRIM 2008). That being said, regional innovation indicators show that the institutional part of Mauricie's innovation expenditures is lower than the provincial average. For example, even though expenditures in university research increased by 46 per cent for the year 2002, this growth places Mauricie second to last out of the ten regions with a university. After a period of rationalization during the second half of the 1990s, the number of professors and researchers has increased in all Quebec regions except for Mauricie, which was unable to maintain the rank it had five years before. Similarly, the number of inventions patented at the United States Patent and Trademark Office in 2006 declined compared to 1996 (MDEIE 2008).

Sectoral Patterns of Networking

Value-added Papers and Associated Technologies

In the Trois-Rivières city-region, pulp and paper companies form a mature sector made up of some large companies producing paper or manufacturing equipment for the international market. Around this core revolve SMEs like machinery manufacturers, suppliers of raw material and products used in the manufacturing of paper, as well as manufacturers of specialized paper products. Table 13.2 presents the networking practices of the companies.

Within the pulp and paper sector (value-added papers), innovation is especially directed towards the improvement of existing products and processes. Incremental innovations are carried out on the products in order to present products with appreciable modification and improvement in terms of appearance or performance. In certain cases, those

Table 13.2 Networking practices of firms in the value-added papers and associated technologies niche (*n* = 4)

Types of innovation	Assets	Intervening actors	Actors' main location
Product improvement	Talent	CEGEP de 3-R, UQTR; high schools; universities (QC)	R
	Know-how	Local equipment manufacturers; CIPP, Centre international de couchage; similar firms	R
Process improvement	Knowledge, R&D	CIPP, Centre international de couchage; PAPRICAN research centre; from the inside (locally and internationally)	R
	Innovation opportunities	Clients, suppliers; from the inside (internationally)	NR
	Material and financial resources	Head office; banks; governments; equipment suppliers	NR

Note: R = region; NR = non-region.

small innovations aim to allow different uses of the product. For the processes, a good part of the innovation relates to the pulp "recipe" which is used for manufacturing and largely influences the characteristics of the paper produced: the ingredients used in the recipe are changed, the mixtures are improved, and new components are introduced. Radical innovations are rare: the production equipment's life cycle is of twenty to twenty-five years and the development of "radically new" papers is also not very frequent. However, firms devote some resources and efforts to fundamental research, which ensures them a follow-up and a minimal participation in projects of this type.

In all the innovation projects discussed by the participants, one observes strong and frequent collaborations with regional R&D and technology transfer centres. The major part of the technical know-how and knowledge and also of the R&D which the firms mobilize in their innovation projects originate within the enterprises or in organizations of the Trois-Rivières city-region. These specific projects (R&D projects, interventions, tests, and processes improvement) but also everyday interactions between the people concerned produce results and innovations which are initially established locally and then transferred inside the companies of the same group and, in certain cases, in firms of the same region and sector and which are not direct competitors. In addition, as some of the companies belong to large groups, part of the tests, research projects, and production of knowledge necessary to innovation are carried out outside the region within other companies belonging to the same group.

In paper transformation factories, the large equipment was replaced by smaller and more "technological" machinery. Formerly, an employee could learn how to operate, repair, and modify the machinery from on-the-job training provided by an older employee. Today, the manufacturing processes require more specialized, more advanced, and technical training. The companies find these talented resources mainly in educational establishments of the region: at UQTR at the university level (scientific training), in CEGEP de Trois-Rivières at the collegiate level (technical training), and in the vocational programs of the region's high schools. In all these establishments, one finds training programs specific to pulp and papers and industrial mechanics which are offered exclusively in the Mauricie region. While offering a specific support to this industry, these training programs also play a role in the building of its strong position at the regional and provincial levels. Many firms of the sector thus consolidate their technological capacity by increasing the education level of their personnel.

The markets for the products manufactured by the companies are international. Thus, as this mature sector is marked above all by incremental innovation, ideas on product innovations come more frequently from the customers and those are almost invariably outside the city-region. With regard to process innovation, ideas on innovations come more frequently from the suppliers, and, with one exception, they are also sourced outside the region. In both cases, the networking practices of the companies reflect their international status as well. However, as we mentioned earlier, all does not occur abroad, since an equipment manufacturer of international status was established in the region. They develop, manufacture, and install the major part of the equipment and machinery in the factories. In fact, high-tech equipment is thus developed and manufactured in the region to meet the needs of local costumers but also of Canadians, North Americans, and, increasingly, Asian ones. Finally, one observes the same phenomenon of extra-regional networking for the acquisition of the financial resources required to innovate. Being international firms, these companies find the money they need from governmental or private institutions which work at the national or international level.

Generally, one observes that the firms of this sector often resort to regional organizations for their innovation projects. They find in regional institutions several of the essential assets they need to innovate. At the same time, firms but also research and technology transfer centres have frequent and regular interactions/collaborations with organizations outside the city-region. There, they find innovation opportunities but also the financial resources and the new knowledge they do not have at hand, but which they need to move ahead in their innovation projects. Within the sector, major firms and research organizations assume a role of "global pipelines" (Bathelt et al. 2004; Owen-Smith and Powell 2004; Scott 2002).

Metal Transformation Technologies

As is the case with pulp and paper, the metal transformation industry is a mature sector whose presence goes back to the first phases of industrialization in the region. The sector counts forty companies which provide nearly 3000 direct jobs. The sector comprises some companies acting at a national or international scale but also includes more specialized SMEs whose products are generally intended for large industrial clients. In our sample, the majority of the companies transform metals used in the manufacturing of other products. Table 13.3 presents the networking practices of these companies.

In the firms we encountered, innovation is generally an improvement made to existing products or processes. The customers of these companies ask for products that, without being radically new, nevertheless have a certain degree of innovation or an improvement in comparison to available products. Since several of these companies develop and manufacture components which are integrated into other products, the innovation often consists in re-examining the characteristics of certain elements to improve performance or functionality or to reduce costs. One of our respondents summarized innovation this way: "The customers ask for improved products and force us to have new ideas." In addition, to make these innovations possible and successful, the companies must introduce new equipment and improve their manufacturing process. Another frequent "source" of process innovation is the need to reduce costs and be more productive.

Table 13.3 Networking practices of firms in the metal transformation niche ($n = 4$)

Types of innovation	Assets	Intervening actors	Actors' main location
Product improvement	Talent	CEGEP de 3-R, high schools, UQTR; universities (QC); intermediaries	R
	Know-how	CCTT in metallurgy (CEGEP de 3-R); industrial partners; CNRC/IRAP (local)	R
Process improvement	Knowledge, R&D	From the inside (local); CCTT in metallurgy (CEGEP de 3-R); UQTR	R
	Innovation opportunities	Clients, suppliers; environmental-industry partners/ firms	NR
	Material and financial resources	From the inside (local); SDÉ (local); banks (regional); governments (regional offices); equipment suppliers	R

Note: R = region; NR = non-region.

In all these innovation projects and without regard to the level of education the employee must have attained, the talent resources have been, for the most part, trained and recruited within the region. The graduates of UQTR, CEGEP de Trois-Rivières, and establishments of the different school boards constitute the principal recruitment pool. Moreover, not only do these companies work with the educational establishments to recruit their labour force, they also use them for the in-service training of their employees.

For the development of the knowledge and know-how necessary to their innovation projects, the firms interviewed work frequently and on a relatively continuous basis with the research and technology transfer centres of the region, mainly UQTR's department of chemistry and the Centre de métallurgie du Québec of the CEGEP de Trois-Rivières.

Within these organizations' settings, the companies carry out tests, trials, and some R&D projects. In the same fashion, firms of this sector collaborate with the engineers of the NRC's regional office, where they seek expertise on production processes and linkages with organizations where they can test their new products. As in the pulp and paper industry, it is mainly their customers who suggest innovation opportunities. In that sense, it is their respective markets, which are primarily national or international, that afford the enterprises their innovation opportunities. Lastly, when the money invested does not come directly from the enterprise itself, the financial resources of the companies mostly come from regional organizations. The firms also have frequent regional interactions with governmental organizations of economic development to obtain technical assistance or financial support.

In short, one can note the strong presence of regional organizations in the innovation projects of this sector's companies. Even more than the companies of the pulp and paper sector, they access the essential assets which they need to innovate in regional institutions. That being said, the sector is not "closed" on its city-region, mainly because it is their national and international suppliers and clients that provide firms with ideas on desirable innovations and also with a significant part of the tacit knowledge they need to conduct those projects. As with the pulp and paper sector, major firms and research organizations play the role of "pipelines" connecting the city-region to the outside.

Furniture Industry

This mature, low-tech, and relatively traditional sector is massively directed towards export, considering that, year in, year out, about 80 per cent of its production goes directly to the United States. Within the Trois-Rivières city-region the sector is structured around two large firms and about fifty subcontractors, who manufacture components which are then assembled by the contractor firm, who ensures the products' completion and commercialization. Also parts of this sector are SMEs which design, manufacture, and market products for the local and national market. The great majority of these companies work in the mid-level and high-end market of kitchen and dining room furniture. Table 13.4 presents the networking practices of the companies encountered in the course of our research. In this sector, innovation mainly consists in re-examining and improving existing products and processes. When it exists, R&D specifically aims at optimizing the industrial processes.

Table 13.4 Networking practices of firms in the furniture industry (*n* = 4)

Types of innovation	Assets	Intervening actors	Actors' main location
Product improvement	Talent	CEGEP de 3-R, CEGEP de Victoriaville; high schools (regional)	R
	Know-how	Suppliers; high schools (regional); CCTT du meuble (Victoriaville); transfer of know-how within the firm	R
Process improvement	Knowledge, R&D	Chaire du meuble UQTR; CCTT du meuble (Victoriaville); suppliers (regional)	R
	Innovation opportunities	QFMA (provincial); trade shows; conferences; readings	NR
	Material and financial resources	Suppliers (national, international); banks (provincial, national); from the inside	NR

Note: R = region; NR = non-region.

In their innovation projects, firms benefit from regional organizations; particularly the Chaire industrielle de recherche sur le meuble at UQTR, FPInnovations' regional office, and the École québécoise du meuble et du bois ouvré at CEGEP de Victoriaville (EQMBO-ENTREPRISES). In this last case, the organization is located outside the Mauricie region, but is still adjacent (less than one hour away from Trois-Rivières). The companies of the sector turn to these organizations for knowledge in innovation management, product development, industrial design, industrial engineering, wood machining, surface treatment, and CNC technologies. With these regional organizations, the companies carry out research, development, and transfer projects which are closely related to their own innovation projects. For the projects more specifically directed towards production process, equipment and raw-material suppliers of (mainly solid or laminated) wood are privileged sources of technical and operational information. Even if some of these suppliers are within the region, the majority of them are located outside Mauricie and operate at the provincial and even national level.

If the regional organizations remain their principal source of knowledge in regard to innovation, it has nevertheless been observed recently that companies of the sector also called upon extra-regional organizations: particularly Laval University's research centres on wood (Chaire industrielle sur les bois d'ingénierie structuraux et d'apparence [CIBISA], FORAC), Québec's Centre de recherche industriel (CRIQ), and FPInnovations' laboratories, which are all located in the Quebec city-region. Given the minimum technological level of the firms in this industry, the majority of its workers have received a basic vocational training in high school. Some of them benefited from a more advanced and technical training at the college level. The companies mostly find their personnel within the region: at UQTR and the CEGEP of Trois-Rivières for what relates to industrial engineering, at the CEGEP de Victoriaville's École nationale du meuble et de l'ébénisterie, and in the high schools of the Trois-Rivières region for what concerns manufacturing as such (cabinet work and serial production). Moreover, all these organizations participate in the continuing education of the sector's employees. However, it is still common in this industry that employees are partly trained on the job by more experienced workers, who transfer to them practical knowledge and know-how.

Like all the companies present in the export markets, the sector's largest firms identify innovation opportunities in interaction with their customers and by carrying out surveys on these exterior markets. These large firms also participate in international furniture and design exhibitions and trade shows of production equipment. SMEs which market their own products, as well as those which subcontract to other companies, also participate in these exhibitions in search of ideas for new innovations. Sectorial associations – provincial (Quebec Furniture Manufacturers Association) or national (Canadian Home Furnishings Alliance) – are largely used by all companies.

The material and financial resources mobilized by firms in their innovation projects mainly come from outside the city-region. On the one hand, the machinery used in the manufacturing process is generally not of regional origin; on the other hand, the financial resources come primarily from financial institutions located in larger metropolitan city-centres (Montreal) and, in small part, from institutions of the region.

All in all, the networking practices of the companies met in the furniture sector give an important place to interactions with organizations of the Trois-Rivières region. As with the metals and pulp and paper sectors, the companies of the furniture sector turn initially and especially towards regional organization to find the innovation assets they partially or totally lack. But as was the case with the previous two sectors, extra-regional firms and support organizations also play a significant role.

Hydrogen and Electro-Technologies

In Mauricie, the hydrogen and electro-technologies niche must be regarded as emergent. The firms are relatively few and they are usually young in age. There are more enterprises in the electro-technologies sub-sector than in the hydrogen one. Moreover, this last niche is articulated mainly around a research infrastructure at UQTR: the Institut de recherche sur l'hydrogène. Several regional economic development agents believe that the hydrogen field has a strong potential for growth, in which they hope to see the region assuming a leadership role. As for the electro-technologies field, it is composed of enterprises which use the knowledge accumulated throughout the years to link with hydroelectric and nuclear projects (Gentilly-2 nuclear generating station): developers and manufacturers of transformers, machinery for power generation, energy systems, location systems, and consulting services for Hydro-Québec's large hydroelectric and nuclear projects. Table 13.5 presents the networking practices of these firms.

The companies of this sector are characterized by important R&D activities which are mainly carried out within the company. These innovation activities mainly consist in developing new production processes or in improving existing products. With their customers, these companies not only construct infrastructures, but also undertake R&D projects whose results are immediately used in actual engineering work. Most of the time, these projects require a great contribution of knowledge, R&D, and know-how and are carried out in tight collaboration with the customers, which are located in Quebec, everywhere in Canada, and internationally. Moreover, the majority of these companies are "branch plants" and are thus supplied with knowledge by the head office or others plants/offices of their group. In this sector, the regional research organizations, particularly the IRH and the LTE, are not used with the same intensity we observed in other sectors. Of course, the companies call upon the services of the researchers of these two organizations, but their external independent source of knowledge remains their customers or other plants of their group, who are mainly located outside the region.

Table 13.5 Networking practices of firms in the hydrogen and electro-technologies niche ($n = 4$)

Types of innovation	Assets	Intervening actors	Actors' main location
Product improvement	Talent	CEGEPs de 3-R, Shawinigan; UQTR (IRH); Hydro-Québec	R
	Know-how	Clients; firms from the same sector; firms from other sectors	NR
New process	Knowledge, R&D	Clients; firms from other sectors UQTR (IRH), Hydro-Québec (LTE)	NR
	Innovation opportunities	Clients	NR
	Material and financial resources	Governments (provincial, national); from the inside	NR

Note: R = region; NR = non-region.

In this sector, most of the talented resources, including the top-level management, were born or were trained in the region. Highly qualified young employees are recruited directly at UQTR and, to a lesser extent, in other Quebec universities. It is observed that this young "talent" circulates from one company to another because of economic fluctuations in the attribution of contracts to those firms. Innovation opportunities are identified within the partnerships set up for the completion of projects or, more frequently, by customers who express well-defined needs. In certain cases, networking with regional development organizations and professional extra-regional associations also plays a role in the identification of innovation opportunities. The material and financial resources needed by the company who wishes to innovate come mainly from the group to which it belongs.

All in all, the networking practices of firms in this sector are characterized by strong extra-regional relationships. Among the five assets, only the recruitment of talented resources is done within the city-region and rest on networking practices at the regional level. For the other "elements," particularly know-how and knowledge, companies find what they need primarily outside the region.

Table 13.6 Networking practices of firms in the industrial bioprocesses niche (*n* = 4)

Types of innovation	Assets	Intervening actors	Actors' main location
New product, product improvement	Talent	CEGEPs de 3-R, Shawinigan), UQTR; universities (QC)	R
	Know-how	Firms from other sectors; government laboratories (QC, Cdn); CNETE (CEGEP de Shawinigan)	NR
New process, process improvement	Knowledge, R&D	Government laboratories (QC, Cdn); UQTR (CIPP); private R&D centres; firms from other sectors	NR
	Innovation opportunities	Clients	NR
	Material and financial resources	banks; Innovatech; governments (provincial, national); CRSNG	NR

Note: R = region; NR = non-region.

Industrial Bioprocesses

This sector is mainly composed of young technological SMEs still at the startup stage. Of all the sectors studied, this is the one most dependent on university spin-offs. Focused on R&D, some of these firms turn innovation into their raison d'être, whereas for the others, the oldest ones in particular, product innovation still occupies the central place in their development strategy. All in all, the bioprocesses sector is still emerging in the Trois-Rivières region. Table 13.6 presents the practices of networking of these firms.

The companies of this regional niche of excellence work mainly in the design and production of new biomolecules, the improvement of the properties of existing biomolecules, the development of new processes that make their manufacturing possible, and, finally, quality-control technologies adapted to the new products/processes. They offer several applications for the sectors of pulp and paper, energy, and the environment. Among the companies interviewed, only one had reached a developmental stage allowing it an appreciable growth by regularly improving its products. The others had rather reached the stage of the pre-commercialization and were missing risk capital to undertake industrial tests or the structured distribution of their products on large markets.

Because they do not possess or cannot afford everything they need, these small firms resort almost constantly to "external help" for their innovation projects. Knowledge, R&D, and know-how often originate from work or collaborations with research and technology transfer centres that are more often than not external to the region. Thus, even if the companies interviewed occasionally work with UQTR's researchers (CIPP and various groups of research) and those of the CNETE at Collège de Shawinigan, they usually work with extra-regional institutions: the NRC's Biotechnology Research Institute (Montreal), INRS–Institut Armand-Frappier (Laval), UQAM, École de technologie supérieure (Montreal), French or Brazilian universities, and so on. These companies count on a strong scientific capital, usually held by their principal leaders, surrounded by a team of scientists who carry out R&D activities within the company. Generally, these talented resources come not only from UQTR but also from the CEGEPs of the region (Trois-Rivières, Collège Laflèche, and Collège Shawinigan) and, to a lesser extent, from other Quebec universities.

Most of the time, firms identify innovation opportunities in collaboration with customers who are mainly located outside the region. Finally, their material and financial resources are not obtained from regional organizations. The majority of the companies we met said they desperately lack risk capital. Being unable to find these resources in the region, they almost all thought of establishing either their head office or the central part of their activities in the geographical and social space of their funding sources or within an industrial agglomeration that resembles them. Concretely, the companies working in the agro-alimentary field thought of moving into the region of Saint-Hyacinthe, near the research centres with which they work and institutions that have a specialized expertise in their domain. The others planned to move to Montreal in order to better integrate themselves into the social networks of financing, research, and transfer in biotechnology.

In fact, the only asset for which intra-regional relations prevail is the recruitment of talented resources. As for the rest, it is mainly outside the region that these firms find what they need, and this distance to their "supply sources" encourages them to consider moving their activities.

Table 13.7 Regional portrait of networking practices

| | Mature sectors in phase with the SRI | | | Emerging sectors out of phase with the SRI | |
	Pulp and paper	Metal transformation	Furniture	Hydrogen and electro-technologies	Industrial bioprocesses
Talent	R	R	R	R	R
Know-how	R	R	R	NR	NR
Knowledge and R&D	R	R	R	NR	NR
Innovation opportunities	NR	NR	NR	NR	NR
Resources	NR	R	NR	NR	NR

Note: R = region; NR = non-region.

Networking Patterns, Performance and Sector – City-Region Alignment

The comparison of the networking practices of the five sectors studied reveals two general and distinct patterns of interaction between a firm and its region (table 13.7). Both appear to be strongly related to the sector to which a firm belongs (patterns of interaction are not the same across all sectors) and also to the history of the city-region's industrialization and the progressive and concomitant construction of an innovation support infrastructure. This observation shows that innovation practices are determined as much by sectoral factors as by the characteristics of the city-region.

Sectoral Characteristics and Performance in the Mature Sectors

The first pattern of interaction is found in the three mature sectors that have been present in the region for a long time: pulp and paper, metal transformation, and furniture. Firms of the three mature sectors have steady relations with the regional organizations of research, technology transfer, and teaching. Through these organizations they regularly access the talent, know-how, and knowledge (including the R&D) which they need in their innovation projects, whether it concerns the improvement of their products or their manufacturing processes. Moreover, it is to the organizations of the region they turn when they encounter day-to-day difficulties. That being said, these firms do not get everything they need from within the city-region. In fact, their material and financial resources as well as innovation opportunities are most frequently drawn from outside the region, a behaviour that corresponds to their status as exporters and large, often multinational, firms. Firm in those sectors also find at least part of the knowledge and know-how they need to undertake their innovation projects outside the region. Working with their clients and suppliers, but also with researchers and technicians in other branch plants or the home office, they access the knowledge and know-how they don't have or can't find locally. As we observed earlier, each sector's major firms and research organizations act as "global pipelines" to access the knowledge and know-how produced and located outside the city-region.

Thus, we observe that the Trois-Rivières city-region is composed of the organizations of research, technology transfer, and support that allow the companies of these three mature sectors to carry out and conclude their innovation projects. In other words, the fact that a firm finds in the city-region the majority of the assets it needs shows that it possesses the expertise, competences, and resources to satisfy its requests and needs. In these three cases, the networking practices indicate that the companies, on the one hand, and the city-region's innovation support organizations and institutions, on the other hand, are to some extent in phase with one another.

Over the course of many decades, the companies of these three sectors ended up occupying a central place in the socio-economic fabric of the region and constituted an "organized" grouping of large companies, SME, suppliers, and private or governmental service organizations which developed in symbiosis with not only one another but also with the regional milieu. The educational, research, and technology transfer system that developed from the 1970s onwards was to some extent articulated with this industrial infrastructure: UQTR and the Collèges de Trois-Rivières and Shawinigan with their specialized or complementary programs in pulp and paper and metallurgy; the creation of research and technology transfer centres in these two fields during the 1980s; the creation of research chairs interested in these fields from the 2000s; and the consolidation and expansion of the existing organizations. All in all, the various levels of government, with the collaboration of the companies, gradually built around the dominant industrial sectors a support infrastructure for innovation which is now abundantly used by the firms and constitutes the "heart" of the region's support infrastructure for innovation.

Nevertheless, the economic performance of these three mature sectors has been quite difficult over the past decade. In the pulp and paper sector alone, the region has lost a little more than 2000 jobs since 2001 with the closing of AbitibiBowater in Shawinigan (Presse canadienne 2007) and repeated downsizing of Kruger's plants in Trois-Rivières (Le Nouvelliste 2009; Rochette 2010a). The metal transformation sector has gone through difficult times with the closing of the Norsk Hydro plant in Bécancour (Radio-Canada Nouvelles 2006) in 2007, the carry-forward of the building site of REC Silicon (Aubry 2010), projects indefinitely postponed at Silicium Bécancour (ibid.), the 15 per cent reduction in working hours for the employees of Aluminerie de Bécancour (Vermot-Desroches 2009), and so on. Finally, since 2005 the furniture industry has seen a significant reduction of its exports to the United States and has consequently reduced its number of employees (Lafrenière 2011). There have been some closings (QMI Agence 2011). However, one furniture firm, Bermex, is growing steadily since 2007 in terms of both sales and employees (Lafrenière 2009).

Keeping in mind that the innovation support structure of the city-region is well equipped and well in tune with the needs of these firms, it is difficult to associate their weak performance with an absence of support organizations attributable to the small size of the city-region. Moreover, since firms have made extensive use of the existing organizations, one cannot point in the direction of a simple mismatch of expertise or activities. In line with the city-region literature, one can turn instead towards the lack of diversity and/or cross-fertilization between sectors for an explanation. Here, the first observation is that the Trois-Rivières city-region does not have the kind of diversity found in a large city-region like Montreal. Of Trois-Rivières we can say that it is specialized in a few complementary sectors that in theory can interact to their mutual benefit. But as far as our results show, there are almost no cross-sectoral relations and, accordingly, no cross-fertilization.

For example, we find no evidence of significant interactions between the pulp and paper and biotechnology sectors. Now, the first one can surely benefit from collaborations with firms and research organization belonging to the second, since its future in industrialized countries is more and more linked to the use of biotechnologies for the development and production of added-value and "intelligent" papers (Forest

Products Association of Canada 2010; Rochette 2010b; Nadeau 2009). One the other hand, pulp and paper firms certainly represent interesting first markets for young biotech firms. In this context, it is tempting to explain the difficulties of those two sectors by pointing to the fact that, the city-region being small, diversity is very restricted (both in quantity and quality) and does not allow the cross-fertilization that is so important for good economic performance. The weak economic performance of both sectors would then be explained by the characteristics of the city-region.

Although inviting, however, this explanation does not take into account all the information gathered in the field. First, even if the diversity found in the Trois-Rivières city-region is far from what can be found in a large city-region, it nevertheless seems to a certain point "sufficient." Pulp and paper firms can find within the city-region a great deal of the biotechnological expertise they need. They don't even have to go outside their sector, because the Centre intégré en pâtes et papiers is a reservoir of biotechnological knowledge and know-how earmarked for pulp and paper products and processes. The relevance of the research centre is quite high and, as we have seen, firms already collaborate with its researchers and technicians. Thus, the diversity those firms need is not necessarily of the degree observed at the macro level in the large city-region. Second, our example shows that there is at least a theoretical possibility that cross-fertilization can occur within a sector instead of between sectors. Here, UQTR and the CEGEP de Trois-Rivières play a central role as "sources of diversity" and producer/supplier of the relevant knowledge and know-how.

If we explore the pulp and paper example, we find that the factors that inhibit fertilization of firms by biotechnological knowledge and know-how are sectoral in nature:

(a) The actual processes and equipment are not well adapted to a "biotechnological turn";
(b) For the production of new intelligent papers, firms need smaller and different plants and also smaller and different equipment;
(c) To replace the actual equipment and reorganize plants would be very costly, and since the recent years have been so difficult, firms do not have the financial capacity to invest that much;
(d) The high value of the Canadian dollar would negatively affect exportation of new value-added/intelligent papers.

Here, the characteristics of the sector rather than those of the city-region explain its difficulties and slow down its renewal. It is interesting to underline that in the case of its three mature sectors, the Trois-Rivières city-region's characteristics seems adequate to open the door for their renewal and redevelopment: innovation-support organizations exist, are relevant, and are working with firms; global pipelines exist; cross-fertilization is possible. Nevertheless, mainly because of sectoral factors, growth is not at a turning point.

Characteristics of the City-Region and Performance in the Emergent Sectors

The second pattern of interaction is observed in the industrial bioprocesses sector and the hydrogen and electro-technologies sector, which are both emergent. The networking practices of these companies are characterized by interactions with economic actors and organizations of research and transfer, which are generally located outside the region. For the purposes of knowledge, know-how, innovation opportunities, or material and financial resources, the partners of these companies are generally from Montreal, Quebec, Canada, or even the United States or Europe. Only the talent resources are mainly recruited within the region. Of course, the companies of these two sectors find within the region some of the assets they need to support their innovation projects: the IRH, the LTE, and the CNETE. However, their practices of networking indicate that those are not sufficient: that is, not sufficient in number, not sufficiently diversified, and not sufficiently relevant. In these two sectors, everything goes on as if the firms and the organizations of research, transfer, and support were out of phase and the support organizations and institutions of the city-region were qualitatively and quantitatively inadequate to support innovation in the companies.

In the industrial bioprocesses sector, the young SMEs have great difficulty in growing and some of them have simply closed. As we said earlier some of the SMEs we met during our study were thinking about moving to a milieu more favourable to their growth. In the hydrogen and electro-technologies sector, a diversity of obstacles inhibits growth. The hydrogen sector has great difficulty in moving beyond the stage of R&D, since IRH at UQTR is the only significant organization and there are just a few firms. In electro-technologies, the closing of the nuclear power plant Gentilly-2 is significantly slowing down the development of firms. In those emergent sectors, the understanding of their difficulties is quite different from what we have put forward concerning the mature ones. Here, the characteristics of the city-region rather than those of the sectors explain the difficulties observed.

The Trois-Rivières city-region is too small and too specialized to be able to offer to the firms and actors of these emergent sectors the kind of milieu where they can find useful and relevant partners to collaborate with. Neither the "industrial atmosphere" of Marshall nor the "local buzz" of Bathelt et al. exists. In fact, the firms of those two sectors are primarily working outside the city-region to find the assets they need to conclude their innovation projects. The same is true for a research centre like IRH, which in the absence of "local" firms turns towards companies located in other city-regions. This observation reminds us that to collaborate within its city-region, an organization needs to have someone relevant to talk too (Drejer and Vinding 2007). From this perspective, it is easy to understand that a small city-region is unable to offer its firms and organizations the same kind of diversified support and potential partnerships one can find in a large one.

All in all, our observations and analysis show that the interaction patterns and propensity to collaborate are dependent on the characteristics of both the industries and the city-region. More specifically, in the case of the latter, the characteristics of the support organizations and institutions play a significant role. Moreover, the interaction patterns and propensity to collaborate evolve as sectors and support institutions evolve (Andersen and Drejer 2008). The interaction between companies and the supporting infrastructure within a city-region has a historical dimension that takes the form of "path dependency," leading to situations where, at a given time, certain firms and certain support organizations are not embedded. Put differently, the interactions within a city-region and the propensity to collaborate on the regional scale vary according to the different sectors. To understand what happens as well as the effects of these behaviours on the firms and the region concerned, it is necessary to take account of the history of the alignment of the interests of the actors and institutions which constitute the city-region (Andersen and Drejer 2008).

Conclusion

Generally, our study of the networking practices of the companies of the Trois-Rivières city-region shows that the embeddedness of these practices within the region is not identical in all the industrial sectors. The "traditional" sectors of the regional economy are historically aligned with the infrastructure of research, transfer, and support of the Trois-Rivières city-region, which was gradually put in place from 1970s onward. It results in a regular collaboration between firms and support organizations, that is, a regional embeddedness of the innovation activities.

In the three mature sectors we observed the presence of both "local buzz" and "global pipelines" with a potential for cross-fertilization. In light of these observations, it is difficult to see the characteristics of the city-region (size and diversity) as explanatory factors for their weak economic performance and challenges for renewal. Instead, our analysis shows that the characteristics of the sectors are at the heart of the difficulties they face.

In the two emergent sectors, the patterns of interaction which lead to innovation indicate an important contribution of extra-regional assets. Although the geographical and social proximity of the partner is also important for them, the companies of these sectors collaborate outside the region because they do not find in their milieu what they need for their innovation projects. In certain cases, they simply do not have a "relevant" regional intermediary with whom to speak. For the leaders of these firms, this situation is really alarming. In the bioprocesses sector, for example, all the people interviewed were thinking of leaving the region. Obviously, the companies of these sectors are partly "disembedded" or less embedded than those of the mature sectors. No "local buzz" – only "pipelines" to go outside the city-region. Here, the characteristics of the city-region appear more determining than the sector for understanding organizations' behaviours and performance.

In sum, our analysis shows that both the sector and the characteristics of place play a significant role in the innovative capacity and performance of firms, and therefore in the economic performance of the city-region to which they belong. Since Trois-Rivières city-region does not have and will have great difficulties in providing[3] all the richness and diversity necessary to support and bring to maturity a large variety of sectors/actors, the characteristics of place are key factors in the difficulties encountered by firms and organizations in emergent sectors. On the other hand, in the mature sectors around which an impressive and relevant sectoral support structure has been built and used by the actors, their difficulties have more to do with sectoral realities than with characteristics of place.

NOTES

1 These 20 interviews were carried out with one or two key members of
each firm over a 9-month period, from September 2007 to May 2008. The
approximate duration of the talks was 90 minutes. A single interview guide
was used for each meeting in order to ensure the validity and comparabil-
ity of the information collected (Yin 2003): the *Interview Guide Theme 1 – The
social nature of innovation*, developed for the ISRN city-region studies. The
main focus of the interview was the history of the firm and of exemplar innovation projects relating primarily to the social and interactive dimension
of innovation activities: the nature of the activities and projects, external
collaborations, recruitment and retention of talented resources, and the diffusion and exchange of information (flux). Those four topics were selected
on the basis of their being key "ingredients" of innovation (St-Pierre and
Trépanier 2011; Tidd, Bessant, and Pavitt 2005). Information collected during interviews was compiled for each niche in a matrix which included the
following categories:

- the nature of the innovating activity (product development or improvement, process development or improvement);
- the asset/ingredient for innovation (talented resources, know-how,
  knowledge and R&D, innovation opportunities, material or financial
  resources);
- identification of the organization through which the company gets, partly
  or entirely, a specific asset/ingredient;
- the regional (R) or extra regional (NR) location of each organization.

2 FPInnovations Wood Products Division is Canada's national wood products research institute. Its role is to support the forest products industry in
optimizing manufacturing processes, extracting higher-value products from
available resources, and meeting customer's expectations for performance,
durability, and affordability. FPInnovations is a partnership currently
involving more than 300 primary, secondary, supplier, and government
members. http://www.forintek.ca/public/eng/E1-about/0.about.html.

3 There is little chance that a transformation of the city-region's innovation
support infrastructure to align it with the needs of the emergent sectors will
take place. First, the relevant supporting organizations already exist outside
Trois-Rivières and it is highly improbable that the provincial and federal
governments will pay for the creation of new ones. Second, the city-region
does not have at his disposal the financial resources such a development
would imply. In this context, it is difficult to see how the institutional and

financial resources necessary for a major and durable reorientation of the city-region's economic structure can become available. For an analysis of the feasibility and difficulties of a "transformative process," see Parker 2010.

REFERENCES

Andersen, P.H., and I. Drejer. 2008. "Systemic innovation in a distributed network: The case of Danish wind turbines, 1972–2007." *Strategic Organization* 6 (1): 13–46. http://dx.doi.org/10.1177/1476127007087152.

Asheim, B., and P. Cooke. 1999. "Local learning and interactive innovation networks in a global economy." In *Making connections: Technological learning and regional economic change*, ed. E.J. Malecki and P. Oinas, 145–78. Aldershot: Ashgate.

Asheim, B., and M. Gertler. 2007. "The geography of innovation, regional innovation systems." In *The Oxford handbook of innovation*, ed. J. Fagerberg, D.C. Mowery, and R.R. Nelson, 291–317. London: Oxford University Press.

Aubry, M. (2010), "Maurice Richard convoque les leaders. " *Le Nouvelliste*, 27 October. http://www.cyberpresse.ca/le-nouvelliste/economie/201010/27/01-4336534-maurice-richard-convoque-les-leaders.php.

Bathelt, H., A. Malmberg, and P. Maskell. 2004. "Clusters and knowledge: Local buzz, global pipelines and the process of knowledge creation." *Progress in Human Geography* 28 (1): 31–56. http://dx.doi.org/10.1191/0309132504 ph469oa.

Beaudry, C., and A. Schiffauerova. 2009. "Who's right, Marshall or Jacobs? The localization versus urbanization debate." *Research Policy* 38 (2): 318–37. http://dx.doi.org/10.1016/j.respol.2008.11.010.

Becheikh, N., R. Landry, and N. Amara. 2006. "Lessons from innovation empirical studies in the manufacturing sector: A systematic review of the literature from 1993–2003." *Technovation* 26 (5–6): 644–64. http://dx.doi.org/10.1016/j.technovation.2005.06.016.

Bloomfield, G.T., and E. Bloomfield. 1994. "'Our prosperity rests upon manufactures': Industry in the central Canada urban system, 1871." *Urban History Review / Revue d'histoire urbaine* 22 (2): 75.

Boschma, R.A. 2005. "Proximity and innovation: A critical assessment." *Regional Studies* 39 (1): 61–74. http://dx.doi.org/10.1080/0034340052000320887.

Calantone, R.J., S.T. Cavusgil, and Y. Zhao. 2002. "Learning orientation, firm innovation capability, and firm performance." *Industrial Marketing Management* 31 (6): 515–24. http://dx.doi.org/10.1016/S0019-8501(01)00203-6.

Carrier, M., and P. Gingras. 2004. "Les villes moyennes: Analyse démographique et économique, 1971–2001. Note de recherche." *Recherches Sociographiques* 45 (3): 569–92. http://dx.doi.org/10.7202/011470ar.

Cooke, P. 2001. "Regional innovation systems, clusters, and the knowledge economy." *Industrial and Corporate Change* 10 (4): 945–74. http://dx.doi.org/10.1093/icc/10.4.945.

Cooke, P., M. Heidenreich, and H.-J. Braczyk. 2004. *Regional innovation systems.* 2nd ed. Trowbridge: Cromwell Press.

Desjardins Études économiques. 2007. *Région administrative de la Mauricie, survol de la situation économique.* Montreal.

Doloreux, D. 2004. "Regional innovation systems in Canada: A comparative study." *Regional Studies* 38 (5): 479–94. http://dx.doi.org/10.1080/0143116042000229267.

Dosi, G. 1988. "Sources, procedures and microeconomic effects of innovation." *Journal of Economic Literature* 26 (3): 1120–71.

Drejer, I., and A.L. Vinding. 2007. "Searching near and far: Determinants of innovative firms' propensity to collaborate across geographical distance." *Industry and Innovation* 14 (3): 259–75. http://dx.doi.org/10.1080/13662710701369205.

Florida, R., C. Mellander, and K. Stolarick. 2008. "Inside the black box of regional development: Human capital, the creative class and tolerance." *Journal of Economic Geography* 8 (5): 615–49. http://dx.doi.org/10.1093/jeg/lbn023.

Forest Products Association of Canada. 2010. *Transforming Canada's forest products industry. Summary of findings from the Future Bio-pathways Project.* http://www.fpac.ca/publications/Biopathways%20ENG.pdf.

Frenken, K., F. Van Oort, and T. Verburg. 2007. "Related variety, unrelated variety and regional economic growth." *Regional Studies* 41 (5): 685–97. http://dx.doi.org/10.1080/00343400601120296.

Isaksen, A., and K. Onsager. 2010. "Regions, networks and innovative performance: The case of knowledge-intensive industries in Norway." *European Urban and Regional Studies* 17 (3): 227–43. http://dx.doi.org/10.1177/0969776409356217.

Lafrenière, M. 2009. "Investissement de plus de 20 millions chez Bermex." *Le Nouvelliste,* 20 November. http://www.cyberpresse.ca/le-nouvelliste/economie/200911/20/01-923703-investissement-de-plus-de-20-millions-chez-bermex.php.

Lafrenière, M. 2011. "Canadel veut maintenir son volume d'activités." *Le Nouvelliste,* 15 June. http://www.cyberpresse.ca/le-nouvelliste/economie/201106/15/01-4409357-canadel-veut-maintenir-son-volume-dactivites.php.

Lazonick, W. 2006. "The innovative firm. " In *The Oxford handbook of innovation,* ed. J. Fagerberg, D.C. Mowery, and R.R. Nelson, 29–55. London: Oxford University Press.

LeMay, J. 2002. "The impact of the Quiet Revolution: The business environment of smaller cities and regions of Québec: 1960–2000." *Québec Studies* 34: 19–30.

Le Nouvelliste. 2009. "Une année horrible pour les pâtes et papiers." *Le Nouvelliste* 11 (December): 13.

Lundvall, B.-A. 1992. *National systems of innovation.* London: Pinter Publishers.

MacKinnon, D., A. Cumbers, and K. Chapman. 2002. "Learning, innovation, and regional development: A critical appraisal of recent debates." *Progress in Human Geography* 26 (3): 293–311. http://dx.doi.org/10.1191/0309132502 ph371ra.

Malerba, F. 2002. "Sectoral systems of innovation and production." *Research Policy* 31 (2): 247–64. http://dx.doi.org/10.1016/S0048-7333(01)00139-1.

MDEIE. 2007. Ministère du Développement Économique, de l'Innovation et de l'Exportation. *Tableau de bord des systèmes régionaux d'innovation du Québec.* Quebec.

MDEIE. 2008. Ministère du Développement Économique, de l'Innovation et de l'Exportation. *Portrait socioéconomique des régions du Québec.* Quebec.

MFEQ. 2013. Ministère des Finances et de l'Économie, *Projet ACCORD.* http://www.economie.gouv.qc.ca/objectifs/creer-liens/projet-accord/page/le-projet-accord-12796/?tx_igaffichagepages_pi1%5Bmode%5D=single&tx_igaffichagepages_pi1%5BbackPid%5D=70&tx_igaffichagepages_pi1%5Bcurrent Cat%5D=&cHash=1481e17f48de6f4d91e68257968b5311.

Nadeau, J.B. 2009. "Patrice Mangin, le magicien du papier: Trois-Rivières vit au rythme du papier et ce pourrait bien être là que naîtront des papiers 'flyés.'" *L'Actualité* 15 (September): 31.

OECD. 2001. *The new economy: Beyond the hype.* Paris: OECD.

OECD. 2006. *Territorial reviews: Competitive cities in the global economy.* Paris: OECD.

OSRIM. 2008. *Tableau de bord du système régional d'innovation de la Mauricie.* Technopole Vallée du Saint-Maurice, Trois-Rivières.

Owen-Smith, J., and W.W. Powell. 2004. "Knowledge networks as channels and conduits: The effects of spillovers in the Boston biotechnology community." *Organization Science* 15 (1): 5–21. http://dx.doi.org/10.1287/orsc.1030.0054.

Parker, R. 2010. "Evolution and change in industrial clusters: An analysis of Hsinchgu and Sophia Antipolis." *European Urban and Regional Studies* 17 (3): 245–60. http://dx.doi.org/10.1177/0969776409358244.

Pittaway, L., M. Robertson, K. Munir, D. Denyer, and A. Neely. 2004. "Networking and innovation: a systematic review of the evidence." *International Journal of Management Reviews* 5–6 (3–4): 137–68. http://dx.doi.org/10.1111/j.1460-8545.2004.00101.x.

Presse canadienne. 2007. "AbitibiBowater ferme des usines." *La Presse,* 29 November. http://lapresseaffaires.cyberpresse.ca/economie/200901/06/01-689243-abitibibowater-ferme-des-usines.php.

QMI Agence. 2011. "Fermeture de Meubles EG en Mauricie." *Argent,* 11 March.

Radio-Canada Nouvelles. 2000. *Trois-Rivières, capitale nationale du chômage.* 7 July. http://www.radio-canada.ca/nouvelles/51/51665.htm.

Radio-Canada Nouvelles. 2006. *Norsk Hydro met la clef sous la porte.* 24 October. http://www.radio-canada.ca/nouvelles/Economie-Affaires/2006/10/24/003-norsk_hydro.shtml.

Rochette, M. 2010a. "Les dix événements de l'année en économie. Rétrospective 2010." *Le Nouvelliste* 27 (December): 3.

Rochette, M. 2010b. "L'industrie de la fôret doit prendre le virage biotechnologique pour survivre." *Le Nouvelliste* 2 (February): 12.

Rochette, M. 2011. "Trois-Rivières: Capitale québécoise du chômage." *Le Nouvelliste*, 4 February. http://www.cyberpresse.ca/le-nouvelliste/economie/201102/04/01-4366919-trois-rivieres-capitale-quebecoise-du-chomage.php.

Safford, S. 2004. *Searching for Silicon Valley in the RustBelt: The evolution of knowledge networks in Akron and Rochester.* Working paper MIT-IPC-04-001, Cambridge, MA, MIT Industrial Performance Center.

Scott, A.J. 2002. "A new map of Hollywood: The production and distribution of American motion pictures." *Regional Studies* 36 (9): 957–75. http://dx.doi.org/10.1080/0034340022000022215.

St-Pierre, J., and M. Trépanier. 2011. "Concomitance de la capacité d'innovation des PME et de la capacité des territoires à les soutenir dans quatre régions du Québec." In *Les performances des territoires. Les politiques locales, remèdes au déclin industriel,* ed. D. Carré and N. Levratto, chap. 5, 191–238. Paris: Éditions Le Manuscrit.

Sternberg, R. 2007. "Entrepreneurship, proximity and regional innovation systems." *Tijdchrift Voor Economische En Sociale Geografie* 98 (5): 652–66. http://dx.doi.org/10.1111/j.1467-9663.2007.00431.x.

Therrien, P. 2005. "City and innovation: Different size, different strategy." *European Planning Studies* 13 (6): 853–77. http://dx.doi.org/10.1080/09654310500187961.

Tidd, J., J. Bessant, and K. Pavitt. 2005. *Managing innovation: Integrating technological, market and organizational change.* 3rd ed. London: Wiley.

Uzzi, B. 1997. "Social structure and competitiveness in interfirm networks: The paradox of embeddedness." *Administrative Science Quarterly* 42 (1): 35–67. http://dx.doi.org/10.2307/2393808.

Vermot-Desroches, P. 2009. "Alcoa nous demande l'impossible." *Le Nouvelliste,* 4 March. http://www.cyberpresse.ca/le-nouvelliste/centre-du-quebec/200903/04/01-833068-alcoa-nous-demande-limpossible.php.

Wolfe, D. 2009. *21st century cities in Canada: The geography of innovation.* Ottawa: Conference Board of Canada.

Yin, R.K. 2003. *Case study research: Design and methods.* 3rd ed. Thousand Oaks, CA: Sage Publications.

PART V

The Global Challenge for Innovation in Canadian City-Regions

14 Related Variety, Knowledge Platforms, and the Challenge for Cities and Regions in the Global Economy

PHILIP COOKE

Over the past decade, two major research projects have been undertaken by members of the Innovation Systems Research Network to investigate the key innovation challenges facing cities and regions in the global economy. The first of these projects focused on the role of local and regional clusters in the Canadian economy, while the second project investigated the contribution of innovation, creativity, and governance to the economic dynamism of city-regions in Canada. From the outset, ISRN research was grounded in some of the key concepts that emerged from regional systems thinking, which grew out of a dawning recognition, in academe and among diverse policy communities, of the need for a fundamental rethinking at the analytical and policy levels of regional systems analysis. The growing acknowledgment of this need for a reorientation of regional policy embodied the reaction to the common problem of the "one size fits all" solutions to spatial asymmetries, whether enterprise zones, science parks, business incubators, or new waterfronts, many of which dilute the authenticity of urban and regional economic strategies with their generic policy "recipes." Regional innovation systems thinking recognized from the beginning (Cooke 1992) the diversity of regional innovation characteristics of business and regional governance competences and capabilities and advocated diverse policy responses. A second influence arose from the practical experience of examining and understanding actual modes of delivering regional innovation policy.

One of the important policy and practice innovations of the 1990s – which was associated with the accelerated demise of the Industrial Age in advanced economies – concerned clusters. These fundamentally *regional* phenomena portended the rise of networked forms of knowledge distribution and flow, clusters forming nodes in global networks. In the first project, ISRN researchers undertook detailed case studies of the dynamics of innovation in twenty-six industrial clusters across Canada, the results of which were summarized in a special issue of *European Planning Studies* (Wolfe 2009a). The second project expanded its focus from the specific dynamics of individual clusters to the broader role that innovation and knowledge flows, talent and creativity, and local governance arrangements contribute to the economic dynamism of city-regions across Canada. This volume reports on the findings of the first of the three themes from the second research project on creative and innovative industry in sixteen Canadian city-regions. One of the key issues addressed in this project was paraphrased by Anne Golden, president and CEO of the Conference Board of Canada as follows. In strategizing city-region growth "should reliance be placed on urban specialisation – as Harvard professor Michael Porter claims? … Or should we trust in diversity – as the late urban expert Jane Jacobs believed?" (Golden 2009).

Readers will be familiar with this hot theoretical topic, which focuses on many urban economic concepts ranging from the hoary old "localization" versus "urban" externalities dichotomy to the more recent literature on "knowledge spillovers," "absorptive capacity," and "path dependence." With respect to externalities and regional growth a key research question has been the extent to which firms in agglomerations benefit from "Romer externalities" of localization or "Jacobs externalities" of urbanization. Specialization and diversification are the key differentiating dynamics with respect to these two perspectives on growth and agglomeration. Specialization has been a mantra of the supply-side, clustering, and cluster policy era.

For a time economists, especially, favoured the "view" of an imagined if improbable power alliance involving the long-dead Alfred Marshall, the happily still alive Kenneth Arrow, and Paul Romer, the younger progenitor of "neoclassical endogenous growth theory." This "MAR" view is the line from which spring Porter (1998) and another Harvard urban economist, Ed Glaeser (2000). Porter, of course, espouses "clusters," while Glaeser claims that specialized labour markets promote most urban growth. Against these eminent names stood, for many decades, Jane Jacobs (1969), the lone advocate of the view that variety was the superior condition for urban prosperity and that quality was best supplied by large cities. The result was an intense academic debate which focused on the juxtaposition of MAR spillovers (or externalities) and Jacobian ones for a period which saw a great rise in the publication of articles on the topic, mainly in the first decade of this millennium.

As stimulating as this debate has been over the past decade, it has often overlooked the insights brought from empirical research by evolutionary economic geographers, showing that urban growth was stronger in localities with "related variety" in their economies. This, of course, appears to vindicate Jacobs as well as fitting well with ideas about networks, interactive learning, and innovation more generally. This chapter briefly touches on some of the key insights and findings from the current ISRN studies of innovation in Canadian cities. It then goes on to show that situating the results of the ISRN research program in the context of a more theoretically informed framework based in regional innovation systems thinking, allied to evolutionary economic geography and development analysis, produces a superior change or transition model. This is particularly so in reference to its basic idea of economic development caused by interactions between elements in regional economies displaying related variety. This contribution is provided as a spur to extend the frameworks that have underpinned the ISRN research program in light of some recent conceptual developments in the field. The goal of this effort is to draw out the implications of this new line of thinking in evolutionary economic geography and regional development for the results of the ISRN research program and suggest new directions for future research.

Innovation in Canadian Cities

In his introduction to this volume, David Wolfe makes a fine attempt to bridge the MAR-Jacobs divide with his differentiation of urban innovativeness according to the city scale. Thus, the Canadian urban system is characterized by specialized local economies with few spillovers and little innovation or creativity (e.g., pulp and paper), small-to-medium-sized cities that are usually relatively specialized, and, often, manufacturing towns, again with little innovativeness or creativity because of low spillovers and metropolitan cities. The first two, although urban, display localization externalities, that is, they are there because of some initial resource endowment that continues to be economically viable. The third is a bit more sophisticated than that, but often externally controlled, with little endogenous innovation or creativity, and so living from localization, probably due to location (near the US manufacturing belt) or labour market (semi-skilled routine jobs) advantages.

The other kind of urban setting, though, is the one that really benefits from urbanization economies, giving it its diversity. In Canada, as perhaps elsewhere, the knowledge economy really benefits from and dynamizes the growth of metropolitan cities. Canada has between three and five metropoles, the biggest in rank order being Toronto, Montreal, and Vancouver with Calgary and Ottawa as possible contenders on some indices. The papers in this volume present a substantial body of material that the ISRN city-region research program has yielded on the clustering of distinctive industries in different scales and kinds of Canadian city. But the three biggest display the most variety (many clusters), growth, and innovativeness or "cognitive-cultural" dimensionality. Wolfe here uses the term "Schumpeterian hubs"[1] to capture this innovative-creative aspect as well as their relative economic powerhouse characteristics. Thus, in terms of urban growth theory, Wolfe, reporting on the preliminary findings of this research, found both MAR and Jacobian spillovers in Canada's urban system, but the Jacobian ones explain growth far better. One notable statistic is that the five aforementioned cities plus the next five accounted for 2 million of Canada's net job growth of 3.1 million from 1995–2005 (Wolfe 2009b).

The evidence from the Canadian experience also suggests that traditional manufacturing industries that are more concentrated in some of the larger, as well as medium-sized, cities of Central Canada are drawing upon the research infrastructure of the knowledge economy and that some of the most dynamic concentrations of industrial firms stretch beyond the boundaries of a single city or metropolitan region. In these circumstances, well-developed clusters have numerous specialist research centres, laboratories, and training facilities in the same region as user-firms. These firms may have been responsible for the evolution of such a territorial knowledge system by sponsoring needed innovations in the institutional knowledge base. A good example of this point is the celebrated automotive-parts cluster in southern Ontario, which encompasses the geographic area extending from the eastern boundary of the Greater Toronto Area west to London and Windsor. The extended cluster involves a strong supplier base of 280 firms in the automotive parts, rubber, and chemicals and steel sectors that draw upon the research results of 89 science and engineering university programs, and 30 steel and materials science research institutes with access to a highly educated and highly skilled workforce drawn from over 4300 engineering graduates per year (Cooke and De Laurentis 2010).

Notice the immense reliance in this very large and relatively medium-to-high-tech cluster upon a massive range of specialist science and engineering knowledge institutes and materials research laboratories within and beyond the university sector. It is in large part through the networks of globally leading researchers and trainers elsewhere that this cluster keeps its leading-edge knowledge capability, which is in turn relayed back into the pedagogy of students and the knowledge content of companies in the cluster. There is evidence of the same phenomenon at work in some of the other cases reported in this volume, especially the western Canadian ones in Saskatoon, Calgary, and Vancouver. Of course, they have their own company-based in-house knowledge capabilities as well, including accessing corporate knowledge from within what, in many cases, are multinational firms. Hence, a top cluster is both locally and globally knowledge-integrated to the intellectual, industry, and international flows of ideas, practice, and solutions available from its relevant "knowledge communities" – sometimes also referred to as "epistemic communities." However, as the next section begins to suggest, it may be more fruitful, in both analytic and policy terms, to conceive of the scientific and knowledge base that has supported the growth of this cluster as a knowledge platform, with the potential to spin off new forms of economic activity in related economic sectors. We now turn to a consideration of this process.

Regional Innovation and Platform-like Related Variety

As empirically rich and theoretically grounded as the individual ISRN case studies of Canadian city-regions have been, there is a concern that they have not yet fully absorbed some of the most significant trends in regional innovation thinking in the last few years, particularly those arising from evolutionary urban economics and Schumpeterian economics. Taken together, these approaches lead to a more agency-centred perspective that draws theoretical inspiration from the fields of evolutionary urban economics (Jacobs 1969) and evolutionary innovation economics (Schumpeter 1975); both Jacobs and Schumpeter have been shown to be important architects of a powerful theoretical framework capable of explaining regional growth and development. The two master concepts or axes of this frame are the Jacobian concept of "diversity" or its evolutionary variant of "variety," especially "related variety" (Frenken et al. 2007), and Schumpeter's concept of "regional innovation" as rendered in Andersen's (2011) magisterial reconstruction and synthesis of Schumpeter's complete political economy.

Two recurring themes in this co-evolutionary spatial analysis are relatedness of industry, by means of which regional growth is assisted, and the implications of path dependence, or the process by which it can be constrained. Exploration of the first is a relatively recent phenomenon, pioneered by Frenken et al. (2007), but already it is a core body of theory and empirical research in evolutionary economic geography (Boschma and Frenken 2006; Boschma 2005; Boschma and Wenting 2007). The main mechanism by which relatedness influences regional growth is through knowledge transfer between firms, one result of which can be innovation. The key agents of such transfer are employees developing their careers by changing jobs in neighbouring areas and new companies being formed by the spin-off process that may also be a vehicle for innovations. Path dependence is a more established concept arising in economic history, particularly the branch interested in the history of innovation (David 1985). It has been analysed fruitfully in the context of evolutionary economic geography and particularly regional development, adaptation, and change. The champions of "relatedness" indicate the pivotal position occupied by the idea of "related variety" in evolutionary economic geography. Comparable to "proximity," it has numerous dimensions, notably the cognitive, social, organizational, institutional, and geographical. Much research effort is exercised in relation to both concepts seeking to assess the relative importance of each in understanding the evolution of agglomerations or clusters, the core problematic of economic geography. In doing this, light is cast on the role of numerous other key process elements of interest to evolutionary economic geography, such as innovation, technology, knowledge spillovers, learning, and the creation of new regional developmental pathways.

Together, these provide a clear insight into the dynamics of regional evolution. From the concept of Jacobs can be deduced explanations of both firm and, more important, cluster mutations by means of cross-pollination of knowledge (knowledge spillovers) that lead to innovation by means of geographically proximate inter-cluster "collisions" of ideas. Arising from Jacobs (1969), evidence is growing that regional evolution and growth are affected by the degree of regional related variety of industry (Boschma and Wenting 2007; Cantwell and Iammarino 2003). Related variety speeds up lateral absorptive capacity between neighbouring sectors and stimulates innovation cross-fertilization via knowledge spillovers. Relatedness arises from research into regional economic growth, where it is found that economies with "related variety" among industries perform better than those without it. This is called the "proximity" effect, superseding the "portfolio" effect from the viewpoint of industrial structure. More related variety means more lateral "absorptive capacity" from related "knowledge spillovers." These can enhance the innovation potential of regions, and evolutionary economic geography research goes further into this relationship, finding the element of "relatedness" within the required variety to be the independent variable. Moving on, "path dependence" at the regional level can explain stability but also system stagnation and inertia (Martin and Sunley 2010). However, in contexts such as that of a regional economy with related variety, path interdependence can be envisaged where two or more economic trajectories may intersect in regional space, conceivably producing unforeseen innovations from their "revealed related variety" or ex post relatedness.

From the work of Schumpeter, we get the mechanism for these mutations, for, as Andersen (2011) shows, against many unreflective conflations of "innovator" and "entrepreneur," they are, for Schumpeter, rigidly distinctive categories. The former "recombines" knowledge, while the latter acts upon and "commercializes" it. These are conceptually distinct skill sets, even though they may occasionally be combined empirically in a single actor. Accordingly, imitative "swarming" (or clustering) is principally an "entrepreneurial event." By these combined process mechanisms, urban and regional evolution, including "branching" to new path creation, occurs. The neo-Schumpeterian and evolutionary economic geography perspectives integrate well in theoretical terms (Boschma and Frenken 2006; Boschma and Wenting 2007). They share many core concepts such as path dependence, learning, search, and selection, even niche ecologies, as choices made by firms and agencies regarding innovation and regional evolution (Andersen 2011; Nuttall 2008). Of particular interest is their shared reliance on the key concept of variety and especially the notion of "related variety."

Consistent with the other key concepts is proximity, which has greater reach than simply its geographical dimension. Proximity can involve cognitive and relational dimensions (Boschma 2005), but geographical proximity still facilitates rapid knowledge transfer through lateral absorptive capacity among entrepreneurs and managers in related industries, assisted by knowledge spillovers and external economies of scope, where cognitive dissonance among sub-sectoral actors is relatively low. In these respects we envisage the rise of regional economic "platforms" of related industry activity, which is best exemplified in observed cases of "green innovation." "Green innovation" is defined as "diverse new and commercial products, technologies and processes which, through improvements in the clean energy supply chain from energy source through to point of consumption and recycling, result in reduction in greenhouse gases" (Cooke 2008).

These early analyses were static, so attention turned to the dynamics of technological relatedness and regional branching (new path creation). This invited discussion of relatedness in the short and long term, one hypothesis being that constructing advantage from related variety only brings short-term advantage. Long term, some wholly new branches are needed to sustain regional growth. This is clearly an open question, warranting deep thought, because at the heart of spatial evolution is a notion of an industrial ecosystem, which means complementarities foster growth while unrelatedness destroys it. As noted in "transversality" analysis of regional innovation and growth, keeping industry conscious of regional relatedness is one of the key tasks of the advanced regional development agency (Cooke 2011).

Thus, regional innovation processes display cross-fertilization among clusters in convergent industries that display "platform-like" related variety (Jacobs 1969). Recall that mutation occurs through high lateral absorptive capacity of knowledge spillovers among related industries and their entrepreneurs. Thus, such external economies realize a form of collective entrepreneurship which is massively assisted by geographic proximity. This avoids the "portfolio" effect experienced in regions with low related variety, low knowledge spillovers, and cognitive dissonance among firms rather than high lateral absorptive capacity. These are clear gains to regional policy on innovation and growth performance, whereby a single innovation might be adapted and applied in, say, six different contexts within the regional development platform.

The insights available from evolutionary economic geography in relation to regional economic growth were outlined above; here they are further elaborated. First, applying the notion of related variety has led to new insights in the externalities literature. Empirical studies tend to show it is not so much regional specialization or regional diversification regarding externalities that induce knowledge spillovers and enhance regional growth, but a regional economy that encompasses related activities in terms of competences (i.e., regions well endowed with related variety). Second, evolutionary economic geography has provided additional insights into the question whether or not extra-regional linkages matter for regional growth. Adopting a relatedness framework, empirical studies on trade patterns tend to show that it is not inflows of knowledge per se that matter for regional growth, but inflows of knowledge that are related (not similar) to the existing knowledge base of regions, as is demonstrated in some of the case studies in this volume, particularly those in Vancouver and Saskatoon. Related flows concern new knowledge that can be understood and exploited and, thus, be transformed in regional growth.

Third, relatedness is now also investigated in network analysis. For instance, studies show that collaborative research projects tend to create more new knowledge when they consist of agents that bring in complementary competences. Fourth, the notion of relatedness enriches the literature on labour mobility, which is often regarded as one of the key mechanisms through which knowledge diffuses. Recent studies show that neither inflows nor outflows of labour are properly assessed if one does not also consider how these knowledge flows match the already existing knowledge base of firms and regions (Boschma and Wenting 2007).

Fifth, relatedness may also show its relevance through entrepreneurship dynamics. Here, Jacobs makes a brief, but important re-entry based on variety, relatedness, and the transversality that may arise in contexts of abundant knowledge spillovers and high lateral absorptive capacity. This is the point at which an innovation mutates through the interaction of these forces as entrepreneurs act upon the knowledge symbiosis effected by inter-cluster, inter-firm, or inter-individual innovators. From this, a new cluster or sector may emerge to prominence, as discussed above, thereby branching, or in a more Schumpeterian context of "creative destruction," breaking with the former regional path dependence. Experienced entrepreneurs (those that have acquired knowledge in related industries), as opposed to spin-off companies, may play a crucial role in the regional diversification process. Generally speaking, longitudinal studies show that the long-term development of regions depends on their ability to diversify into new sectors while building on their current knowledge base.

This ability to diversify into new sectors is particularly relevant for the discovery of cumulative knowledge and innovation, which has been traditional for sectors and even clusters, although clusters may be precisely "transitional" forms, and combinative knowledge and innovation dynamics typical of the emergent and evolving "platform" knowledge flows model. This, it will be recalled, is based on "related variety" of inter-industry knowledge spillovers and lateral absorptive capacity among firms. Whereas intra-corporate spatial divisions of labour placed routine assembly industry at geographical peripheries and management headquarters in core regions, knowledge dynamics under knowledge economy conditions are multi-locational and distributed, and innovation is more "open" because cognate to norms associated with public "open science" than in the older, "closed innovation" model. Accordingly, regional governance moves away from the localized "container" model of knowledge geography associated with clustering towards distributed knowledge platforms with pronounced "global antennae."

The concept of industries coexisting in a regional "platform" as a basis for mobilizing regional evolution connects directly to the related-variety argument of the previous section. Neither over-diversified nor over-specialized, and with opportunities present for revealed relatedness in "new combinations" of innovation at interfaces between industries, the accomplished regional economy works with agility and flexibility to meet increasingly user-driven demand. That is not to say that innovation does not continue to be an interactive process between user and producer; rather, it recognizes that innovation studies in the past, perhaps echoing aspects of the practice of innovative businesses, has been overly "productivist." That is, during the years of excess firms competed on the basis of disruptive innovation (Christensen 1997).

So, we see the beginnings of the difference between a cluster and a platform. A cluster is a specialized concentration of business and innovation expertise and support in a localized setting. A platform is a more complex combination of clusters and non-cluster industry organizations such as large corporations that operate in fields that display "related variety." A key reason for the platform being a good place for firms of all sizes to find themselves is that it offers greater potential for innovation than the cluster. Firms can suffer "lock-in" when confined in the specialized conditions of a cluster. This can be cognitive, where they know their neighbours' expertise very well and the opportunities for learning new things from them are very low. It can be functional, in that they see themselves as having a very narrow expertise in a single focus area. Finally, lock-in can be political, where stepping outside cluster boundaries, functionally or geographically, may mean that the cluster support is weakened (Grabher 1993). By contrast, as we shall see, the platform model celebrates the difference and cross-fertilization of ideas and practices among firms in the same or different industries.

This relatedness works because of two important, subsidiary concepts. These are, first, "absorptive capacity" and, second, "knowledge spillovers." In regional economic development terms, absorptive capacity is lateral, whereas in industrial economics it is vertical. Lateral "absorptive capacity" means that entrepreneurs in adjoining or "revealed relatedness" industries can understand each other's business models and focus and apply tacit knowledge or even "routines" from one business type or model to their own. In this way innovations might cross-fertilize and migrate from one industry to a related or revealed-related one. The clearest evidence of this phenomenon emerges from the Calgary case study, but there are hints of it in some of the others, such as the growing strength of Toronto's cognitive cultural economy and the lateral spillovers occurring among the creative, design, and artistic sectors of the Montreal economy. The policy pay-off from this perspective is that regional development "platforms" need to be facilitated that reproduce these conditions. The means by which such cross-fertilizations occur rely upon "knowledge spill-overs" – external economies that spill over accidentally from firms located in geographical proximity that have the absorptive capacity to translate such tacit knowledge into explicit, codified, usable, and repeatable knowledge in a new business context, as may be happening in the case of Calgary.

Where a regional economy is over-diversified, there are few knowledge spillovers and little absorptive capacity except of the generic kind that was promoting, for example, the virtues of outsourcing to "supply chains" in a context of "lean production." Such generic knowledge is by no means useless, but it does not offer specific opportunities for novelty since it is available to all competitor firms. Equally, where it is over-specialized, everyone is so familiar with the fundamentals that knowledge spillovers are ubiquitous, but absorptive capacity absorbs less and less novelty accordingly. Michael Porter's example of the alloy golf-club head cluster in Carlsbad, California, is an example of such an over-specialized, by now not especially innovative sub-sector dominated by Calloway, the firm that once conceived innovative opportunity from aerospace materials to revolutionize the last bastion of wood in the drivers of that Royal & Ancient game (Porter 1998).

The onset of "Phase 2 Globalisation" (Cooke 2005), in which not only routine production and services but knowledge exploration, examination, and exploitation are increasingly outsourced to China, India, and other Asian countries more generally, means essentially one thing: these countries, with slight exceptions, will remain for a long time at best incremental innovators but in the main, like Japan before them, mainly imitators. Schumpeter (1975) predicted this mimicry effect that "swarms" in clusters as imitators pile into the market following the spawning of innovation. The West, especially the United States, will remain a significant, possibly overwhelming source of disruptive, even radical, innovation. The former is technology destroying, the latter competence destroying. Hence, it is incumbent on Western countries, including Canada, to stimulate much "related variety" in their economies, particularly of a thematic, rather than the old-fashioned, sectoral kind: thus, security software, bioenergy, and human-computer interfaces with high-grade networking capabilities rather than the stale old ICT, biotechnology, nanotechnology trilogy highlighted in most national and sometimes regional science and technology prospectuses.

Conclusions

A key goal of this chapter was to suggest the directions that future research on cities and regions in Canada should take. While the ISRN case studies have added considerably to our knowledge of the innovation dynamics in city-regions of widely varying size and competences, they do not fully address how these cities and regions can adapt to the challenge of an evolving post-recession global economy. The chapter demonstrates how these challenges can be overcome by the adoption of an evolutionary economic geography approach that is rooted in Schumpeterian dynamics and related-variety concepts, both closely allied to the neighbouring concepts of path dependence and proximity (geographical and relational). Where these phenomena converge sectorally and geographically, we found the notion of regional platforms useful because the concept captures the multi-cluster manner in which "cluster mutation" among related-variety industries actually occurs in such settings. Evolutionary mutation occurs as entrepreneurs take knowledge from their own and their firm's path-dependent evolution in one sector and find ways in combination with network partners from related but distinctive industry clusters to form a new or emergent cluster built from these knowledge convergences. Such skills in the labour market are thus crucial to such regional innovation and economic development.

Clusters and platforms are not in contradiction; rather, they are both complementary and evolutionary in that, ideally, clusters may evolve into platforms. Rather than being over-specialized and locked in to one main category of knowledge and expertise, cluster firms are able to recombine their knowledge with firms and clusters that are different. In this way, they may take advantage of the related variety among companies that is the essence of imagining and implementing innovation.

Most innovation comes from existing knowledge rather than from new, although new knowledge can be more important in science-based innovation, as in biotechnology. But even there, finding new uses for established discoveries, including new uses for pre-existing drugs, is not unknown. This makes innovation less daunting for SMEs if they are able to discuss possibilities with other firms rather than with university researchers. It also makes the role of urban centres as the critical sites for related-knowledge platforms all the more pressing.

It may be that the ending of the twenty-five-year neo-liberal experiment, coinciding with the onset of the knowledge economy as a main new metropolitan employer, and the further haemorrhaging of manufacturing jobs to Asian locations, requires a revitalization of urban government, not just governance. To re-balance urban economies towards new kinds of manufacturing, requiring both creativity and innovation in, for example, eco-innovation, renewable, and recycling, the Schumpeterian hubs are potentially key and powerful actors through a range of public policies, including their procurement policies. They can bring about substantial parts of their future by judicious procurement strategies. Actually, big Canadian cities such as Toronto and Vancouver have honourable records in both green production and consumption. This is mainly due to political actions of metropolitan governments, notably as customers of eco-innovations. Now may be the time for old-style urban political leadership to give way to the skills of "urban orchestration" of the many, not the few, interests in the complex mosaic that is, today, the modern city.

NOTE

1 A term coined by Pierre Veltz. See "Big cities and the global economy," keynote address at the Leverhulme International Symposium, "The Resurgent City," London School of Economics, 19–21 April 2004. http://www.veltz.fr/pierre_veltz/articles/pierre_veltz_article_resurgent_city_london_school_of_economics.html.

REFERENCES

Andersen, E.S. 2011. *Joseph A. Schumpeter: A theory of social and economic evolution*. London: Palgrave Macmillan.

Boschma, R. 2005. "Proximity and innovation: A critical assessment." *Regional Studies* 39 (1): 61–74. http://dx.doi.org/10.1080/0034340052000320887.

Boschma, R., and K. Frenken. 2006. "Why is economic geography not an evolutionary science? Towards an evolutionary economic geography." *Journal of Economic Geography* 6 (3): 273–302. http://dx.doi.org/10.1093/jeg/lbi022.

Boschma, R., and R. Wenting. 2007. "The spatial evolution of the British automobile industry: Does location matter?" *Industrial and Corporate Change* 16 (2): 213–38. http://dx.doi.org/10.1093/icc/dtm004.

Cantwell, J., and S. Iammarino. 2003. *Multinational corporations and European regional systems of innovation.* London: Routledge.

Christensen, C. 1997. *The innovator's dilemma.* Boston: Harvard Business School Press.

Cooke, P. 1992. "Regional innovation systems: Competitive regulation in the new Europe." *Geoforum* 23 (3): 365–82. http://dx.doi.org/10.1016/0016-7185 (92)90048-9.

Cooke, P. 2005. "Regionally asymmetric knowledge capabilities and open innovation: Exploring 'Globalisation 2' – a new model of industry organisation." *Research Policy* 34 (8): 1128–49. http://dx.doi.org/10.1016/ j.respol.2004.12.005.

Cooke, P. 2008. "Cleantech and an analysis of the platform nature of life sciences: Further reflections upon platform policies." *European Planning Studies* 16 (3): 375–93. http://dx.doi.org/10.1080/09654310801939672.

Cooke, P. 2011. "Transversality and regional innovation platforms." In *The handbook of regional innovation and growth*, ed. P. Cooke, B. Asheim, R. Boschma, R. Martin, D. Schwartz, and F. Tödtling, 303–14. Cheltenham: Edward Elgar.

Cooke, P., and C. De Laurentis. 2010. "The Matrix: Evolving policies for platform knowledge flows." In *Platforms of innovation: Dynamics of new industrial knowledge flows*, ed. P. Cooke, C. De Laurentis, S. MacNeill, and C. Collinge, 311–60. Cheltenham: Edward Elgar.

David, P. 1985. "Clio and the economics of QWERTY." *American Economic Review* 75: 332–7.

Frenken, K., F. van Oort, and T. Verburg. 2007. "Related variety, unrelated variety and regional economic growth." *Regional Studies* 41 (5): 685–97. http:// dx.doi.org/10.1080/00343400601120296.

Glaeser, E. 2000. "The new economics of urban and regional growth." In *The Oxford handbook of economic geography*, ed. G.L. Clark, M.P. Feldman, and M.S. Gertler, 83–98. Oxford: Oxford University Press.

Golden, A. 2009. Introduction to D.A. Wolfe, *21st century cities in Canada: The geography of innovation.* Ottawa: Conference Board of Canada.

Grabher, G., ed. 1993. *The embedded firm.* London: Routledge.

Jacobs, J. 1969. *The economy of cities.* New York: Random House.

Martin, R.L., and P. Sunley. 2010. "The place of path dependence in an evolutionary perspective on the economic landscape." In *Handbook of evolutionary economic geography*, ed. R. Boschma and R.L. Martin, 62–92. Cheltenham: Edward Elgar.

Nuttall, C. 2008. "Silicon feels the power of the sun." *Financial Times*, 25 March, 12.

Porter, M. 1998. *On competition*. Boston: Harvard Business School Press.

Schumpeter, J. 1975. *Capitalism, socialism and democracy*. New York: Harper Torchbooks.

Wolfe, D.A. 2009a. "Introduction: Embedded clusters in a global economy." *European Planning Studies* 17 (2): 179–87. http://dx.doi.org/10.1080/096543 10802553407.

Wolfe, D.A. 2009b. *21st century cities in Canada: The geography of innovation*. Ottawa: Conference Board of Canada.

Contributors

Harald Bathelt holds the Canada Research Chair in Innovation and Governance in the Department of Political Science, University of Toronto. He is also cross-appointed in the Department of Geography and Program in Planning, University of Toronto and Zijiang Visiting Professor at East China Normal University, Shanghai. He has published several books and articles on clusters, innovation systems and knowledge creation, political economy, industrial restructuring, globalization, and regional policy and governance.

Yves Bourgeois is director of the Urban and Community Studies Institute, University of New Brunswick, Saint John. A graduate of UCLA (PhD), Edinburgh, Oxford, and Moncton universities, he writes on the linkages between the city and cultural identity, innovation, and growth.

Neil Bradford teaches political science at Huron University College, Western University. His research interests include place-based public policy, urban and community economic development, and multi-level governance.

Philip Cooke is University Research Professor in regional economic development, and founding director (1993) of the Centre for Advanced Studies, University of Wales, Cardiff. He is the editor of *European Planning Studies*, a monthly journal devoted to European urban and regional governance, innovation, and development issues. In 2003 he was elected Academician of the UK Academy of Social Sciences. He is a long-standing member of the Research Advisory Committee of ISRN in Canada and the Swedish CIND and CIRCLE research centres.

Rosemarie Dallaire, is an adviser for innovation and SMEs with Innovation et Développement économique Trois-Rivières.

Charles H. Davis holds the Edward S. Rogers Sr. Research Chair in Media Management and Entrepreneurship at Ryerson University, where he teaches in the School of Radio and Television Arts' Master of Arts in Media Production program and in Ryerson and York Universities' joint graduate program in Communication and Culture. His research interests include new product development, labour, and the political economy of media.

Pierre-Marc Gosselin is a professor in the Department of Sociology and Anthopology, Université d'Ottawa.

Jill L. Grant is professor of planning at Dalhousie University in Halifax, Nova Scotia. Her current research interests include development trends in Canadian suburbs, health and the built environment, and the factors that attract musicians and artists to Halifax. She is the author or editor of four books and dozens of articles.

J. Adam Holbrook is an adjunct professor and Associate Director of the Centre for Policy Research on Science and Technology (CPROST) at Simon Fraser University in Vancouver, BC. He was trained as a physicist and electrical engineer, and is a registered professional engineer in Ontario and BC. He started his career as a satellite engineer at Telesat, after which he spent twenty years in the federal public service in several S&T policy positions. At CPROST his research, consulting, and teaching activities centre on the analysis and impact of science, technology, and innovation in the public sector and private sector, both in Canada and abroad.

Juan-Luis Klein is head of the Research Centre for Social Innovation (Centre de recherche sur les innovations sociales, CRISES) and professor in the Department of Geography at the Université du Québec à Montréal. He is in charge of the Geographie contemporaine series published by the Presses de l'Université du Québec and has authored or co-authored several books and articles on the topic of regional development, local initiatives, and social innovation.

Cooper H. Langford is Faculty Professor in Communication and Culture at the University of Calgary. His previous appointments include vice-president for research, University of Calgary, and director of Physical and Mathematical Sciences at NSERC. His research focuses on innovation and knowledge flows.

Ben Li has diverse interests as a member of the University of Calgary InnoLab, as a policy and communication analyst at the Legislative Assembly of Alberta, and as a social media consultant, holding degrees in biology, political science, and open-source innovation.

Nicholas Mills holds a Rogers Doctoral Fellowship in Communication and Culture at Ryerson University. His research interests include business innovation and the political economy of media industries.

Andrew Munro is a PhD candidate at the Institute for the History and Philosophy of Science and Technology, University of Toronto. He is interested in and has published about clusters, innovation systems, and university-industry technology transfer. His dissertation examines cross-regional knowledge transfer in the medical and agricultural biotechnology sectors of Toronto and Guelph, Ontario.

Jen Nelles is an adjunct assistant professor at Hunter College (CUNY) in the urban affairs and planning department. Her research focuses on metropolitan governance and regional economic development. She has led projects in Europe and North America and is currently based in New York, NY. She is the author of *Comparative Metropolitan Policy: Governing beyond Local Boundaries in the Imagined Metropolis* (Routledge, 2012).

Peter W.B. Phillips, an international political economist, is professor of public policy in the Johnson Shoyama Graduate School of Public Policy at the University of Saskatchewan. He undertakes research on governing transformative innovation, including biotechnology regulation and policy, innovation systems, intellectual property, supply-chain management, and trade policy.

Cami Ryan is a former post-doctoral fellow with the University of Calgary and is now a professional research associate with the Department of Bioresource Policy, Business and Economics, College of Agriculture and Bioresources, at the University of Saskatchewan.

Gregory M. Spencer, PhD, is the manager of Local IDEAs (Indicator Database for Economic Analysis) at the Munk School of Global Affairs at the University of Toronto. Greg has published on the subjects of creativity and innovation in city-regions, industrial clusters, and rural economic development. Previous to undertaking his doctorate Greg worked from 1999 to 2003 at the Local Futures Group, an economic development consultancy based in London, England.

Diane-Gabrielle Tremblay is Canada Research Chair on the socio-organizational challenges of the knowledge economy, professor of labour economics and human resources management at the Télé-université of the Université du Québec, and director of the CURA (Community-University Research Alliance on work-life articulation over the life course).

Michel Trépanier is a professor at the Institut national de la recherche scientifique (INRS UCS) in Montreal and the Institut de recherche sur les PME, Université du Québec à Trois-Rivières (UQTR).

Peter Warrian is Senior Research Fellow at the Munk School of Global Affairs, University of Toronto. He is Canada's leading academic expert on the steel industry. His current research is on knowledge networks, supply chains, and engineering labour markets.

Graeme Webb is a doctoral student in the School of Communication at Simon Fraser University. His research interests combine political sociology with communication theory, specifically examining how communication technologies can facilitate a participatory governance paradigm at the city-region level.

Brian Wixted (PhD) is a research fellow with the Centre for Policy Research on Science and Technology (CPROST) at Simon Fraser University. His research interests include the international geography of sectoral innovation systems and the governance and strategic environment of publicly funded research systems.

David A. Wolfe is the Royal Bank Chair in Public and Economic Policy and director of the Program on Globalization and Regional Innovation Systems at the University of Toronto. He has been national coordinator of the Innovation Systems Research Network since its inception and the principal investigator on its two major collaborative research projects.